The Insect and the Image

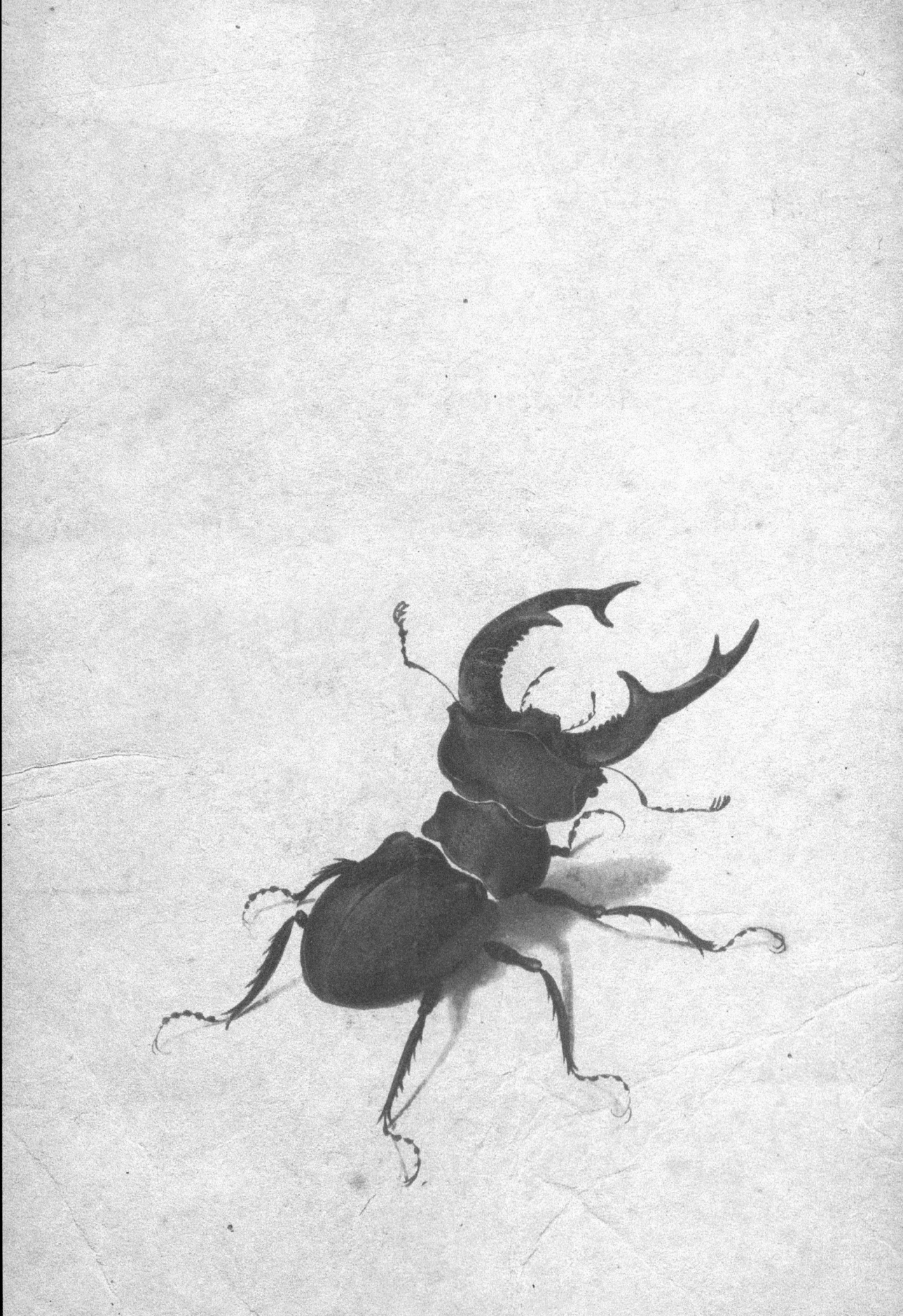

THE INSECT AND THE IMAGE

*Visualizing Nature in
Early Modern Europe, 1500–1700*

JANICE NERI

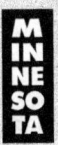

University of Minnesota Press

Minneapolis

London

Frontispiece: Albrecht Dürer, *Stag Beetle,* 1505 (detail). Watercolor and gouache. The J. Paul Getty Museum, Los Angeles.

An earlier version of chapter 1 was previously published as "From Insect to Icon: Joris Hoefnagel and the 'Screened Objects' of the Natural World," in *Ways of Knowing: Ten Interdisciplinary Essays,* ed. Mary Lindemann, Studies in Central European Histories (Boston: Brill Academic Publishers, 2004). An earlier version of chapter 4 was previously published as "Between Observation and Image: Representations of Insects in Robert Hooke's *Micrographia,*" in *The Art of Natural History: Illustrated Treatises and Botanical Paintings, 1400–1850,* ed. Amy R. W. Meyers and Therese O'Malley, Studies in the History of Art (Washington, D.C.: National Gallery of Art, 2008).

Published by the University of Minnesota Press
111 Third Avenue South, Suite 290
Minneapolis, MN 55401–2520
http://www.upress.umn.edu

Library of Congress Cataloging-in-Publication Data

Neri, Janice.
  The insect and the image : visualizing nature in early modern Europe, 1500–1700 / Janice Neri.
    p. cm.
  Includes bibliographical references and index.
  ISBN 978-0-8166-6764-2 (hc : alk. paper)
  ISBN 978-0-8166-6765-9 (pb : alk. paper)
1. Insects in art. 2. Art, European—Themes, motives. I. Title.
  N7668.I56N47 2011
  704.9'43257094—dc23

                          2011031739

Printed in the United States of America on acid-free paper

The University of Minnesota is an equal-opportunity educator and employer.

18 17 16 15 14 13 12 11   10 9 8 7 6 5 4 3 2 1

For Ted and Abingdon

CONTENTS

*I*n 1475 a woodcut illustration of insects appeared in Konrad von Megenberg's *Buch der Natur,* the earliest natural history book in the German language (Figure I.1). By present-day standards of scientific illustration, and more importantly by the image-making standards that would be established by the end of the sixteenth century, this woodcut image can only be described as schematic or crude. Many different types of insects are jumbled together in a composition that awkwardly wavers between two- and three-dimensional space. The individual insects lack detail and precise description, and they would quickly frustrate any attempt to identify their genus or species—despite the inviting and friendly smiles some of them proffer the viewer.

At first glance the Megenberg woodcut seems far removed from the exquisitely rendered images of insects of the sixteenth and seventeenth centuries, many of which are the subject of this book. But as a capsule of knowledge about the insect world as it stood in 1475, before the explosion of interest in insects that would occur in the next century, the woodcut tells us a great deal about the contours of this knowledge and the constraints and boundaries that would shape later practitioners' approaches to understanding this segment of the natural world. With the exception of beetles and dragonflies, most of the early modern categories of insects are present in the Megenberg image: bees,

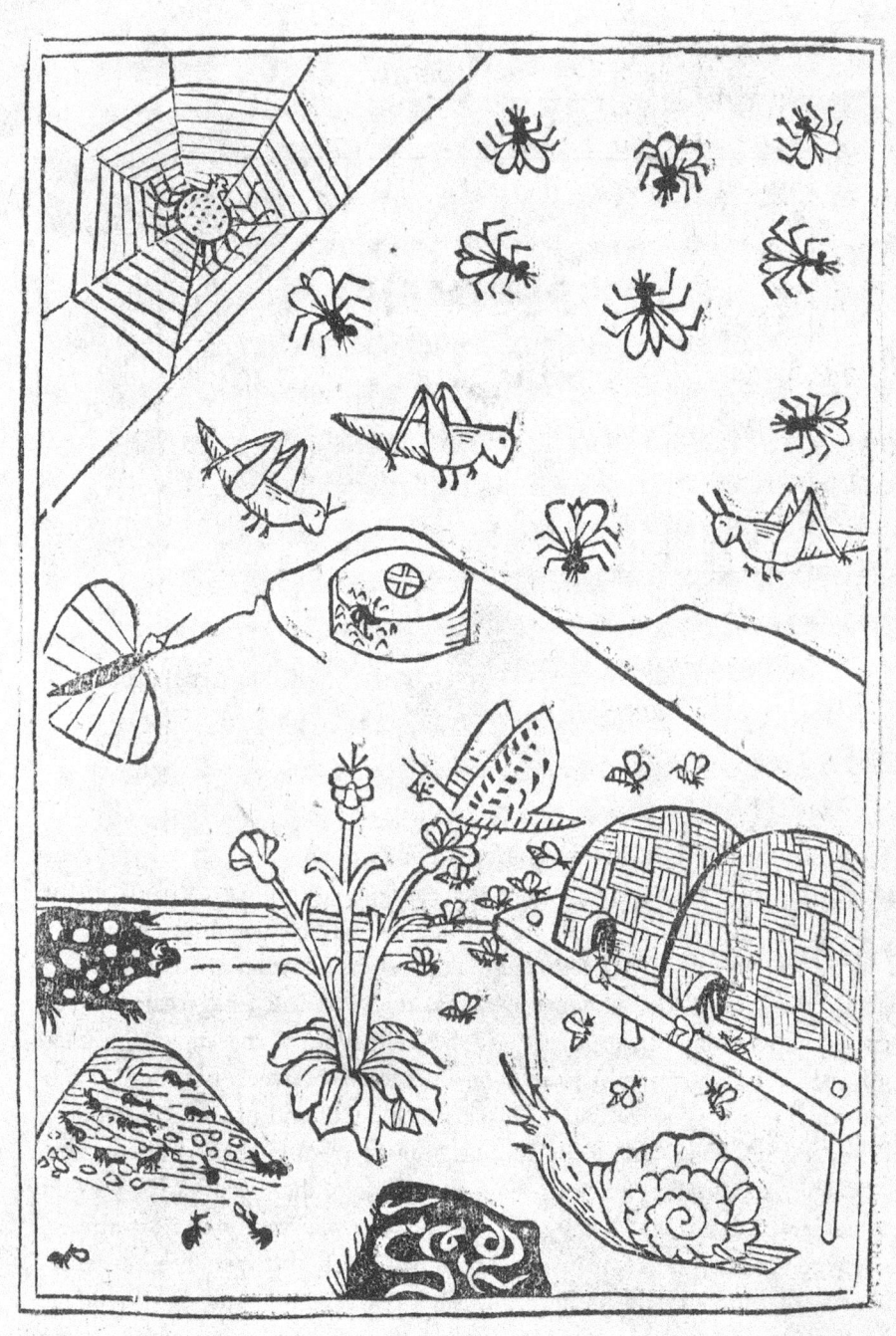

Figure 1.1. Konrad von Megenberg, *Buch der Natur* (Augsburg, 1475). Reproduced by permission of The Huntington Library, San Marino, California.

spiders, ants, grasshoppers and locusts, and flies—a category that encompassed houseflies as well as butterflies and moths. The beehives and the circular box containing silkworms indicate the economic benefits of studying insects and their life cycles, topics that would be pursued and greatly expanded upon in the publications and practices of later investigators. The flowering plant at the center of the image and the cross-section view of an ant colony point to the importance that artists and naturalists would place on studying the relationships between insects and their environment. Finally, the artist's attention to the patterning of the wings of the butterfly and the moth alerts us to a key element in the study and representation of insects in the sixteenth and seventeenth centuries: an emphasis on external appearance and forms, especially those that were colorful, intricate, or unusually delicate. The woodcut also reminds us that in the early modern period the term "insect" encompassed many creatures found crawling on the ground and flying in the air, as demonstrated by the presence in it of a snail, a frog, and the fossilized remains of worms.[1] Above all, this woodcut reflects the effort to understand the insect world through the process of making a visual image.

Understanding the natural world through image making is one of the distinguishing characteristics of the study of insects in the sixteenth and seventeenth centuries, and it is the focus of the five different moments in the history of insect illustration that I present in *The Insect and the Image*. If knowledge about insects could be summed up in a single woodcut in 1475, by the end of the seventeenth century this knowledge had expanded to fill entire books, many of them extensively illustrated. Between the years 1500 and 1700, Europeans began to turn their attention to insects and their life cycles as a new subject of study, and the strange, alien forms of insects and their unusual habits fascinated naturalists, collectors, artists, and the other practitioners who began to collect and display them in a variety of settings and media. In this book I trace the processes by which insects were constituted as subjects through making and using visual images. At the same time that insects emerged as new subjects, they also served as vehicles for the construction of artistic and scientific personae. In transforming insects into both subjects and objects, image makers and other early modern practitioners made use of a visual strategy I call "specimen logic," which allowed them to construct themselves as the gatekeepers to a strange and fascinating new world.

Insects appeared in a number of visual contexts prior to the sixteenth century, most frequently in the margins of illuminated manuscripts where they appeared along with images of other small objects such as flowers, shells, and medals. The shift of insects from the margins to the center was signaled by Albrecht Dürer's depiction of a single stag beetle in 1505 (Figure I.2). In this iconic drawing, Dürer lavished attention upon a creature that had previously, even within his own oeuvre, only appeared in the margins of images.[2] It was not only the act of choosing an insect as subject matter for an independent drawing that was notable in Dürer's image of the stag beetle but also the way in which he depicted this subject. The beetle is placed in the center of the page, and there are no other objects or elements to distract the viewer's attention from the details of its form. The viewer's investigation of the beetle reenacts the artist's careful observations of the insect's body, which is painstakingly worked in watercolor and gouache. With the stag beetle and other drawings from this period, Dürer was instrumental in establishing the genre of the nature study, notably through its distinctive visual form as defined by a concentration on individual specimens.[3] This "specimen logic" would dominate the approach to depicting insects throughout the early modern period.

The visual technique of presenting an isolated object against a blank background was the foundation of the sixteenth-century nature study, and as such it is central to many of the visual images discussed in this book. Although specimen logic was not limited to insects, as a way of understanding the natural

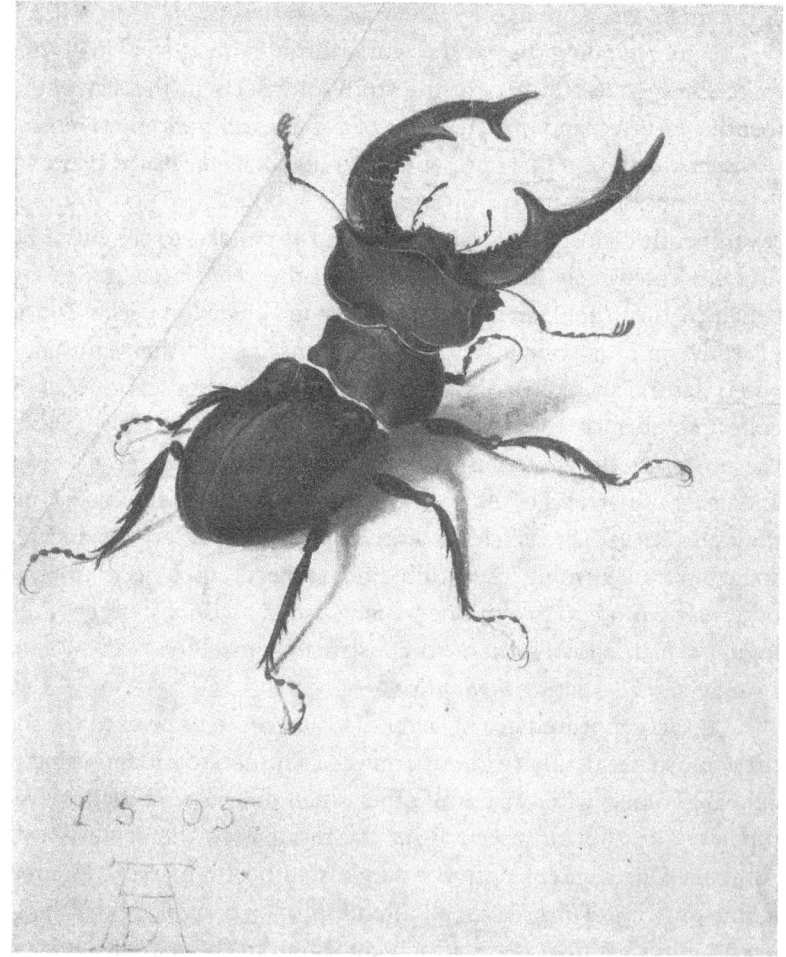

Figure I.2.
Albrecht Dürer, *Stag Beetle,* 1505. Watercolor and gouache. Upper-left corner added, with tip of left antenna painted in by a later hand. The J. Paul Getty Museum, Los Angeles.

world as a succession of isolated objects it had a far-reaching influence on image makers who took insects as their subject matter. Specimen logic turns nature into object by decontextualizing select creatures and items—that is, by removing them from their habitats, environments, and settings. Conversely, only those creatures and items that can be depicted or displayed as objects, those that possess clearly defined edges or contours and whose surfaces are visually distinct, are suited to the aims of specimen logic. Insects—or rather, certain types of insects—meet the criteria of specimen logic and were thus well suited to the broader impulse to visualize nature as a collection of objects. Specimen logic produces images that are stubbornly insistent on the materiality of their subject matter. Dürer's stag beetle and its many descendants entice us with their glistening surfaces and impossibly delicate structures, inviting us to imagine them as inhabitants of a timeless space of display, and to imagine ourselves as their possessors. The artists and other image makers discussed in this book deployed specimen logic to explore their new subject and to associate themselves with a mode of picturing in which the ability to create a highly detailed image was a sign of artistic talent and a keenly observant eye.

Specimen logic was also part of the commodification of nature that was central to global trade and commerce during the early modern period. Images of insects took part in this global trade, albeit on a small scale, but are nonetheless important for understanding how networks of economic exchange factored into the construction of knowledge of nature. This book therefore contributes to the growing literature treating the convergence of art, science, and commerce in the early modern period.[4] These interdisciplinary studies emphasize the ways that material practices and ideas surrounding the study and representation of nature intersected with global movements of peoples, plants, and other objects. In the field of art history, taking such a perspective has opened up new ways of thinking about the social, cultural, and material contexts of familiar images. The art historian Julie Hochstrasser, for example, has argued that the exotic food-stuffs pictured in seventeenth-century Dutch still life paintings mask the historical realities of global trade, especially the violence visited by Europeans upon the inhabitants of the Spice Islands.[5] As with the still life paintings discussed by Hochstrasser, the images of insects presented in this book are often admired for their seemingly timeless beauty, but it is my intention to uncover some of the unseen social, cultural, and historical processes at work in their production.

Insects as Subjects

The multiple sites where insects emerged as subjects reflect the rich visual and material culture of studying nature in early modern Europe. Insects took on a new status as exotic rarities fit for display in settings ranging from princely cabinets to lavishly illuminated manuscripts, still life paintings, and natural history collections. It is for this reason that this book addresses images that would be considered "amateur," as well as in some cases images made by anonymous or unidentified artists, alongside the more skillful works made by professional artists.[6] All of these types of imagery are important for understanding how insects were constructed as subjects through visual images, and the ways that artists actively participated in establishing the contents and parameters of the natural world for early modern Europeans. One of the central arguments of this book is that the choices artists made about how and what to include in their images of the natural world helped to constitute key ideas about nature as exotic, controllable, and ultimately subject to the movements of global trade and capitalism. This vision of nature was created by both skilled and unskilled hands, in elite and nonelite settings. The majority of the images presented herein can be understood as part of the dominant framework for understanding nature in early modern Europe, centered on the idea of the natural world as composed of beautiful, exotic, and collectable objects. Several key counterexamples, however, provide important glimpses into moments when this framework was being formed and what types of natural subjects were not compatible with specimen logic.

As surprising as it may be to readers for whom insects seem to be a marginal or excessively narrow topic, this book examines only a selection of the available materials pertaining to insects in the early modern period. The insect has long been a "member of the universe of animal metaphors,"[7] and the endless variety of insect forms and habits has provided much material for exploring its religious, social, cultural, and symbolic associations. Bees in particular have been of interest to commentators concerned with exploring the parallels between insect and human societies, and in early modern England silkworms joined bees as examples in the discourse surrounding labor and labor relations.[8] In eighteenth-century France, writers of natural history books for elite audiences focused on the study of insects as a point of entry for improving the character of aristocrats through the study of nature.[9] Insects also took on heavy symbolic meaning in the early modern period, with beetles and butterflies often serving as symbols for Christ and the resurrection, or as references to the Greek myth of Psyche and the soul.[10] Critical studies of insects by literary scholars and historians have generally focused on the insect as a metaphor for human relations, or for the relationship between the individual and society.[11] *The Insect and the Image* differs from these studies and approaches in two ways. First, it is concerned with examining the emergence of the insect as subject matter in the last half of the sixteenth century, the period that established the conditions of visibility upon which later musings on insects were founded. Second, the developments traced in this book tell the story of increasingly decontextualized images and objects, that is, of insects removed from their habitats and entered into networks of circulation and exchange. Although the networks through which specimens and insects traveled were most certainly made up of a web of social relations, the construction of the insect as subject was founded upon its capacity to be removed from context and setting.[12] This book offers an account of the emergence of the insect as an independent subject in visual images, a process that ultimately depended on constructing the insect as an exotic, alien inhabitant of a nonhuman world.

The Insect and the Image is not, therefore, a comprehensive history of insect illustration in the sixteenth and seventeenth centuries, nor is it a book of "firsts." The artists, image makers, and other practitioners who are the subject of this book made fascinating and important contributions to the study of insects, but their significance for the purposes of this volume does not derive solely from their merit as the first to discover, describe, or publish on a particular insect or phenomenon. Such distinctions can often be difficult to determine, but even in cases where precedence is relatively clear this aspect is not usually the most interesting one of these figures' activities. The first illustrated book on insects, for example, was Ulisse Aldrovandi's *De animalibus insectis,* published in 1602. This achievement is complicated, however, by Thomas Moffet's book on the same topic, *Theatrum insectorum,* published posthumously in 1634 but finished in 1590. Moving beyond the criteria of first discovery or publication allows us to explore the questions that arise when considering that both men were working on their separate studies of insects during the same time and devising remarkably similar methods despite their probable ignorance of each other's research. Similarly, Joris Hoefnagel was not the first artist to paint insects; Robert Hooke was not the first to observe and represent insects with the aid of the microscope; and Maria Sibylla Merian was not the first to depict the stages of insect metamorphosis. Early seventeenth-century still life painters who included insects in their compositions copied and recopied the same insects in numerous paintings. *The Insect and the Image* shows that the establishment of insects as subject matter was not a linear process but was instead a phenomenon that took place across a wide range of practitioners, contexts, and geographical locations. This book is therefore organized chronologically and thematically to give the reader a sense of the diversity of the material and the dispersed nature of the practitioners. As such, not all early modern European makers of insect imagery appear in this study.[13] The images and image makers that I examine were selected based on their capacity to help understand the central concerns of my work in this volume, among which are issues of accuracy and observation in visual images, the construction of insects as the inhabitants of an exotic and mysterious "natural world," and the parallel construction of artistic and scientific personae around insects.

The interdisciplinary approach that I use to examine these developments places this book at the methodological intersection of the history of science, art history, and visual culture. In my concern with uncovering the conditions under which it was possible for insects to emerge as subjects in the early modern period, I firmly ground my work in the archaeological approach to studying the past as theorized by Michel Foucault. Apart from the issues surrounding the usefulness of the idea of *epistemes,* discussed in further detail below, Foucault's ideas regarding discourse and power relations have been enormously influential on the development of the theories of visuality upon which this study is based. As with theories of discourse, theories of visuality seek to understand the conditions under which vision operates, the processes by which subjects become visible, and the conditions of visibility.[14] In *The Insect and the Image* I draw on these theories to explore the conditions and constraints that were at play when a new subject was formed. In the case of insects, this subject was constituted at the intersection of discourse and visuality. In my consideration of material practices, I also draw on Jonathan Crary's formulation of the relationships between vision, visuality, bodies, and technology. Crary uses the camera obscura as a model for "premodern" vision, "a site at which a discursive formation intersects with material practices."[15] While I do not share Crary's idea of a sharp break between modern and premodern modes of vision, I am indebted to this important theoretical shift from relying solely on visual images for understanding vision and visuality to an understanding of representation as an interaction between visual and material practices.[16]

This book is intended to contribute to the field of study variously labeled visual culture or visual studies, a discipline that is distinguished by its concern with images, materials, and practices that lie outside of the traditional scope of art history. Svetlana Alpers in her groundbreaking book *The Art of Describing: Dutch Art in the Seventeenth Century* argued that Dutch art should be understood as a "mode of picturing" that had parallels with mapmaking, emblems, natural history, and optical devices such as the camera obscura and the microscope. Alpers's book had important repercussions for the study of the visual culture of early modern Europe, especially the prominence given to alternative modes of image making and the integration of fine arts materials with nonart objects and images.[17] More recent studies of art and science in early modern Europe have demonstrated a variety of approaches to understanding early modern European representations of the natural world and have reached conclusions ranging from the ultimate failure of images as a means of knowing nature to the paradoxical imbrication of science and the imagination.[18]

While a number of studies of art and science focus on the continuities or disparities between early modern and modern science, in *The Insect and the Image* I am not primarily concerned with articulating the roots of modern science. A central tenet of much of the recent scholarship in the history of early modern science is its emphasis on the social and cultural construction of knowledge and its incorporation of ideas, texts, and practices of figures working outside of what has come to be known as the "mainstream" venues of the university and the scientific society. In this way, the history of early modern science has a great deal in common with the approach of visual culture and visual studies. Pamela Smith's *The Body of the Artisan: Art and Experience in the Scientific Revolution,* for example, explores the ways that bodily engagement with material objects and processes contributed to the formation of knowledge of the natural world and is therefore an essential part of the developments leading to the Scientific Revolution. Smith has also argued for using the early modern concept of *ars* to understand the relationship between art and science, since it "possessed a much broader connotation of practice and experience and was used in the case of the mechanical arts to refer to the work of the human hand."[19] This broadened definition of art to include the ideas and practices of artisans and other practitioners who made and used visual images is central to my approach in this book, but I also seek to preserve the distinction between artist and artisan in order to attend to the ways that the construction of the professional identity of the artist factored into the formation of insects as subject matter.

Art versus Science

The emergence of insects as subject matter, and as subjects in visual images, took place within the broader framework of the revolutionary developments taking place in Renaissance art, in particular the increased interest in naturalistic representation and the depiction of three-dimensional space. As has often been pointed out, these new developments in art coincided with the new ways of studying and observing the natural world that would eventually come to be grouped under the umbrella of the Scientific Revolution. Careful observation and accuracy were highly valued in both of these areas of inquiry, and there are numerous examples from the period that point to the shared desire for precision in art and science.[20] Much important research has come out of studies that attend to the correspondence between the concerns of art and science, but reliance on this correspondence as the sole explanatory model for early modern European images of the natural world can preclude asking (and answering) more complex questions about the particular qualities of visual images and how they function. The larger questions that I seek to explore in *The Insect and the Image* are about the ways that images work in terms of how they were produced, used, and understood. Why did people become interested in insects, and why did visual representations of insects take this particular form at this particular time? Why was "precision" valued and what did it mean? Why did specimen logic dominate, how did it come to be accepted, and what were its implications and consequences?

Issues surrounding observation, accuracy, and precision as they pertain to art and science in the early modern period have been discussed extensively in relation to botanical illustration of the sixteenth century.[21] The rise of botany and natural history during this period was accompanied by debates among botanists and others about the value of illustrations, which resulted in several clear statements about the use of images in the study of nature. The renowned botanist Carolus Clusius described his method of working with an artist on the illustrations for his book on Spanish and Portuguese plants. Clusius explained that although he did not make the illustrations himself, he closely supervised their production: "Having found an industrious and diligent artist, I had the images of plants depicted on wood blocks, and often I was beside the artist to indicate those aspects that had to be carefully observed when expressing the forms of dried plants."[22] Leonard Fuchs gave a more detailed explication of the method of making the illustrations for his *De historia stirpium* of 1542:

> As for the pictures themselves, every single one of them portrays the lines and appearance of the living plant. We were especially careful that they should be absolutely correct, and we have devoted the greatest diligence that every plant should be depicted with its own roots, stalks, leaves, flowers, seeds, and fruits. Over and over again, we have purposely and deliberately avoided the obliteration of the natural form of the plants lest they be obscured by shading and other artifices that painters sometimes employ to win artistic glory. And we have not allowed the craftsmen so to indulge their whims as to cause the drawing not to correspond accurately to the truth.[23]

Both Clusius and Fuchs emphasize the need to supervise artists and artisans in order to prevent errors, or even worse, flights of fancy that would confuse or obfuscate a viewer's understanding of the plant. Clusius's attempt to control his artist by sitting beside him parallels Fuchs's stern efforts to keep his artists from pursuing "artistic glory." In both of these accounts, working with artists is presented as a heroic struggle on the part of the botanist, one in which the botanist triumphs over the devious, vainglorious, or incompetent artist. This characterization of the relationship between the artist and the botanist as an ongoing conflict between the concerns of "art" and the aims of "science" would have profound consequences for both early modern practitioners and modern historiography. In the seventeenth century, for example, Robert Hooke echoed the concerns of Clusius and Fuchs in the preface to his *Micrographia* by stating in an often-quoted passage that he endeavored to present nature with a "sincere hand and faithful eye."[24] The subtext of Hooke's statement is his assurance to the reader that he too has waged battle in the arena of art and science, and that the values of science have prevailed over those of art. From this point of view, "art" serves "science" in communicating its discoveries, but art can hinder the clear expression of these discoveries by indulging in "unscientific" pursuits that belong to the realm of aesthetics.

There is no question that artists (like naturalists) made errors that needed to be corrected, and that misunderstandings between naturalists and artists took place. But what has not been sufficiently understood is the way that the opposition between "art" and "science" was a product of rhetoric designed to serve the purposes of specific individuals. The study of nature in early modern Europe was an intensely social activity, and although lone investigators could and did venture into the field to collect specimens and illustrate them, observing and investigating the natural world was rarely a completely solitary enterprise. The process of organizing, describing, representing, and displaying nature was more often than not done in collaboration with others in professional and domestic settings such as museums, classrooms, and artists' workshops. In some instances these collaborations were conducted across time and space through the medium of print, letters, or the exchange of specimens and images. Because the investigation of nature was a social activity, power dynamics were inevitably at play. The passages from Clusius and Fuchs quoted above are examples of the way that the "competition" between art and science could be deployed to construct the persona of the botanist as a reliable and indispensable source of information about nature. In both cases the artist or artisan is construed as potentially unreliable and in need of the guidance of a trained expert. The persona of the naturalist as the arbiter of accuracy thus depends upon the presence of a figure who is unreliable, and as such the artist becomes the other against whom the naturalist forms his own identity and stakes claim to his territory.[25]

What I argue in *The Insect and the Image* is that the relationship between observations and images is much more complex than the dichotomy of art versus science suggests, and that the concepts of accuracy and precision that were central to the study of nature in the early modern period were contingent upon particular social and material contexts. In returning to the example of Hooke, who constructed his own persona as a reliable observer in tandem with creating startling visual images of insects, it is important to understand that these activities took place as part of his efforts to negotiate the social terrain of the Royal Society and his uncertain status within that group. Similarly, the Flemish manuscript illuminator Joris Hoefnagel used insects as a venue for displaying his extraordinary artistic talent and as a way of fashioning himself as the heir to Dürer. The imaginary and "real" insects in his late sixteenth-century illuminated manuscripts engage in a visual dialogue with his predecessors and contemporaries, thereby establishing him as the equal of Dürer—and perhaps even suggesting that he surpassed the master. The relationship between "art" and "science" is not understood here as a struggle or conflict, and the central questions addressed in this book are not founded on the idea that one is subservient to the other. Instead, both are approached as practices relating to the project of describing the natural world, a project in which visualizing the limits and contents of what was to be included in the category of "nature" was under constant construction in a variety of sites and contexts. At the same time that ideas about nature were being explored, the image makers discussed in this book used these developments as an opportunity to forge professional identities for themselves. The personae they crafted aimed above all else to assure their audiences that they were trustworthy. Once this was established, they could be relied upon as sources for a variety of activities ranging from shaping courtly tastes to witnessing experiments.

Images and Objects, Images as Objects

For Hoefnagel and Hooke, and for the other image makers whose works are discussed in this book, the process of creating a visual image entailed not simply the direct transference of information but a transformation and translation of observations into images. Once these observations took visual form, they could then travel along a number of different paths. Exchanges of images and specimens between individual collectors are therefore an important part of the story I tell here, but I conceptualize broadly the idea of exchange to include not only physical movements but also virtual interactions and visual practices. As the diverse settings and materials studied in this book show, the process of transforming and translating observations into knowledge took many forms. At the end of the sixteenth century, the naturalists Ulisse Aldrovandi and Thomas Moffet utilized practices of physically cutting and pasting images in order to manage and control the vast amount of information they collected about insects and to forge this knowledge into the medium of the printed book.[26] A century later, Maria Sibylla Merian used the same techniques of virtually cutting and pasting in order to fix the forms of the insects and to arrange and rearrange them into compositions for her own very different book on the insects of Surinam. The transformative process of image making was also at play in depictions of insects in early Flemish still life painting at the turn of the seventeenth century. Still life artists contributed to the construction of insects as subject matter within this new genre of painting by pairing them with shells and coins—items with established pedigrees as exotic and collectible objects—thereby effecting a virtual transformation of insects into shells.

Images that employ specimen logic often use formal techniques associated with naturalism to create the illusion of three-dimensionality. Perspective, foreshortening, shadows, and modeling are the strategies most often used by artists to evoke the presence of an object occupying a three-dimensional space. Brian Ogilvie has pointed out that techniques such as these could be problematic within the realm of botanical illustration, and that some sixteenth-century botanists preferred images in which plant forms were flattened in the manner of pressed specimens.[27] Such preferences were not always evinced by naturalists who studied insects, although certain insects such as butterflies and moths were well suited to depiction as flattened specimens. For other insects, however, an emphasis on three-dimensional form was preferred. One of the recurring issues in this book is the tension between two-dimensional images and three-dimensional objects, a border that is constantly traversed by the insect as it was constructed as both illustration and specimen. The Megenberg woodcut offers another early visual example of this tension between two- and three-dimensionality—it is especially notable in the top half of the image, where the surface that the flies rest upon can be perceived as both a flat expanse and the wall of a room. It would make sense to explain this shift in perspective to the artist's lack of skill if it were not for the fact that similar juxtapositions often occur in the work of highly skilled artists such as Joris Hoefnagel and Maria Sibylla Merian. Rather than being a question of skill, then, I argue that such instabilities instead pertain to the fluid relationship between images and objects.[28] For many early modern practitioners who worked with insects, images and specimens were often interchangeable objects that circulated within the same spaces of display and exchange, equally at home in drawings and specimen trays. In *The Insect and the Image* I thus endeavor to treat visual images *as* objects in order to explore the ways that insects blurred the line between image and object.

The Question of Order

Early modern European students of nature often engaged in ordering the objects they collected, the texts they wrote, and the images they created. The development of organizational schemes and systems for ordering the natural world was a central concern for sixteenth- and seventeenth-century practitioners as well as for modern-day historians of science. Although the question of order is therefore of great importance to those who studied nature in early modern Europe—as well as to those of us who study those practitioners—one of my aims in this book is to explore the myriad other issues at play when an artist or image maker decided to make a visual image with the natural world as its subject. The diverse range of people and practices that I treat in *The Insect and the Image* is intended to provide a sense of the complexity of these visual practices and the ways that they often engaged with questions other than those pertaining to the ordering of the natural world. Collecting, exchanging, buying, selling, displaying, and crafting nature were activities that often intersected with the question of order, but ordering was not always the primary concern driving these efforts.

Nonetheless, any exploration of attitudes toward the natural world in early modern Europe warrants a consideration of the ordering of nature, and the ideas expressed by one of the most influential, if vexing, works on the topic, Michel Foucault's *The Order of Things*. In this book Foucault puts forward numerous intriguing and contradictory theories about the ways that knowledge about the natural world was formed in the Renaissance and how this knowledge of nature underwent a major shift during what he calls the Classical age (the mid-seventeenth century through the eighteenth century). Central to Foucault's ideas about these changes is the concept of the *episteme*, which is loosely defined as the practices and ideas that make up the dominant mindset of a period. For Foucault, the Renaissance episteme and its associated approach to ordering nature were characterized by networks of correspondence in which similitude, resemblance, and surface appearances were the qualities through which connections between objects and other entities were made. In contrast, the ordering of nature that took place in the Classical age focused on internal qualities and involved a radical paring down of the field of inquiry by eliminating connections and removing objects from these earlier networks.[29] While Foucault has been criticized for drawing too rigid a distinction between the Renaissance and Classical epistemes, and while the usefulness of the concept of episteme has been questioned and debated extensively, Foucault's Renaissance and Classical epistemes have served as important touchstones for numerous studies of the history of natural history, scientific illustration, and other related areas. David Freedberg, for example, has argued persuasively that the activities of Federico Cesi and the members of the Academy of the Linceans and their commissioning and accumulation of thousands of drawings of natural subjects reflect a desire to order nature that incorporated elements of the Classical episteme, in particular the drive toward abstraction and geometry.

Apart from the problem of epistemes, Foucault's *The Order of Things* poses another set of problems for students of early modern images of the natural world. Although vision, visibility, resemblance, and similitude are central to Foucault's thinking, in *The Order of Things* he does not actually offer a coherent theory of visual images. Foucault's discussions of resemblance and similitude often sound very much like they would be useful for thinking about visual images that utilize techniques of naturalism and realism, but in fact he has little to say about natural history illustration or other types of images that took part in "specimen logic."[30] This is not surprising, given that Foucault's primary concern was with *les mots* (the words).[31] Foucault's focus on words over images in the eighteenth century is also not surprising given the divergent history of natural history texts as compared with that of natural history illustrations. Beginning in the sixteenth century, natural history illustrations and similar images began to partake of specimen logic by removing objects from their contexts and placing them against the blank space of a page for the viewer's inspection. In contrast, natural history texts from the same period wove dense, often impenetrable connections between objects and their histories. The famed Lincean bee, the first illustration of an insect made with the use of a microscope, is an example of this disjunct between text and image (Figure I.3). On this engraved broadsheet issued by the members of the Academy of the Linceans in 1625 to celebrate the jubilee year of Pope Urban VII, three views of a magnified bee occupy an uncluttered space and encourage a prolonged, leisurely inspection by the viewer. In contrast, the *Apiarium,* a broadsheet issued in connection with this image, consists of blocks of text that require focused effort for the reader to work through (Figure I.4). Indeed, as Freedberg remarks, "A more daunting and forbidding-looking sheet than the *Apiarium* could hardly be imagined."[32] Both image and text contain an abundance of detail, but it is the quality of this information and the type of response it engenders that distinguishes the specimen logic of images from descriptions of specimens in texts.

This book is divided into two parts. Part 1 examines the emergence of insects as subject matter and the crafting of personae around insects in illuminated manuscripts, natural history illustration, and early still life painting during the period circa 1580 to 1620. Chapter 1 attends to the divergent history of texts and images as it relates to Foucault's epistemes and the question of order by examining works by Joris Hoefnagel and the techniques that the artist used to organize his approach to imaging insects. I use the concept of "screening" that Foucault developed in reference to the natural history texts of the Classical age to understand Hoefnagel's selections of insects. I argue that Hoefnagel's screening techniques enabled him to reduce the insect world to a very small selection while at the same time conveying the appearance of far-ranging inclusiveness. Thus, part of my intent in chapter 1 is to show that Foucault draws too sharp a distinction between the Renaissance and Classical epistemes, and that both ways of thinking about nature are present in Hoefnagel's work. Closer attention to visual images such as Hoefnagel's (while not "natural history" per se, but having much in common with the natural history illustration of the period) yields a more complex picture of how the natural world was conceptualized in the late sixteenth century.

Joris Hoefnagel screened nature to make insects accessible as precious objects and sophisticated entertainments for courtly audiences, and to make himself into an equally treasured figure within such circles. At the same time that Hoefnagel was fashioning insects for the court, other practitioners were working with insects to visualize nature as a similarly vast storehouse of wonder and to ensure their roles as trusted organizers of that storehouse. The vibrant communities of artists, artisans, and naturalists that coalesced around the study of insects in the field of natural history are the subject of chapter 2. The naturalists Thomas Moffet and Ulisse Aldrovandi also screened insects to make them accessible as objects, but their aims were to render nature comprehensible for the medium of print. Unlike Hoefnagel, these naturalists worked collaboratively, although not with one another. Their research on insects and the illustrated books that were the product of this research were built through the visual and material practices of circulation and exchange. Each of them developed techniques of virtually cutting and pasting to fit insects into the

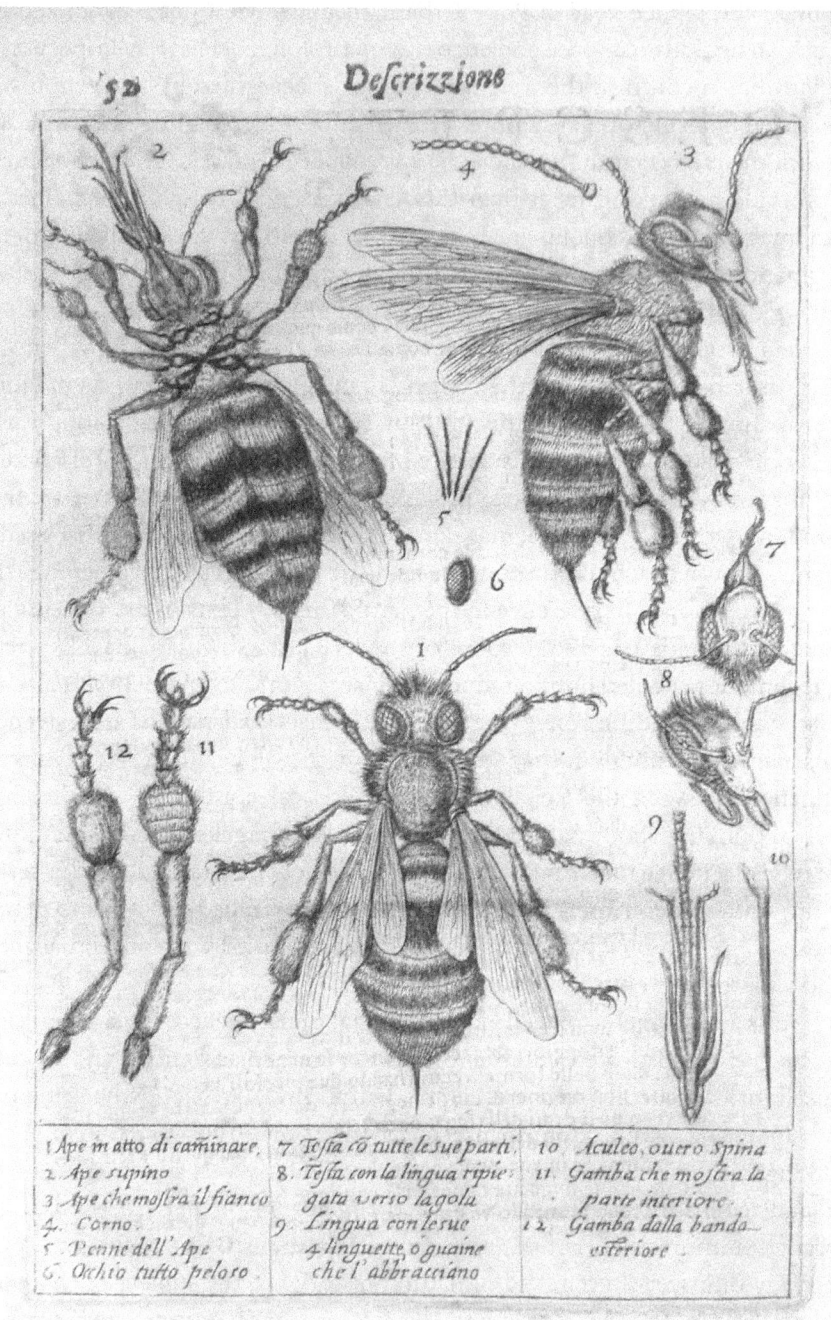

Figure I.3. Lincean bee from Francesco
Stelluti, *Persio tradotto* (Rome, 1630).
Courtesy of History of Science Collections,
University of Oklahoma Libraries.

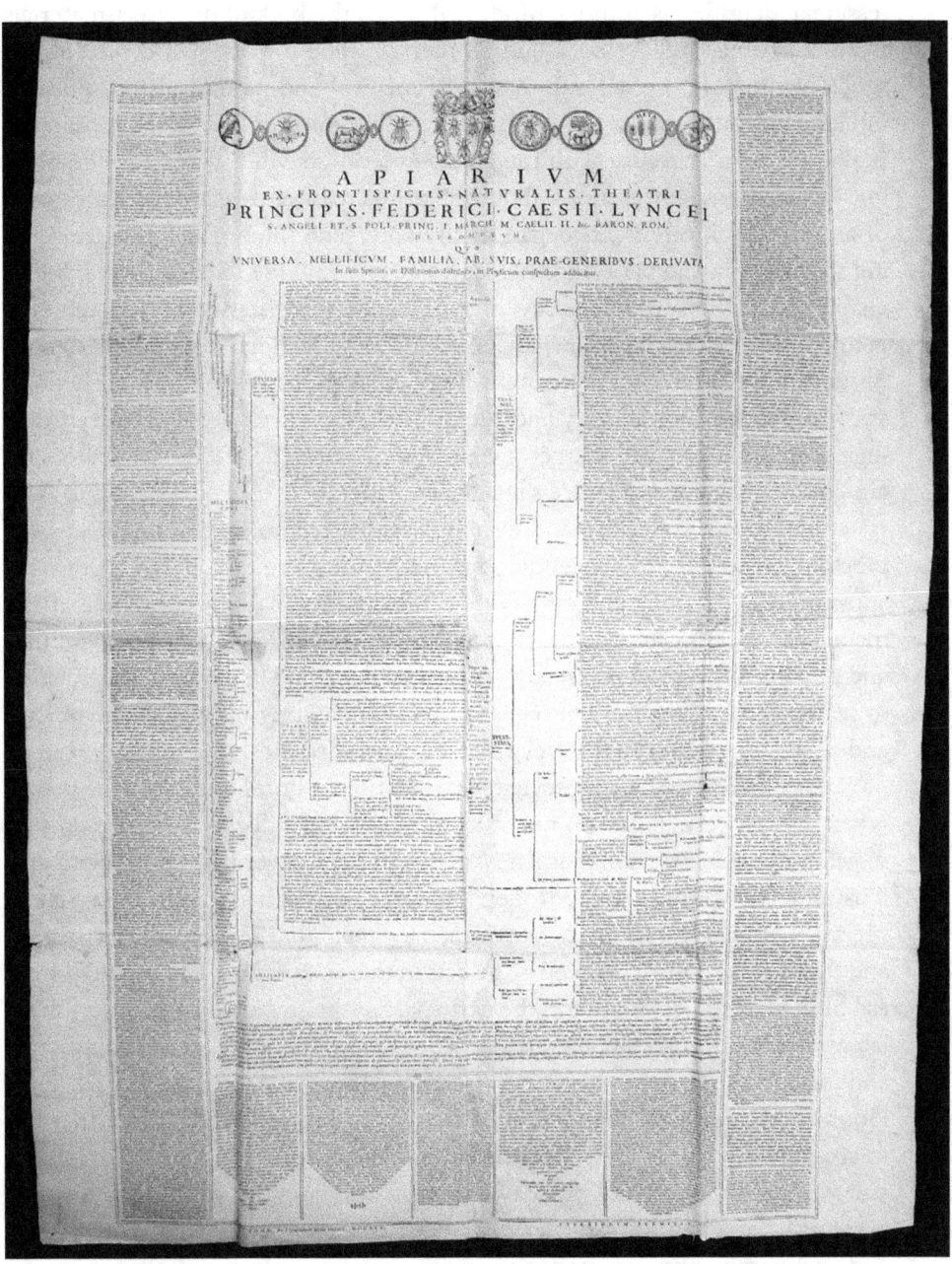

Figure 14. Federico Cesi, Francesco Stelluti, and others, *Apiarum,* 1625. Broadsheet. Courtesy of History of Science Collections, University of Oklahoma Libraries.

printed medium, and in doing so they faced related issues regarding their professional identities as authors. Moffet inherited the bulk of his insect materials from illustrious predecessors, and he had to argue for his own contributions while negotiating the delicate terrain of medical and natural history practice in late sixteenth-century London. Aldrovandi's collaborative work took place within his own museum with the artists and others who were in his employ, but he faced similar questions about the primacy of his research when rumors of another book on insects began to circulate. The third visual practice relating to insects in the period from 1580 to 1620 is still life painting, the topic of chapter 3. At the turn of the seventeenth century, artists working in the new genre of still life decided to include insects among the floral bouquets and other rarities they included in their compositions. The audiences for these early floral still lifes were mostly courtly and elite, unlike later seventeenth-century paintings that had a broader market appeal. Thus, artists active in the new genre aimed to satisfy their audience's tastes for rarities and collectables and the associated culture of collection and display. Still life painters presented insects as precious objects as Hoefnagel did, but they dealt more directly with practices surrounding collecting and displaying three-dimensional objects. These artists made sure to associate the newly popular insect with objects whose cultural and monetary value were already well established, and in doing so they helped to secure the status of both insects and still life paintings as collectable objects and of still life painters as purveyors of these new rarities.

Part 2 shifts to the later seventeenth century in order to explore the period when insects were no longer new subjects but were still exotic objects around which professional personae and new visual strategies were formed. Two major illustrated books on insects, Robert Hooke's *Micrographia* and Maria Sibylla Merian's *Metamorphosis insectorum Surinamensium,* are the focus of this section and the "new worlds" occupied by insects in this later period. After insects were established as subject matter in the early part of the seventeenth century, these later practitioners used approaches incorporating technology, global commerce, and the decorative arts to visualize insects and themselves in new ways. Robert Hooke's *Micrographia,* the subject of chapter 4, begins with the idea that the microscope reveals a "new world" to viewers, and the chapter explores how and why insects were the primary occupants of this new world. Drawing on previously unattributed sketches, I show that Hooke's early work with insects and the microscope was undertaken with several unknown collaborators. I argue that the record of these early collaborations with others on observing and representing insects was excised from the final version of the book, and that Hooke's carefully constructed images of the microworld were directed at convincing his readers and the members of the Royal Society that his work was solitary, and therefore more trustworthy. Chapter 5 explores images made by Maria Sibylla Merian, for whom insects were the primary means by which she marketed nature from the New World to European collectors and consumers. Like Robert Hooke, Merian's professional status was uncertain and she too needed to convince her audiences that her observations were trustworthy. In Merian's need to market her drawings, specimens, and the book on Surinam to an international community of elite collectors, she used visual strategies that combined the conventions of natural history illustration and specimen logic with her background in needlework and decorative arts design. For both Hooke and Merian, images of insects were the primary vehicles around which they constructed their personae as observers, translators, and mediators of the natural world.

The broader themes and concerns of *The Insect and the Image* are based upon questions generated in the field of history of science, while the answers that I hope to offer are based on the type of close visual analysis that is rooted in art history. This book is intended to contribute to models for rethinking the analytical framework of "art and science" in order to move beyond disciplinary boundaries and fixed categories. A growing body of scholarship has shown that alternative concepts, such as "image" and "knowledge," may be more fruitful for understanding the meaning and function of objects and practices that fall outside of the traditional definitions of art and science.[33] This is not to say that art and science are no longer relevant or useful terms—indeed, they remain correct and necessary for describing many artifacts and activities. But my goal is to uncouple the "art and science" formulation in order to better understand how these different modes of knowing the natural world came to be linked during the early modern period. As I hope *The Insect and the Image* will show, moving beyond "art and science" is a development that reflects the maturing of the field and a rich diversity of subject matter that can no longer be contained under a single category of analysis. The visions of nature presented in this book show that a diverse range of theoretical and methodological approaches is needed to understand the ways that artists and other image makers contributed to the construction of nature in early modern Europe.

I

INSECTS AS OBJECTS AND INSECTS AS SUBJECTS

Establishing Conventions for Illustrating Insects

ONE

JORIS HOEFNAGEL'S IMAGINARY INSECTS

Inventing an Artistic Identity

During the late fifteenth and early sixteenth centuries, insects emerged as objects of study for artists, naturalists, and other practitioners as part of the rising interest in classifying, collecting, and representing the natural world in early modern Europe. The Flemish artist Joris Hoefnagel (1542–1601), whose exquisite miniatures were much sought after by northern and central European collectors, directed much of his extraordinary artistic talent toward the depiction of insects. Most of Hoefnagel's insect imagery is contained in three of the artist's major works: the two illuminated manuscripts *Mira calligraphiae monumenta* and *Ignis* (part of the four-volume *Four Elements* series of manuscripts), and the *Archetypa studiaque patris Georgii Hoefnagelii . . .* , a series of copperplate engravings that the artist published with his son Jacob in 1592.[1] Although Hoefnagel was not the first early modern artist to depict insects, he was the first to concentrate so extensively on them, and thus it was necessary for him to devise techniques for managing this vast and potentially limitless subject matter. Hoefnagel utilized an array of image-making strategies to narrow the types and forms of the insects he represented, and by doing so he was able to achieve the appearance of encyclopedic coverage. Placing such strict limitations on his subject matter allowed Hoefnagel to explore the artist's role in manipulating "natural" appearances by interweaving fantastic and naturalistic modes of visual representation. Hoefnagel's insects were made to appeal to the highly refined sensibilities of late sixteenth-century courtly audiences, in particular the taste for the esoteric and the bizarre that is usually associated with mannerism, and as such they positioned him as the gatekeeper to a world filled with jewel-like treasures and curiosities. Hoefnagel's insect illustrations were perfectly calibrated to satisfy his audience's appetite for witty, sophisticated amusements while also establishing his identity as a figure who held the key to unraveling these tantalizing visual games. Like the "lusus naturae," or jokes of nature, that were also popular during this time, Hoefnagel's insects were examples of playful and exotic natural forms.[2] But unlike jokes of nature, these insects were entirely the work of Hoefnagel, and they helped to show that Hoefnagel's creative talents rivaled those of nature itself.

The visual strategies Hoefnagel utilized for limiting and controlling the insect world were complex and varied. One concept that is useful for understanding Hoefnagel's visual strategies is the idea of "screening" introduced by Michel Foucault in reference to the writing of natural history in the eighteenth century. While Hoefnagel's activities cannot be categorized as "natural history" in the sense that it was practiced in the eighteenth century, his illustrations certainly have much in common with natural history illustration as it was developing in the sixteenth century. Both made use of "specimen logic," and screening is a unique aspect of Hoefnagel's engagement with this mode of visual representation. According to Foucault, the flourishing of natural history in the Classical age—the period from the mid-seventeenth through eighteenth centuries—was achieved through strict control of the parameters of the natural world. Naturalists of the Classical age were able to devise a means of standardizing their observations by limiting them to a narrow range of structures and data, and in turn they were able to communicate those observations to a wide network of practitioners. Foucault writes that "natural history did not become possible because men looked harder and more closely. One might say, strictly speaking, that the Classical age used its ingenuity, if not to see as little as possible, at least to restrict deliberately the area of its experience."[3] Natural history texts "screened" certain aspects of nature in order to make others visible, and this screening process defined both the scope of natural history and the character of the description—in other words, *what* could be described and *how* it could be described. "This area," Foucault continues, "much more than the receptivity and attention at last being granted to things themselves, defines natural history's condition of possibility, and the appearance of its screened objects: lines, surfaces, forms, reliefs."[4] Since Foucault's primary concern was with articulating the structures and assumptions governing the *writing* of natural history, he did not consider the role or function of visual images in the history of natural history.[5] However, Foucault's notions of restricted areas of experience and "screened objects" is very useful for understanding certain aspects of sixteenth-century images of the natural world. In many ways, Hoefnagel's insects can be understood as screened objects, and the concept is key to understanding the choices that he made in establishing the parameters of the insect world.[6] In making his decisions about which insects to depict, Hoefnagel chose "if not to see as little as possible, at least to restrict deliberately" the range of insects he would consider. This deliberate narrowing, or screening, of his experience was a key element in his ability to convey the impression that the insect world was vast and varied.

The criteria Hoefnagel used to limit the scope and range of his investigations allowed him to engage in extremely creative elaborations and variations within the parameters he set for himself. Hoefnagel's insect imagery developed out of the broader artistic context in which he worked, in particular the two closely related artistic traditions of late medieval manuscript illumination and sixteenth-century nature painting. These artistic traditions provided Hoefnagel with visual precedents as well as strategies for constructing an exquisitely rendered world of insects that would be associated with him throughout the early modern period. Hoefnagel used these artistic traditions not only to visualize a new world of insects but also to establish himself as the supreme master of that world.

Insect Imagery and the Dürer Renaissance

During the last quarter of the sixteenth century, many northern and central European collectors were captivated by the work of the German artist Albrecht Dürer (1471–1528), and they were especially fascinated with the artist's nature studies. This renewed interest in Dürer took place within the wider context of the formation of princely collections, and thus resulted in intense competition for the artist's original works as well as the commissioning of works from contemporary artists in the "Dürer style." The "Dürer Renaissance," as this revival has come to be known, was fueled by the interests of collectors such as Archduke Ferdinand of Tyrol, Willibald Imhoff the Elder, Emperor Rudolf II and others.[7] In this vein Hoefnagel's illuminated manuscripts and cabinet miniatures were also highly prized by these collectors, along with works by other artists specializing in similar depictions of the natural world. Hoefnagel was a favorite artist of Rudolf II, and his works were among the many pieces by northern European and Italian artists that formed a part of the emperor's *kunstkammer* (chamber of art and wonders). Hoefnagel entered Rudolf's service at some time during the early 1590s, and he completed several of his major works, including *Mira calligraphiae monumenta*, as commissions for the emperor.

It was in the context of this revival of interest in the work of Dürer that Hoefnagel developed and perfected his corpus of insect imagery. The insect that held prime importance within this corpus was the stag beetle, the subject around which Hoefnagel established his artistic mastery of insects through a visual dialogue with Dürer. Dürer's drawing of a stag beetle (Figure I.2) enjoyed great popularity among sixteenth-century artists and patrons, and the beginning of the revival of interest in Dürer's nature studies has been dated to a copy from 1574 of the drawing by Hans Hoffmann.[8] Hoffmann worked for Rudolf II in Prague from 1585 until his death, and he had previously lived in Nuremberg where he completed many Dürer-style nature studies for the collection of Willibald Imhoff. Many copies of Dürer's *Stag Beetle* were made during the late sixteenth century, and it continued to be copied and incorporated into compositions by artists well into the seventeenth century.[9] Dürer's *Stag Beetle* reflects a major milestone in the representation of insects in early modern Europe, and the drawing came to function as a prototype for artists of the Dürer Renaissance. Unlike earlier illustrations of insects in the margins of illuminated books of hours, Dürer made the insect the sole subject of his composition by presenting it alone in the center of the page. With the *Stag Beetle* drawing and his other nature studies, Dürer literally and figuratively moved insects, plants, and other elements of the natural world from the margins to the center. Furthermore, Dürer represented this subject matter in a visual style designed to facilitate his audience's comprehension of the natural world as composed of individual, discrete objects that could be isolated from one another. Apart from the artist's monogram, the date, and the beetle's shadow, Dürer's *Stag Beetle* image contains no contextual elements or other visual information to distract from the viewer's understanding of the insect as an object or specimen. Such images offered sixteenth-century audiences a powerful model for thinking about plants, animals, insects, and other *naturalia* as objects and eventually as specimens, and thus they played a central conceptual role in collecting, exchanging, and displaying the natural world in early modern Europe. Visual images such as Dürer's *Stag Beetle* also served as a structural framework for understanding and representing the natural world for artists of the Dürer Renaissance such as Hoefnagel and Hoffmann. In order to better understand nature, it was necessary for these artists not only to study and observe actual plants and insects but also to study Dürer. However, copying Dürer's prototype image was only the first step toward mastery of the genre. In order to truly establish himself as Dürer's equal, Hoefnagel engaged in an active dialogue with the master's images by altering the prototype and adding his own interpretations to the inventory of established forms.

Hoefnagel's first extended treatment of the insect world appears in the *Ignis* volume of his *Four Elements* series of illuminated manuscripts. Created by Hoefnagel between 1575 and 1582 while working in Munich at the court of Albert IV, the four volumes in the series were intended to present a compendium of the entire known animal kingdom. The art historian Lee Hendrix has shown that Hoefnagel modeled many of his illustrations in the *Aqua, Terra,* and *Aier* volumes of the series on unpublished drawings by other artists and on published sources such as Gessner's *Historia animalium*.[10] However, with the *Ignis* volume and its focus on insects Hoefnagel had few models to follow other than Dürer's. Hoefnagel adopted the sparse, decontextualized format of the *Stag Beetle* drawing for all of the folios in *Ignis,* whereas in the other volumes of the *Four Elements* the artist usually included spatial markers and landscape elements. Hendrix notes that *Ignis* possesses a markedly different visual character and effect than do the other volumes of the *Four Elements:* "The blank vellum surrounding the insects contributes to the effect of visual immediacy not only by accommodating the moving eye, but also by jolting it with luminous whiteness, which isolates the specimens from one another, as pure visual data."[11] In deciding to employ this visual strategy for representing insects, Hoefnagel was not simply depicting specimens but rather "specimenizing" insects.

Dürer's prototype *Stag Beetle* drawing thus provided a model for the *Ignis* volume, but it also posed a challenge to Hoefnagel. The image served as a screen through which Hoefnagel formulated his approach to the insect world by helping him to pare down his subject matter to that already treated by Dürer while also making use of the pared-down visual style of his predecessor. But in order to establish himself as the master of the insect world, Hoefnagel not only had to display artistic skills equal to those of Dürer (by producing a flawless copy of the *Stag Beetle*) but also he needed to produce an image that would be admired and copied as much as Dürer's iconic beetle. Again, it was the stag beetle that served as the means for Hoefnagel to pursue these ambitions. Hoefnagel's version of Dürer's *Stag Beetle* appears on folio 5 of *Ignis* (Figure 1.1), in which he has expertly captured the stance and attitude of Dürer's beetle by mimicking the exaggerated curve of the mandibles and the inquisitive rearing head of the original. One of the central concerns of Hoefnagel's art was the production of vibrant, lifelike appearances, and in his illustrations of insects Hoefnagel follows Dürer in creating a lively, vigorous creature. In addition, Hoefnagel displays his own artistic virtuosity and acute powers of observation by making a number of "improvements" to the prototype. Hoefnagel's treatment of the beetle's shadow is particularly sensitive; whereas Dürer's beetle casts a single shadow, there are many fine gradations present in Hoefnagel's version. These gradations help to communicate the relative distances between the insect's various body parts and the surface of the page; the legs seem to be closer to the surface and therefore cast sharper, darker shadows, while the raised head is more distant and thus throws a softer and vaguer series of shadows. Hoefnagel's efforts to surpass Dürer are most notable in the "corrections" he made to the structure of the insect's body. Where Dürer has left the spaces between the head, thorax, and abdomen empty, Hoefnagel has created organic connections. Hoefnagel most likely consulted a specimen in making his drawing: *Lucanus cervus* is native to Europe, and although its habitat of deciduous

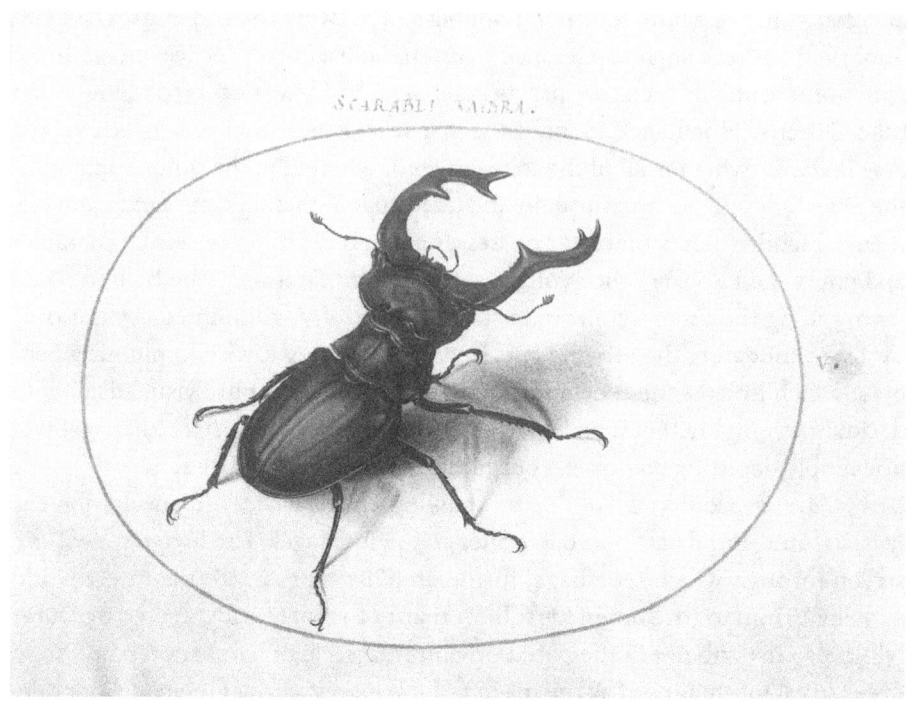

Figure 1.1. Joris Hoefnagel, *Animalia Rationalia et Insecta (Ignis)*, plate V, c. 1575–80. Gift of Mrs. Lessing J. Rosenwald. Courtesy of the Board of Trustees, National Gallery of Art, Washington.

woodlands and forests is much reduced at present, in the early modern period it was extensive and specimens would have been readily available.

It is important to point out here that in creating his version of the Dürer *Stag Beetle* Hoefnagel did not simply "fill in the blanks." The Hoefnagel version must instead be understood as a reinterpretation, since the gaps in the body of Dürer's original beetle do not signify deficiencies in observation or artistry but are integral to the overall lifelike effect of the image. Hoefnagel's stag beetle retains traces of these gaps: on the left side of the insect's body, between the thorax and the abdomen, the artist has included a very small area of white space. This small space is counterbalanced by the slight overlap between thorax and abdomen on the right side. These miniscule adjustments that take place within the framework of the prototype intensify the animated quality of Dürer's image by creating a subtle turning motion within the insect's body. Hoefnagel thus retains the function of the spaces between the body parts while assimilating them into his own refined style of representation. Hoefnagel's adjustments also help us to better understand the gaps in Dürer's beetle not as mistakes but as consistent with an overall approach to visualizing the natural world. The gaps in Dürer's *Stag Beetle* have led some scholars to go so far as to doubt its authenticity, based on arguments that such deficiencies are uncharacteristic of Dürer's usual penchant for accuracy and close observation.[12] But the term "accuracy" is relative to whether one is discussing early modern European images of the natural world or scientific illustrations from other periods, and such a judgment is highly dependent on considerations of the context and function of the image.[13] Instead, the gaps between the body parts of Dürer's *Stag Beetle* should be understood in relation to the early modern impulse to isolate and separate nature into small, manageable units as a means of studying, representing, and creating knowledge of the natural world. The image reflects the artist's desire to see nature in terms of its component parts, and the gaps between the body parts show that such a concern extended into the very fabric of the insect's body. Hoefnagel's later illustrations of insects in *Mira calligraphiae monumenta* form an extended meditation on this theme by exploring the ambiguous relationships between lifelike appearance and accurate observation.

Having established his mastery of the Dürer prototype in folio 5 of *Ignis,* Hoefnagel then presented viewers with another interpretation of the insect in folio 7 (Figure 1.2). This is one of Hoefnagel's most well known insect images, an open-wing stag beetle. In the image the artist opens up the insect's body in order to offer viewers a glimpse of its interior structures. The hard, armorlike plates of the wing case are spread apart to emphasize the beetle's wings, which have been rendered with exceeding care—the thick sinewy veins stand out against the thin transparent membrane, eliciting astonishment from the viewer simultaneously at the delicate structures of the insect's body and at Hoefnagel's consummate display of artistic virtuosity. To make these structures visible it would surely have been necessary to pin a dead specimen's body into the position, yet once again Hoefnagel infuses his subject with life. His expert placement and rendering of the shadows contribute to the impression that the upper half of the insect's body rises up from the surface of the page, a position echoing that of the upraised head of the Dürer prototype. Hoefnagel's open-wing stag beetle image became almost as well known as Dürer's *Stag Beetle;* it was widely disseminated through Hoefnagel's *Archetypa* engravings as well as copied extensively throughout the seventeenth century. The open-wing stag beetle is both an homage to and a variation upon the Dürer prototype, but it was not intended to supplant it. Hoefnagel's placement of his version of the insect directly after Dürer's in the *Ignis* volume indicates that the artist meant for the images to be viewed in relation to one another.[14] The grouping of the two images together is a gesture on Hoefnagel's part toward his artistic lineage and serves to facilitate favorable comparisons between his work and that of his predecessor. The two images are also an example of the way in which Dürer's imagery structured Hoefnagel's approach to studying and visualizing the natural world. Although Hoefnagel's open-wing stag beetle represents a radical

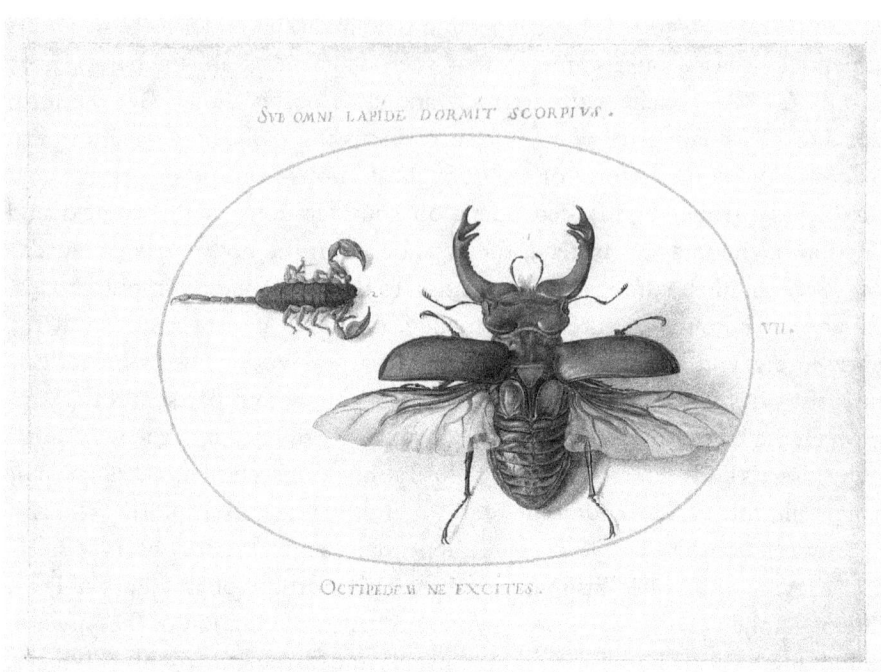

Figure 1.2. Joris Hoefnagel, *Stag Beetle with Open Wings*, c. 1575–80. Originally in *Ignis* volume of the *Four Elements* manuscript series. Berlin, Kupferstichkabinett. Bildarchiv Preussischer Kulturbesitz / Art Resource, NY.

departure from the prototype, this innovation took place within a framework established by Dürer. For Hoefnagel and other artists of the Dürer Renaissance, the motivation for studying nature stemmed in part from the desire to produce flawless imitations of Dürer's images. In this context, the artist's knowledge and mastery of nature progressed in tandem with his knowledge and mastery of image-making practices. Hoefnagel's interpretation of Dürer's *Stag Beetle* and his dramatic transformation of the prototype image also functioned as powerful instruments for forging his own artistic identity within the genre. With the two stag beetle images, Hoefnagel established himself not only as a skilled copyist of Dürer but also as an artist who had equally original contributions to make to the field of sixteenth-century nature painting.

From the Margins to the Center: Imaginary Insects

Sixteenth-century nature painting, in particular Dürer's iconic stag beetle drawing, functioned as a screen for Hoefnagel in developing strategies for representing insects. Late medieval manuscript painting served as another screen for

Hoefnagel for organizing his approach to the insect world and for establishing himself as a uniquely talented observer and presenter of insects. Hoefnagel adopted techniques for depicting insects, small objects, and other animals that were used in the border illustrations of illuminated manuscripts. He also tended to limit his choices of insects to those that were well suited to these techniques. The Book of Hours of Catherine of Cleves offers one example of the emphasis on brightly colored objects, unusual patterns, intricate structures, and other visually interesting features that is characteristic of the border illustrations of late medieval illuminated manuscripts. In this manuscript and others, the insects, flowers, and other minute items that appear in the borders are depicted as discrete, isolated entities, with clear, crisp edges marking a distinct separation between the object and its background.[15] Butterflies and moths were well suited to this mode of representation, and the techniques used in the Book of Hours of Catherine of Cleves draw attention to their colorful and intricate wing patterning (Figure 1.3). Two methods of positioning the bodies of butterflies and moths are on display here: the flat view and the profile view. In the flat view, the insect is depicted with its wings spread to their maximum extension to allow the greatest exposure of the wings—the feature of the insect's body that most interested the artist and the audience. In the profile view the insect's wings are drawn up away from its body, with both sets of wings visible, thereby offering a view of both the dorsal and ventral sides. Although only moths lie flat when at rest, the flat position was often used to depict butterflies, as seen in the center-left insect in Figure 1.3. As such, the flat position does not represent the butterfly's "natural" appearance and would only be possible through pinning the wings of a dead specimen.[16] Hoefnagel made extensive use of both flat and profile views for representing butterflies, moths, and other winged insects, and he subtly manipulated these conventions to transform flat, dead specimens into seemingly living beings. Compared with the butterflies and moths in the Book of Hours of Catherine of Cleves, Hoefnagel's flat insects are imbued with a sense of life and movement, as in the case of the large hornet shown at the top of plate 4, part 5 of the *Archetypa* (Figure 1.4). The placement of the hornet at an angle at the top of the page produces the impression of an insect hovering over the composition to inspect the assemblage of objects, rather than seeming to be a dead specimen pinned to the surface of the page. This interest in producing lifelike impressions was a central feature of Hoefnagel's approach to representing the natural world, and it played an important role in his illustrations of insects.

Hoefnagel used nature painting and manuscript illumination as two differnt types of screens through which he approached insects and devised criteria for choosing them. These criteria for choosing insects were on full display in the first section of Hoefnagel's *Mira calligraphiae monumenta,* the lavishly illustrated manuscript that represents the culmination of the artist's repertoire of

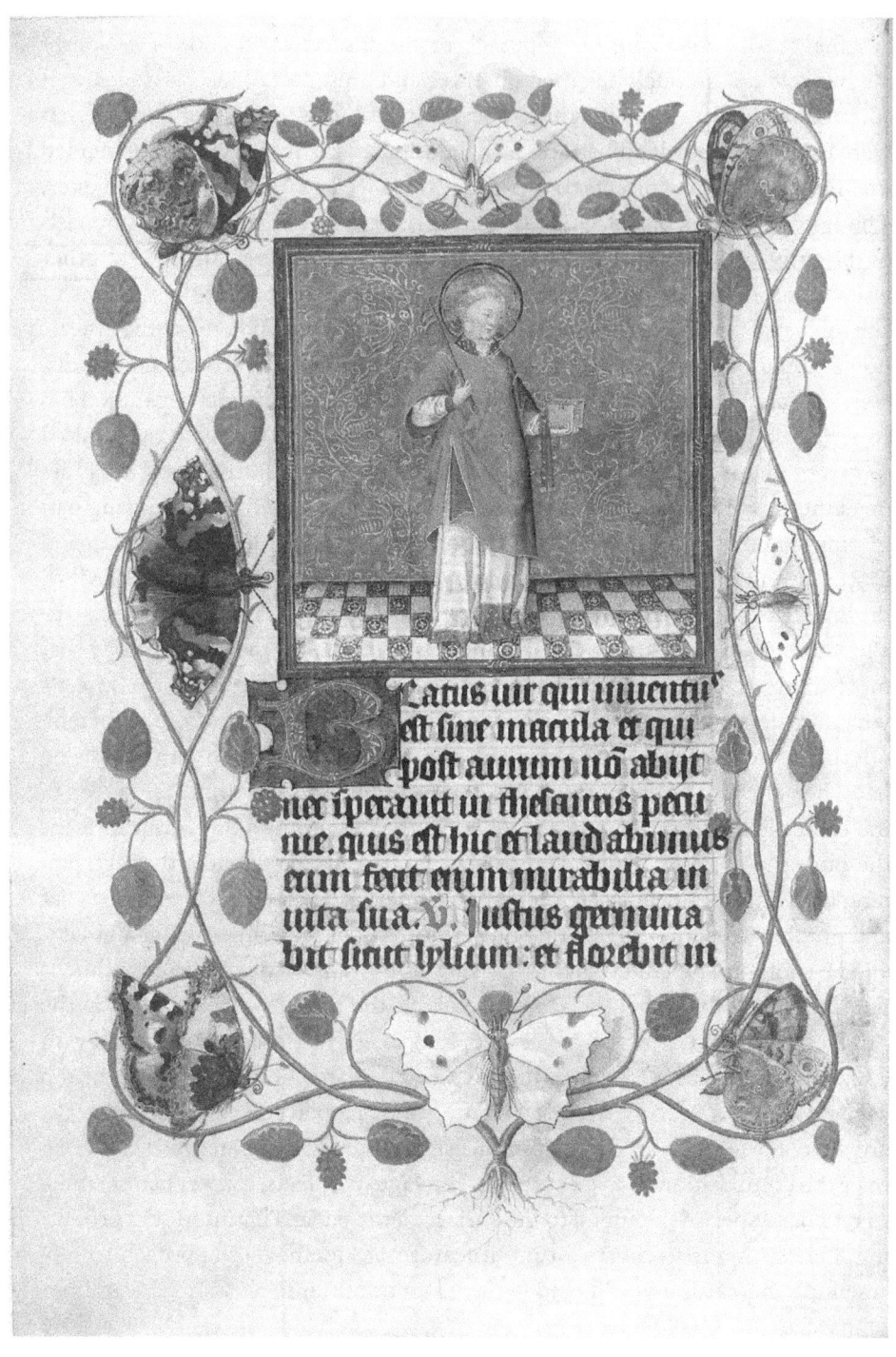

Figure 1.3. Hours of Catherine of Cleves,
c. 1440, page 268. The Pierpont Morgan
Library, New York. Purchased on the Belle
da Costa Greene Fund with the assistance of
the Fellows, 1963. MS M. 917.

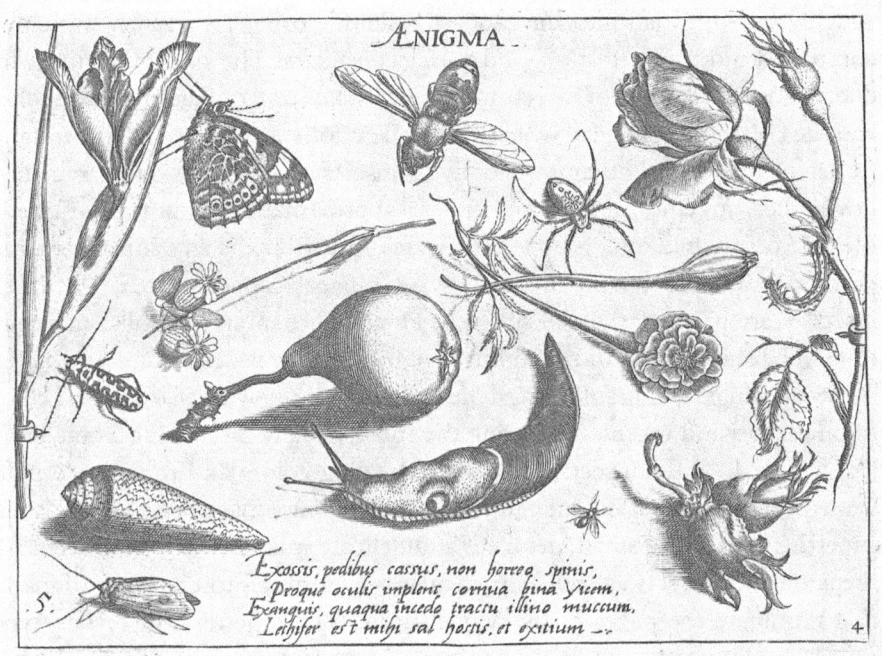

Figure 14. Jacob and Joris Hoefnagel, *Archetypa studiaque patris Georgii Hoefnagelii* . . . (Frankfurt, 1592), part 4, plate 5. Research Library, Getty Research Institute, Los Angeles.

insect forms. The manuscript, also known as the Model Book of Calligraphy, was begun in 1561 by Georg Bocskay, who was commissioned to create the extraordinary examples of calligraphic script by Ferdinand I, grandfather of Rudolf II. Rudolf inherited the manuscript and then later commissioned Hoefnagel to illuminate it. The illustrations were completed during the last decade of Hoefnagel's life while he was living in Vienna. The images in the Model Book present a grand assortment of nature's small but wondrous productions, and in them insects and small animals play a prominent role in "capturing nature's animate quality."[17]

The Model Book images appear to convey an almost encyclopedic coverage of the great variety and diversity of the insect world. Intricately painted and elaborately colored insects and snails cavort among luminous flowers, gleaming shells, and sumptuous fruit, encountering the occasional frog or lizard in the course of their exploratory activities. However, the effect of a bountiful variety of insects is very much the product of Hoefnagel's screening techniques. In the Model Book, Hoefnagel favored certain types of insects over others, notably preferring winged insects and beetles. Of approximately 140 total insects, roughly one-third fall into the modern scientific order *Lepidoptera,* which encompasses moths, butterflies, and their caterpillars. The two other most frequently depicted types of insects are the *Odonata* (dragonflies and damselflies) and the *Coleoptera* (beetles), which each appear fourteen times and fifteen times respectively. Various other types of flying insects such as houseflies, crane flies, hover flies, mayflies, and wasps appear approximately twenty times in the Model Book, while spiders appear six times, grasshoppers and ants three times each, and a scorpion, millipede, and a water bug each appear once.[18] All of these insects were particularly well suited to Hoefnagel's interests, skills, and ambitions. The shiny shells and alien forms of beetles, the iridescent bodies and latticework wings of dragonflies, and the endlessly varied wing patterns and colors of butterflies and moths are among the most visually interesting forms in the insect world. These insects, with their dazzling colors and minute, complex forms serve as the perfect medium for displaying Hoefnagel's artistic skills. His expertise at creating small, detailed, and delicate images was well matched to the subject of insects as well as particularly well suited to the types of insects that dominate the pages of the Model Book. The repetition of certain types of insects allowed Hoefnagel to sharpen and perfect his skills on a finite set of insect forms.

An example of the alliance between the form and the content of Hoefnagel's insect imagery appears in folio 15 of the Model Book (Figure 1.5), in which Bocskay's calligraphic rendering of a chalice is surrounded by plant and insect forms that mirror the script. The rounded blooms of the roses and their tapering stems echo the cup and stem of the chalice, while the flat, horizontal line of the trompe l'oeil slit in the page parallels the bottom line of text, completing the ribbon of empty space that Hoefnagel has fashioned around the text. In the upper-right corner of the page a wasp is perched upon the leaves of a rose bud, its impossibly slender legs forming a perfect fit with the outstretched tendrils of the plant. The insect's body provides yet another elaboration upon the visual motifs of the script in the way that it reflects the forms of the gold initial on the opposite side of the page. The insect's abdomen, wings, and spiky legs evoke the intricately intertwined curves and angles of the pen strokes, with the curve of its abdomen corresponding to the letter's lower stroke and its wings corresponding to the elongated loops of the upper strokes. Another example of Hoefnagel directing his investigation of the natural world toward forms that responded to his particular talents and to the compositional demands of the Model Book can be found on folio 43 (Figure 1.6). Here, the serifs of Bocskay's calligraphy provide another visual motif around which Hoefnagel organized both the composition and the content of the image. The curve of the horn of the rhinoceros beetle is echoed in the taut petals of the lily, a connection that is intensified by showing the outer, bracketlike petals of the flower in sharp profile. The splits in the skin of the pomegranate in the lower-right corner, like the

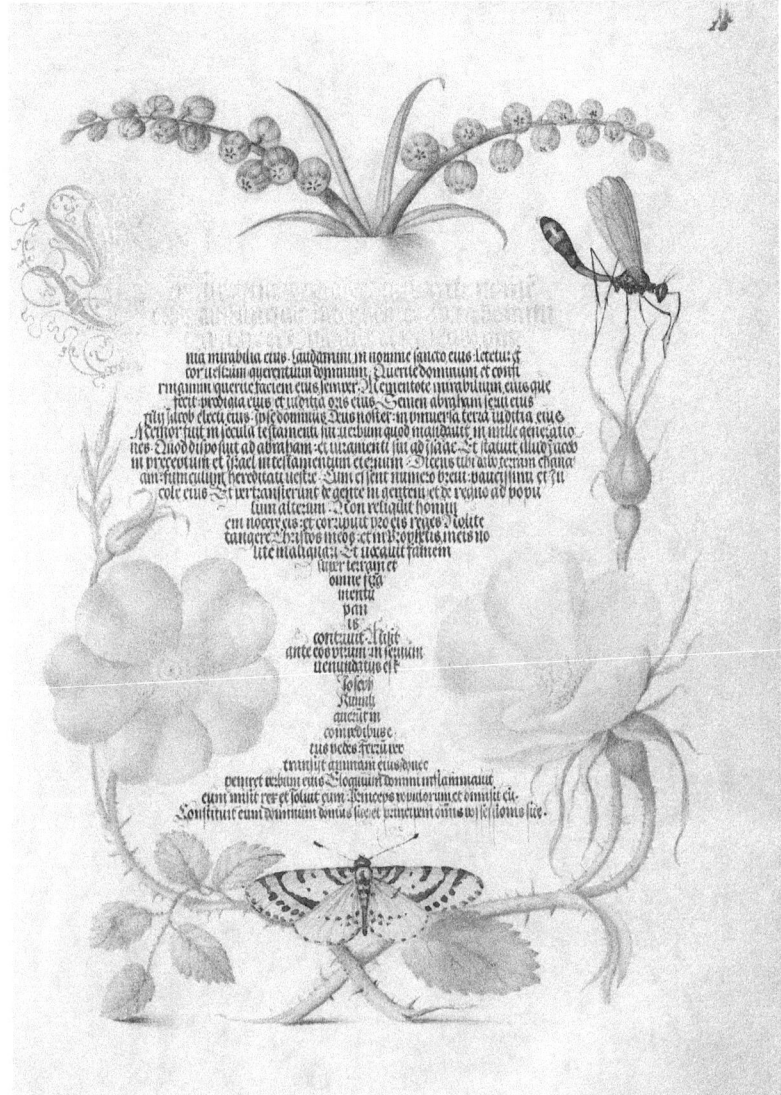

"cut letters" of the calligraphy, are a further elaboration on this visual theme.[19] The illustrations in folios 15 and 43 show how concerns shaped by the visual character of the script influenced both Hoefnagel's choice of subject matter and his method of depicting that subject matter.

For Hoefnagel, the visual motifs suggested by the calligraphy act as another screen through which to examine and organize the natural world. Screening processes such as these allowed Hoefnagel to control the potentially unlimited amount of information about insects by parsing it into small, manageable units. Hoefnagel restricted his choice of insects to those possessing colorful surfaces and unusual wing patterns, such as butterflies, moths, and dragonflies. Also favored were insects with hard, crisp edges that contrast sharply with

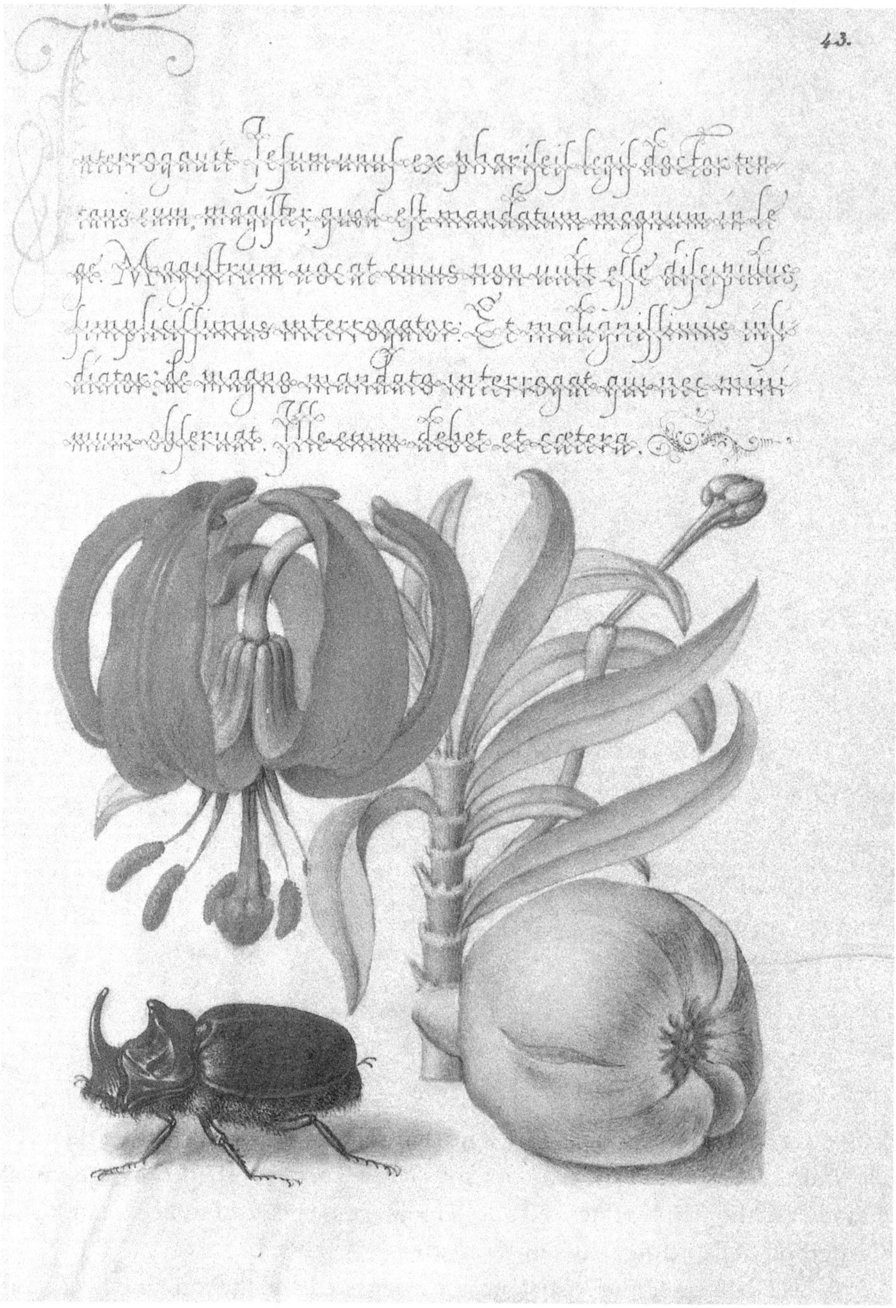

Figure 1.6. Joris Hoefnagel and Georg
Bocskay, *Scarlet Turk's Cap, Rhinoceros Beetle,
and Pomegranate,* 1561–62; illumination
added, 1591–96. The J. Paul Getty Museum,
Los Angeles, Ms. 20, fol. 43.

the smooth white background of the page, such as beetles. Presumably many other insect forms such as cockroaches, earwigs, bed bugs, and various types of insect larvae were readily available for Hoefnagel to study, yet these less visually or aesthetically appealing insects do not appear in the pages of the Model Book or in other works by the artist. Placing such limitations on the insect world allowed Hoefnagel to explore a wide variety of forms within particular categories of insects. Rather than acting as a limitation on the artist's ingenuity, they allowed him to exercise a great deal of artistic freedom and creativity, both within individual compositions, as seen in the folios from the Model Book discussed above, and within the construction of the content of the insect categories.

Hoefnagel's active engagement with screening the natural world is also evident in another set of insect groupings he devised for the Model Book: the real and the imaginary. Forty-three of the insects pictured in the Model Book are imaginary creatures, deliberate fabrications on the part of the artist that were designed to deceive an unknowing audience. These imaginary insects are rendered with the same exacting attention to detail as are the other insects appearing in the Model Book. Drawn with exceeding care, their richly hued bodies gleam and glisten, their wings are delicate and airy, and their stances and attitudes are lively and inquisitive. To the uninitiated observer, the imaginary insects are entirely indistinguishable from insects occurring in nature. In contrast to the fabricated insects, none of the plants Hoefnagel depicts on the pages of the Model Book are fictional.[20] In constructing the imaginary insects Hoefnagel freely arranged parts derived from real insects. For example, the creature pictured in the upper-right corner of folio 118 (Figure 1.7) displays characteristics of two different butterfly families, *Lycaenidae* and *Satyridae*. In folio 15 (Figure 1.5), the insect whose graceful form echoes that of the initial is not a wasp but instead has been identified by modern scholars as an "imaginary wasplike insect."[21] Many other insects in the Model Book cannot be identified at all and are simply designated as "unidentifiable"; other insects resemble known species but are shown in impossible positions or with the wrong number of body parts, such as those described as demonstrating "inaccurately shown copulation."[22] Hoefnagel's "imaginary," "unidentifiable," and "inaccurate" insects coexist with his identifiable butterflies, moths, beetles, and dragonflies, and after taking the fictional and borderline creatures into account, the number of "real" insects in the Model Book is relatively low. Butterflies and moths remain the dominant type, with twelve real examples out of twenty-two total. Only one of the twenty-three caterpillars in the Model Book can be positively identified, though several resemble known species. Of the ten beetles shown in the manuscript, two are unidentifiable and four are inaccurate or questionable; among the twelve dragonflies and damselflies three are imaginary, two are unidentifiable or questionable, and two are inaccurate.

The imaginary and unidentifiable insects do not, however, indicate weak powers of observation or flawed artistic skills on Hoefnagel's part. Like the spaces between the segments of Dürer's *Stag Beetle*, they are examples of the artist's conceptualization of the natural world as a series of autonomous but intricately related parts. In separating individual insect parts from one another and arranging them into new forms, Hoefnagel transformed insects into precious objects akin to those displayed and exchanged in cabinets and kunstkammers. Although Hoefnagel's imaginary insects succeed in convincing the unsuspecting viewer that they are exact transcriptions of nature, these illusory fantasies are better understood as examples of an early modern European artist's active role in constructing nature through the creation of visual images. As the keeper of the secret that these seemingly real insects were his own invention, Hoefnagel positioned himself as a master whose artistic skills and intellect commanded the respect of his audience. Imaginary insects offered an additional layer of appreciation and enjoyment to those viewers who were in the know, and those who were let in on the joke would have been few in number. In this way, the imaginary insects can be thought of as a visual corollary to the kind of complex entertainments offered by other sixteenth-century court figures. The earlier court artist Giuseppe Arcimboldo, for example, painted delightful composite heads that contained many levels of meaning for selected viewers.[23] With both Arcimboldo and Hoefnagel, the artist held the key to these additional levels of meaning. Viewers who did not know what they were looking at risked seeming foolish to those in the know, and Hoefnagel thus enhanced his own status by manipulating his insects, his viewers, and their experience.

The Imaginary and the Real as Modes of Representation

In addition to offering sophisticated and witty entertainment, Hoefnagel's imaginary insects offer the viewer an extended commentary on the ambiguous relationship between representation and reality as it pertains to images of the natural world. Imaginary insects also constituted a link between the images in the first section of the manuscript—the folios that comprise the Model Book—and the images in the second section, a Constructed Alphabet featuring each letter on a single folio in both Roman upper-case and Gothic lower-case script. As with the Model Book section of the manuscript, Rudolf II commissioned Hoefnagel to create illustrations for the Constructed Alphabet, and the two sections were most likely bound together by a joint decision of the artist and the emperor.[24] Hoefnagel employed two distinct but interrelated modes of representation for the two different sections of the manuscript. While he worked in a strictly naturalistic mode in the Model Book, for the Constructed Alphabet he did not seek to convince the viewer that the depicted objects were "real." Instead, the illustrations in this section employ a "fantastic" mode. Lighthearted

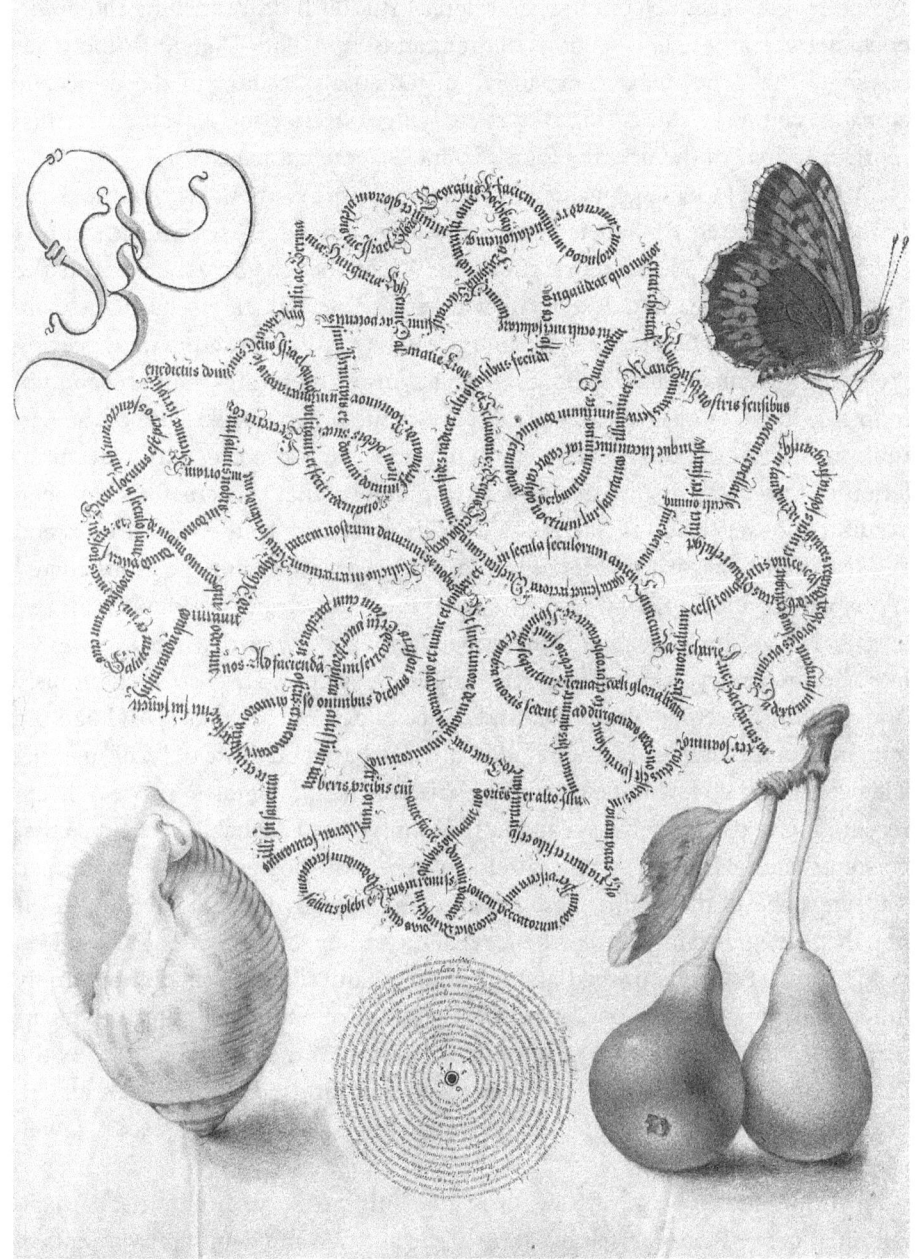

Figure 17. Joris Hoefnagel and Georg
Bócskay, *Butterfly, Marine Mollusk, and Pear,*
1561–62; illumination added, 1591–96.
The J. Paul Getty Museum, Los Angeles,
Ms. 20, fol. 118.

and humorous, they consist of ornamental figures, hybrid creatures, and assorted grotesqueries such as vases, incense burners, scrollwork, and masks, with many symbolic references celebrating the reign of Rudolf II. In describing the differences between these two sections of the manuscript, Thea Vignau-Wilberg has observed that "the relaxed, expansive ornamental program of the miniscule alphabet comes as something of a revivifying jolt after the extremely refined representations of the first 129 folia of *Mira calligraphiae monumenta.*"[25]

Although Hoefnagel included some naturalistic elements within the pages of the Constructed Alphabet, these are always embedded within the context of the fantastic mode of representation. While the imaginary insects in the Model Book are meant to deceive the viewer, the naturalistic insects in the Constructed Alphabet function as decorative fantasy with no pretext of appearing "real." For example, the imaginary wasplike insect teetering on the rosebud of folio 15 convinces a viewer of its reality by virtue of Hoefnagel's use of the naturalistic mode—the careful delineation of its delicate structures and an accumulation of fine detail assure an unsuspecting viewer that this creature has been accurately observed by the artist, as does its proximity to the real flowers and the real moth that it accompanies. Similar wasps appear in the Constructed Alphabet, as in folio 142v (Figure 1.8), but these are clearly fictional creatures, as signaled by their beaklike facial appendages and stylized wings. However, like the imaginary wasp of folio 15 (Figure 1.5) they are also meticulously drawn, with glistening abdomens and delicate legs and antennae. And like the imaginary wasp of folio 15, their bodies are also based on those of "real" insects. This overt and covert intermixture of naturalistic and fantastic modes of representation in the Model Book and in the Constructed Alphabet remind us that concepts such as accuracy and careful observation possess highly contextual meanings within this realm. The message conveyed by Hoefnagel's imagery is not that viewers cannot trust their eyes but rather that knowledge based on vision must always be probed and questioned, and that images do not simply function as transparent conduits of observations of nature, but are active participants in the construction of a particular vision of the natural world. As the creator of this world, Hoefnagel controlled and directed the viewer's experience toward those aspects that reveal nature to be a collection of exotic, jewel-like treasures.

Although the wasp of folio 15 is a fanciful product of the artist's imagination, it nevertheless perfectly captures the attitude and dainty movements of a "real" wasp. In fact, compared to its counterpart on the same page—a magpie moth, one of the few real insects in the manuscript—the imaginary wasp is imbued with life and vitality. The moth has more in common with its artistic ancestors, the butterflies and moths of the border illustration of the Book of Hours of Catherine of Cleves (Figure 1.3), in particular the specimen pictured in the center of the lower edge of the page. Both moths are shown flat against

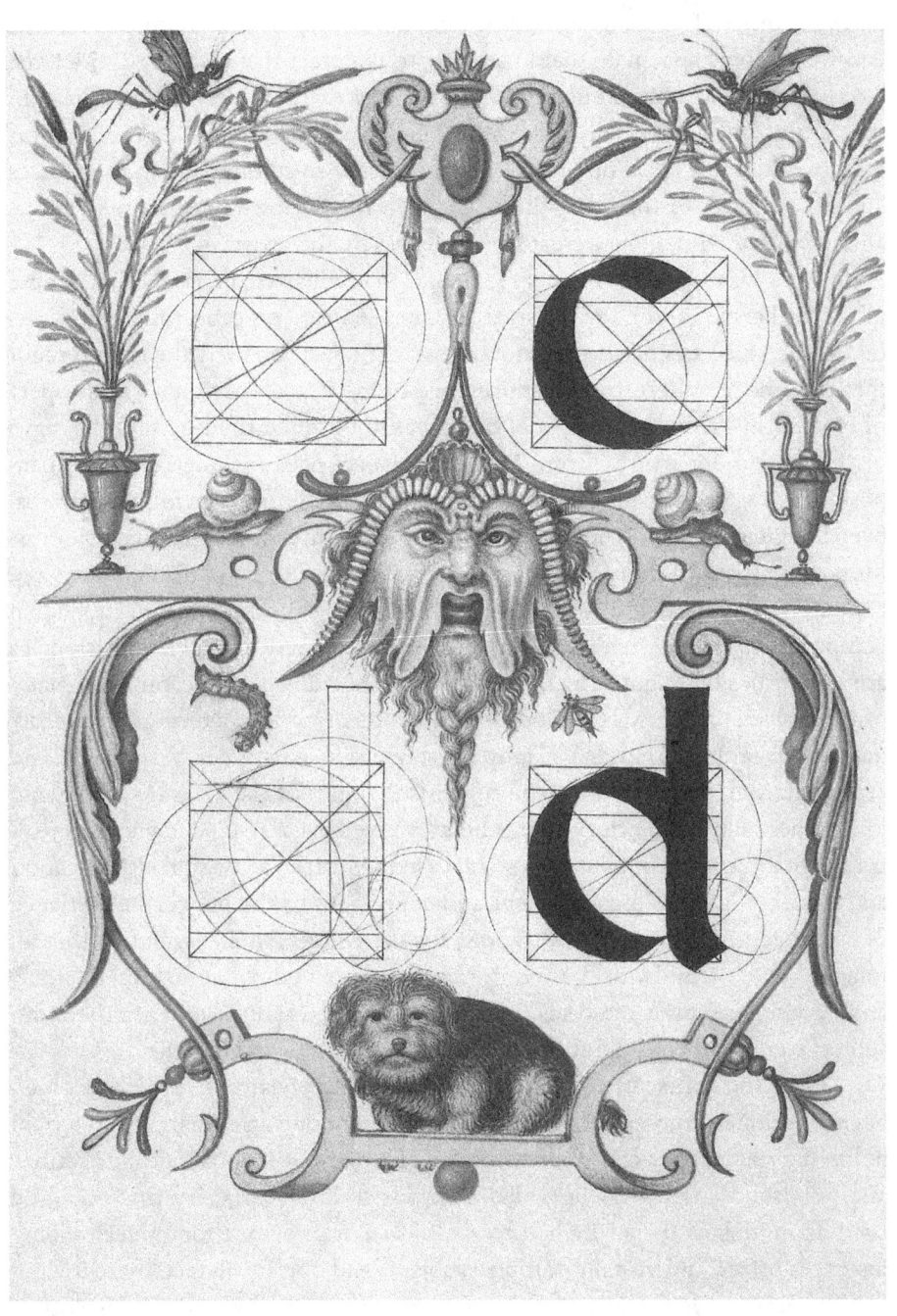

Figure 1.8. Joris Hoefnagel, *Guide for Constructing the Letters c and d,* c. 1591–96. The J. Paul Getty Museum, Los Angeles, Ms. 20, fol. 142v.

the surface of the page, framed by the tendrils of the flowering plant that encircles the border. The magpie moth is one of Hoefnagel's less animated insects, and we can only speculate that in seeking to illustrate this real insect, the artist faced the same problem encountered by many early modern investigators of the natural world, especially those who sought to study insects: in order to thoroughly examine a specimen it was first necessary to subdue and in most cases kill the creature. In this case, the real specimen that exists in nature is not as animated as the imaginary creature fabricated by the artist.

With his imaginary insects Hoefnagel was able to exercise a great degree of artistic license, and he was also able to concentrate on capturing the character of living insects since he was not necessarily constrained by the requirement that the image adhere to the appearance of an actual specimen. This aspect of Hoefnagel's work in the Model Book was not limited to insects. Other signs of the artist's departures from "natural" appearances include the odd juxtapositions of scale that sometimes occur in the Model Book, as in folio 15. Wasps, even imaginary ones, do not usually reach sizes comparable to full rose blooms. Moreover, spatial relationships are often not continuous, with collisions between areas of two- and three-dimensional space occurring within individual compositions.[26] Lee Hendrix has pointed out that these spatial inconsistencies are sometimes highlighted by the oddly shaped shadows that accompany many of the objects in the Model Book.[27] In manipulating and altering supposedly natural appearances, Hoefnagel amplifies the distance between observation and representation, and in doing so he suggests that part of the pleasure of viewing images lies in knowing that what is being viewed is *not* real. The viewer's realization that the wasp of folio 15 is not a wasp but instead a "wasplike" fabrication adds another layer of astonishment and enjoyment to the viewer's experience of the image. Hoefnagel's illustrations for the Model Book remind us that all images of the natural world, even those that are exceedingly exact, meticulous, and precise, are interpretations and translations of "nature" and are therefore subject to myriad considerations of context, audience, and function.

Over the course of the sixteenth century, European artists such as Hoefnagel developed conventions for the visual representation of insects that would be used by artists for centuries to come. Hoefnagel's manuscript illustrations and published engravings provided a corpus of insect imagery that could be used as models and that helped to establish a framework for understanding insects as prized and visually complex subjects and objects. In fact, these images were crucial tools for artists and other practitioners engaged in developing the emerging concept of "the insect world." Image makers such as Hoefnagel devised screening processes that allowed them to parse the insect world into small units that could be easily comprehended and reproduced. Such activities suggest that image makers in the sixteenth century were engaged in efforts to limit the scope of the natural world similar to those that Foucault sees operating in eighteenth-century natural history texts. Hoefnagel's insect imagery in the Model Book was shaped by the image-making traditions of late medieval manuscript illumination and sixteenth-century nature painting, and his choices of objects were in part a response to the visual qualities of Bocskay's script. These choices were also shaped by the artist's particular skills and his efforts to parse the natural world into a manageable and finite number of forms, while at the same time establishing himself as the master of its content and appearance.

Convention and Unconvention

It is important to remember that Hoefnagel's screening processes were developed and directed toward a specific audience in a particular time and place, and that the artistic persona he cultivated was also dependent upon and directed toward this elite courtly environment. The *Mira calligraphiae monumenta* manuscript was created for the delectation of two emperors, and the illustrations supplied by Hoefnagel were meant to appeal to the sophisticated tastes of Rudolf II. The screening methods Hoefnagel employed to organize and describe nature within the pages of the manuscript were developed in the context of the Rudolfine *kunstkammer,* and reflect the passion for rare, precious, and finely wrought objects made of costly and exotic materials that governed the formation of such collections. However, this approach to the insect world was not appropriate for all situations and all contexts, as can be seen in two examples by artists engaged in the representation of insects from the New World. One of these, a drawing of fireflies and a fly from circa 1585 (Figure 1.9), was completed by John White, an English artist who served as the governor of the ill-fated Roanoke settlement in North Carolina. White's drawings of the flora, fauna, and indigenous inhabitants of North America would eventually form the basis of the popular series of engravings *America,* which was published in Frankfurt by Theodor de Bry in 1590. The drawing shows White's familiarity with European conventions for the illustration of insects: three fireflies—or three views of one firefly—are presented in an elegant symmetrical arrangement, each lying flat against the surface of the page with the center specimen opened up to display its wings in the manner of Hoefnagel's open-wing stag beetle. Another insect, a type of fly, occupies the lower part of the drawing.[28]

Although White has masterfully depicted the shape and structure of the insects' bodies and wings, the image does not convey the characteristics of these insects that were most important to European audiences, including White himself. The firefly fascinated Europeans who traveled to the New World because of its ability to produce light, and although this phenomenon could be seen by White it could not be represented visually using the conventions of specimen logic. Instead White communicated this information through the accompanying inscription: "A flye which in the night semeth a flame of fyer." Similarly,

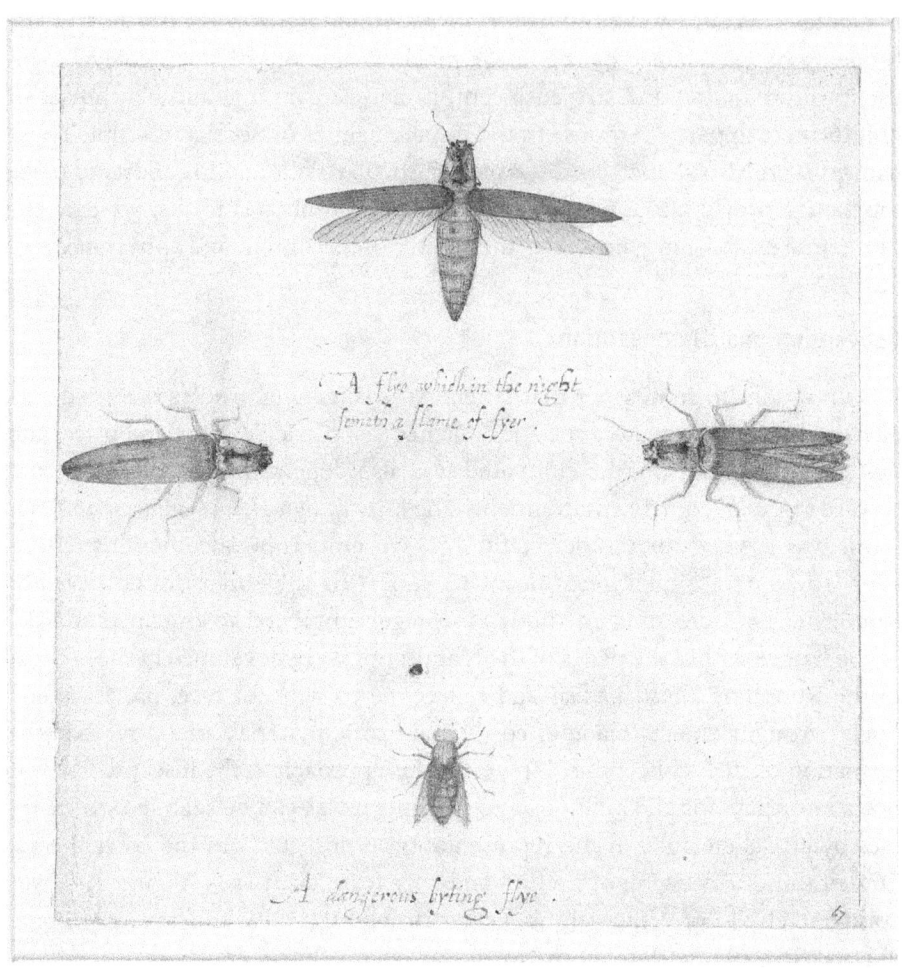

"A flye which in the night semeth a flame of fyer"

"A dangerous byting flye"

Figure 1.9. John White, *Studies of Fireflies and a Gadfly,* c. 1585. Copyright Trustees of the British Museum.

important information about the other insect shown in White's drawing can only be gleaned by reading the text, which describes it as a "dangerous byting flye." Compare White's representation of the biting fly to the illustration of mosquitoes found in the volume titled *Histoire naturelle des Indes,* also known as the Drake Manuscript (Figure 1.10).[29] The Drake Manuscript contains watercolor drawings and written descriptions of West Indian plants, animals, and peoples, and it is believed to have been composed by one or more sixteenth-century French artists and scribes who accompanied Francis Drake on one of his voyages of the 1580s.[30] The illustrations in this manuscript show that the anonymous artists were familiar with European conventions for representing insects and plants, since many of these appear as isolated objects against the

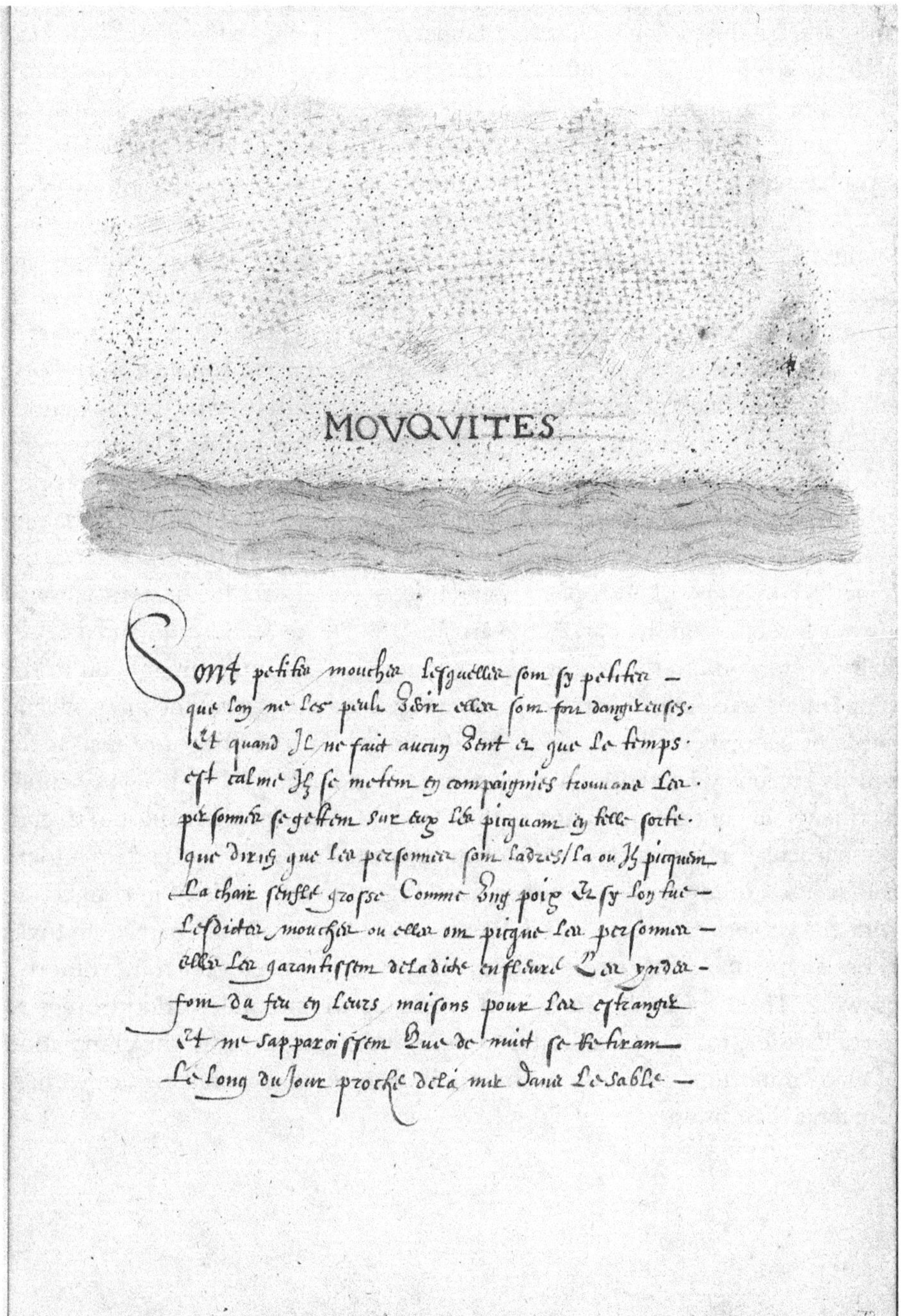

MOVQVITES

Sont petites mouches lesquelles som sy petites —
que lon ne les peult veoir elles som fort dangereuses.
Et quand Il ne fait aucun vent et que le temps
est calme Ilz se mettent en compaignies trouuans leur
personne se gettent sur eulx les picquam en telle sorte
que dirois que les personnes sont ladres/la ou Ilz picquent
La chair senfle grosse Comme vng poig Ie sy soit tue —
Lesdictes mouches ou elles om picque les personnes
elles les garantissem delad dicte enfleure. Et se vndra —
fom da feu en leurs maisons pour les estrangler
Et ne sapparoissem Que de nuict se retiram —
Le long du Jour proche dela mer dans le sable —

Figure 1.10. Anonymous artist,
"Mouquites," *Histoire naturelle des Indes,*
c. 1596, folio 72 recto. The Pierpont Morgan
Library, New York. Bequest; Clara S. Peck;
1983. MA 3900.

blank surface of the page. Unlike White or Hoefnagel, however, the artists responsible for illustrating the Drake Manuscript had only a rudimentary mastery of these techniques. As an illustration of the form and structure of a mosquito's body, the drawing labeled "Mouquites" can only be considered a dismal failure.[31] Rather than presenting a single specimen for the viewer's inspection, the artist has represented a swarm of mosquitoes as a mass of undifferentiated dots of ink. But unlike White's drawing of the biting fly, the drawing in the Drake Manuscript conveys crucial information about the group behavior and appearance of the insect, as well as its habitat. Such information would have been extremely useful to a person hoping to avoid a potentially dangerous swarm of biting insects, as would be the additional information supplied in the text regarding the effects of the insects' bite and the methods used by the indigenous inhabitants to prevent the insects from entering their homes. The screening processes that allowed White to represent the firefly and the biting fly as precious objects removed from their contexts also prevented him from making visible other aspects of their appearance and behavior. On the other hand, the imperfect mastery of European conventions for illustrating insects allowed greater flexibility on the part of the artist of the Drake Manuscript and a better ability to respond to the representational needs of the situation at hand.[32] The anonymous artist created an image of mosquitoes that suited the interests and needs of his or her audience, just as Hoefnagel created images of insects for courtly patrons who expected to be surprised and delighted by his witty entertainments. In contrast, the artist of the Drake Manuscript was expected to convey particular information for a patron interested more in the settlement and colonization of foreign lands rather than entertainment, at least in connection with the subject of insects. While Drake was most likely the patron and was himself a member of elite court circles, this was not the relevant context for this drawing. These examples show that screening the natural world in order to make it visible for early modern European audiences involved the elimination of many important, interesting, and useful features and characteristics. Limits, too, have their limits.

TWO

CUTTING AND PASTING NATURE INTO PRINT

Ulisse Aldrovandi's and Thomas Moffet's Images of Insects

*D*uring the last quarter of the sixteenth century, insects were a rich subject around which the many and varied interests of naturalists, physicians, artists, and other practitioners coalesced. It was during these years that two naturalists, working separately, began assembling the materials that would lead to two major publications on insects. Ulisse Aldrovandi's *De animalibus insectis* of 1602 and Thomas Moffet's *Insectorum sive minimorum animalium theatrum,*[1] published posthumously in 1634, were the first illustrated books on insects to appear in Europe. Although the published versions of these works were separated by a period of over thirty years, the bulk of the preparatory work for both was conducted during the 1580s and 1590s. Aldrovandi and Moffet did not work together, and they may not even have been aware of each other's efforts, but they shared similar educational and professional backgrounds and developed parallel approaches to organizing information, specimens, and images relating to insects. Before they turned their attention to the subject of insects, Thomas Moffet and Ulisse Aldrovandi worked and trained as physicians and were actively involved in the dramatic transformations taking place in medicine during the second half of the sixteenth century. These medical activities would have a strong influence on the way that each of them approached their later projects on insects. This chapter explores the reasons why insects were of interest to Aldrovandi and Moffet, along with others in their circle, while also examining the methods and techniques they developed for understanding and conceptualizing this new subject.

In the same way that Joris Hoefnagel's manuscript illuminations helped to define the parameters of the insect world for the artists in his circle and their courtly patrons, Moffet's and Aldrovandi's books also played key roles in constituting this new field of inquiry. Insects were collectable objects for Moffet and Aldrovandi and were certainly appreciated for their visual and aesthetic appeal, but for them the interest in insects also stemmed from their engagement with the material practices of medicine, museums, and natural history. As I will show in this chapter, one of the ways that early modern European naturalists such as Aldrovandi and Moffet rendered nature comprehensible to themselves and to others was by creating and circulating images. For Aldrovandi, Moffet, and their associates, images and collections of images came to function as virtual specimens and virtual specimen cabinets. In addition, these artists and naturalists bridged the temporal and geographic expanses that separated them by exchanging images and specimens, and through these activities they formed virtual communities devoted to studying insects. In this chapter I argue that an important strategy devised by these naturalists was "cutting and pasting," a technique they used to construct a vision of the insect world that could fit the confines of the printed page. I examine the original drawings and printed illustrations of *De animalibus insectis* and *Theatrum insectorum* and the circumstances of their production in order to demonstrate the importance of image making for naturalists studying insects during the later sixteenth century and to show how images and image-making practices played a central role in the construction of nature in early modern Europe.

Both Ulisse Aldrovandi and Thomas Moffet used insects as a subject around which they constructed their personae as trusted organizers and distillers of the information they collected about the natural world. As each of them struggled to put their versions of the insect world into print, they grappled with the emerging issue of priority of publication. In sensing that this new topic would be of great interest to naturalists and other audiences, each also seems to have felt an urgency to be the first to publish an illustrated book on insects. Thus, the new subject matter offered the potential for recognition as the preeminent authority on insects, but it also carried with it the risk of coming in second. Moffet faced the additional challenge of his book having been the product of communal research, thereby making it impossible for him to present himself as the sole source of information on insects. Although this did not prevent him from making the case for himself and his original contributions, his arguments centered on the financial burden he took on with the project and his organizational efforts.

Medical Practice and Natural History in the Late Sixteenth Century

In Italy during the last half of the sixteenth century, professional medical education at universities increasingly emphasized training in *materia medica,* the preparation of medical remedies. Based on the works of the first-century medical
writer Dioscorides, *materia medica* involved the study of plants, as well as animals and minerals, to determine their healing properties. Although the works
of Dioscorides had been in continuous use by Europeans prior to the sixteenth
century, the renewed interest in Dioscorides during the early modern period
was part of the broader humanist effort to recover and study ancient texts and
learning. Ulisse Aldrovandi was immersed in the humanist movement through
his studies of law, philosophy, and mathematics at the University of Bologna,
and in 1551 he developed a friendship with the renowned Pisan physician,
teacher, and botanist Luca Ghini, under whose influence he became interested
in medicine and botany. Aldrovandi received a degree in medicine from the
University of Bologna in 1551, and in 1554 he began teaching medicine at the
university. There he gave lectures on Dioscorides and other ancient authors
and worked to incorporate into the curriculum the teachings of Galen on pharmacy. In 1561 Aldrovandi was appointed professor of natural history, at which
time he gave a new series of lectures on *materia medica,* and in 1568 a botanical
garden was established in Bologna at his suggestion. Aldrovandi used both the
botanical garden and his natural history museum as sites for training medical
students in the preparation of medicinal simples, plant-based medical preparations. While university-trained physicians in the early sixteenth century could
limit themselves to theoretical learning and leave the preparation of medicinal
simples to apothecaries and other medical practitioners, by the late sixteenth
century this was no longer the case. Aldrovandi's teaching and other activities
were aimed at practical hands-on training. As Paula Findlen has shown, the close
connection between medicine and natural history in Aldrovandi's practice came
about due to increased expectations that physicians be trained in practical applications based on firsthand observation: "Like many collectors of nature, Aldrovandi perceived his primary goal to be the reform of *materia medica,* the formal
study of nature as a medical entity prescribed by Galen, Dioscorides, and medieval commentators such as Avicenna. The appearance of museums of natural
history in Italy was closely tied to the transformation of the medical profession."[2]

Thomas Moffet was also deeply concerned with the reform of medicine,
but his focus was very different from Aldrovandi's concentration on *materia
medica.* Moffet was an adherent of Paracelsian chemical medicine, to which
he was first introduced while studying medicine at the University of Basel with
Felix Plater and Theodor Zwinger in 1578. In the early part of the sixteenth century the medical reformer, alchemist, and mystic Paracelsus (1493–1541) had
argued vehemently for incorporating new ideas and theories into the practice

of medicine. While lecturing at the University of Basel in 1527, Paracelsus put forward the controversial idea that his own experience in medicine was superior to those of the ancients. Paracelsus also emphasized chemistry as the key to understanding both nature and medicine.[3] Chemistry had the potential to reveal hidden secrets of nature by means of a system of correspondence and analogy. As Allen Debus notes, "Primary observations in nature would lead directly to further observations in the chemical laboratory or to analogies based upon them."[4] Paracelsus's ideas were far ranging and often contradictory, and this led to his followers developing a wide variety of opinions about his work in the last quarter of the sixteenth century, when his theories became the subject of great debate among physicians. Because his followers had many versions of Paracelsian theory from which to choose, they were not necessarily bound to accept all of his ideas. Moffet, for example, limited his Paracelsian practice to using chemistry for medicinal applications, and he does not seem to have extended these studies to an interest in alchemical transmutation, one of the most popular aspects of Paracelsus's ideas. In 1584 Moffet published a book promoting chemical medicine, *De jure et praestantia chemicorum medicamentorum*. Structured as a dialogue between two physicians, Moffet's tract reflected his understanding of the Paracelsian principles of salt, sulfur, and mercury, but he argued that these principles should be learned through experiment rather than accepting them based on blind belief.[5]

Moffet was a strong advocate of gaining knowledge though experiment and observation, but he did not urge a rejection of ancient authority, as did some Paracelsians. Taking such a stance would have represented a challenge to medical authority, and there were heated arguments and controversies within some parts of the medical community regarding these topics, notably in France. In England, however, ideas about chemical medicine did not result in such polemical debates, and Moffet's approach was typical of attitudes toward Paracelsian chemical medicine in England. For the most part, English practitioners had adopted a compromise position by the early 1600s by accepting and incorporating new chemical remedies that had been proven to be effective, but few, as Debus states, "concerned themselves with the deeper and more occult aspects of Paracelsian thought."[6] Problems with chemical medicine arose when there were concerns or accusations that they were being prepared or prescribed by those who were not qualified to do so. Chemistry had thus gained a bad reputation, especially among some members of the College of Physicians, an organization with which Moffet had an uneasy relationship. Chemical medicine was, however, sufficiently well respected by members of the College that it was considered appropriate for inclusion in their major publishing project of the 1580s, a pharmacopoeia that was intended to provide apothecaries and physicians in England with a standard reference work on the preparation of medicines. One section of this proposed book was to be devoted to chemical medicines and supervised by a committee of three physicians that included Thomas Moffet. It has been argued that Moffet would have been the person who introduced the idea of including chemical medicine into the pharmacopoeia project, since he was the only member of the College at that time who had received his medical education on the Continent and who possessed an understanding of chemical theory as it pertained to medicine.[7]

The works of Paracelsus and his followers were not the only sources on chemical medicine for physicians and other medical practitioners in England during this period. Numerous books on the art of distillation appeared in the sixteenth century, among them Conrad Gessner's *Euonymus,* published posthumously in 1569 and appearing in an English translation in 1576 as *The newe jewell of health.* Gessner's book contained recipes and remedies for numerous ailments, many of them "borrowed out of a written book," as in the case of a recipe for "oyle of scorpions distilled," as well as that for "an Oyle of antes egges, and the Nettle distylled together, with which the kidneys and bladder anoynted, provoketh speedily urine, this borrowed out of Leonellus."[8] Unlike Paracelsus, Gessner did not reject the learning of the ancients but instead sought to integrate chemistry with ancient knowledge, and this approach was in keeping with English attitudes toward chemical medicine. Although Moffet was an advocate of Paracelsian chemical medicine, he admired Gessner's approach and his works. In *De jure et praestantia chemicorum medicamentorum* the character Chemista gives a long list of persons who used chemical remedies, with special mention to "that upright Gessner of blessed memory."[9] Moffet followed the format of Gessner's book in his own book, *Health's Improvement,* published posthumously in 1655. *Health's Improvement* is a compilation of information on nutrition and diet, and it borrows directly from Gessner and numerous other writers. The book also includes an extensive list of English birds and fish and a discussion of their culinary uses and properties, but in its coverage of over one hundred birds *Health's Improvement* is also recognized as an important work of natural history.[10]

Moffet's other major contribution to natural history was the *Theatrum insectorum,* a project that also originated with Gessner, as I will discuss in further detail below. Gessner's published and unpublished materials thus provided Moffet with a model for integrating ancient knowledge with modern research and methods, and for both Gessner and Moffet—and for Aldrovandi—medicine and natural history were closely interrelated through a shared set of practices, texts, and ideas. For these naturalists, physical engagement with the materials of medicine and natural history were central to the ways that they formed their knowledge of the natural world. In devising techniques and methods for studying the new topic of insects, both Aldrovandi and Moffet drew from their training in medicine, specifically the traditions of compiling recipes, remedies, historical information, and other data from a wide range of published and unpublished sources. Insects proved to be well suited to the compiler's impulse because they were small and durable enough to circulate within the internal and external networks that formed the core of these practitioners' intellectual communities and working methods. In addition, both Aldrovandi and Moffet advocated for reform in medical education and were eager to pursue innovative approaches to the teaching and practice of medicine. This willingness to take on new topics must have contributed to their interest in seeking out the relatively unknown world of insects. Both figures must also have possessed a strong mixture of curiosity and ambition to sustain them through the many challenges and difficulties they faced in bringing their research on insects into print.

Ulisse Aldrovandi: Cutting and Pasting the Virtual Specimen Cabinet

Ulisse Aldrovandi's private museum in Bologna was known and admired by scholars throughout sixteenth-century Europe.[11] Naturalists, noblemen, and other distinguished visitors traveled to his museum to view his renowned collection of plants, animals, fossils, books, manuscripts, and other natural and artificial objects. Aldrovandi's collecting was encyclopedic in its scope, aiming to amass and display all of nature in one space. Findlen argues that this was a common imperative among sixteenth-century collectors, who thought of nature as something that could be owned and displayed alongside other objects: "Within the wider matrix of collecting, the possession of nature figured prominently. Along with art, antiquities and exotica, nature was deemed a desirable object to own . . . From the imaginary to the exotic to the ordinary, the museum was designed to represent nature as a continuum."[12] Visual images were an important component of Aldrovandi's museum, and a significant amount of his energies were directed toward amassing a large collection of drawings of natural subjects. Many of these drawings were made by artists hired by Aldrovandi, and they worked closely with him in his museum. He also brought artists with him on botanizing and collecting trips in order to sketch plants and other objects in the field.[13] These artists were also commissioned to produce the many woodcuts Aldrovandi used to illustrate his massive natural history publication projects. Giuseppe Olmi estimates that as many as eight thousand drawings and paintings were produced in Aldrovandi's workshop during his lifetime, and illustrations continued to be added to the collection after his death.[14] More than simply documenting and preserving the contents of the museum, the drawings functioned as dynamic working documents and were important tools by which Aldrovandi organized and shaped the vision of the natural world he aimed to contain within the walls of his museum.

Most of the original drawings for *De animalibus insectis* are gathered in Tomo VII of the *Tavole di animali,* the seven bound volumes that contain Aldrovandi's collection of drawings of animals. Tomo VII consists of 132 pages,

51 of which depict insects, while other drawings present a wide range of objects and animals, including mice, shells, fossils, fish, frogs, lizards, snakes, and coins.[15] Documentary evidence shows that Tomo VII appeared in its present form and arrangement during Aldrovandi's lifetime, as indicated by the descriptions of its contents and entries in Aldrovandi's notebooks from the 1590s.[16] Most of the insects in Tomo VII were drawn by an artist from Frankfurt named Cornelius Schwindt (1566–1632), who worked for Aldrovandi in Bologna between 1590 and 1594. Schwindt was also commissioned to cut most of the blocks for the woodcut images in De animalibus insectis as well as for De ornithologia, Aldrovandi's three-volume publication on birds published between 1599 and 1603. Little is known about Schwindt's life before and after his stay in Bologna, although he wrote to Aldrovandi at least once after leaving Italy.[17] Several different drawing styles are present in Tomo VII, and drawings seem to have been added by different artists over the course of several years, but the majority of the insect drawings are in Schwindt's hand. The number of insects depicted on each page of Tomo VII ranges from two or three per page to twenty or thirty. Schwindt drew individual insects in horizontal rows, presenting them either in profile or from above and flattened against the page, and most are life-size. Added below many insects are Latin inscriptions, usually providing the name of the general type of insect (e.g., "fly") along with descriptive information about its physical appearance (e.g., "entirely green"). The inscriptions were written by several different people, but none seem to have been written by Aldrovandi himself. One of Aldrovandi's associates, "fratre Gregorio Cappucino," is mentioned in Aldrovandi's notebooks as having written many of the inscriptions in August 1592.[18]

The presence of multiple hands in the drawings and inscriptions reflects the division of labor in Aldrovandi's workshop and the way in which the project of creating the volume of drawings progressed in gradual and overlapping stages. Tomo VII was a constantly evolving work in progress, and it was used variously as a source for models for woodcuts, a reference work, and an organizational tool. Visual images served many purposes for early modern European naturalists and botanists, and one of the most important functions of drawings was to record color, which tended to fade quickly in many plants and animals after they died. In addition to preserving information about the appearance of specimens, colored drawings were also essential for use as models in producing hand-colored editions of illustrated natural history and botanical books. Thus, images could stand in directly for specimens that were old, damaged, or entirely missing.[19] Drawings certainly functioned in this way for Aldrovandi, but in the context of his museum and publication projects, the significance of the insect drawings does not lie solely in their ability to serve as replacements for specimens. Aldrovandi's volume of insect drawings offered the naturalist and his associates a means of accessing and interacting with the collection in ways that could not be achieved through working with specimens alone. The drawings paralleled, but did not duplicate, the function of the naturalist's collection of insect specimens, and in this way the insect drawings came to serve as a virtual collection that was contiguous with the physical collection housed in Aldrovandi's museum.

One of the ways in which Aldrovandi's bound volume of insect drawings functioned as a virtual specimen cabinet was in its capacity to present multiple views of the same specimen. On page 76 of Tomo VII (Figure 2.1), for example, Schwindt has depicted several moths in multiple views. In the first row, dorsal and ventral views of a black, white, and red moth are shown, and the same technique is used to present the larger striped moth below it. Schwindt uses a different technique for the small, pale-brown butterfly shown in the third row. As with the two larger moths above, the insect is shown in two views. But here Schwindt presents the moth with two different positionings of its wings—closed and open. Although no mounting pins are visible, the view of the moth with open wings most likely depicts the appearance of a prepared specimen while the closed-wing view shows the living insect at rest. In his museum, Aldrovandi would have displayed his insect specimens in drawers or tabletops, anchored to a mounting surface with glue or pins. Displaying specimens in this manner would have allowed viewers to see only one side of any given insect, and as such presenting the back and the front of an insect or showing it with its wings open and closed could only be done in a specimen cabinet if Aldrovandi possessed more than one specimen of the same type of insect. As has been shown with the examples above, however, in a drawing it was possible to present multiple views of the same specimen.[20]

Drawings could also be used to create groupings that were not possible with specimens alone. A visual image could convey views of a specimen as it appeared at different points in time, a technique that was particularly useful for showing the stages of an insect's life cycle. On page 16 of Tomo VII, Aldrovandi and Schwindt present the life cycle of a white-and-black moth.[21] The series of images appears on the second row of the page, and it can be "read" from left to right, beginning with the caterpillar and ending with the adult moth. Two views of the pupa (or two different examples of pupae) are presented in the center of the row. This series of images actually presents the life cycles of two different species, since Aldrovandi and Schwindt have also included the eggs and adults of what appears to be a parasitic wasp to the left of the pupae. These images not only present the stages of the moth's metamorphosis, but they also hint at the dynamic and complex relationship between a parasite and its host. Such relationships could be difficult, if not impossible, to describe using specimens alone, given the difficulties of preserving caterpillars, eggs, and pupae. As these examples show, Aldrovandi was not unaware of the processes of insect metamorphosis, and he had an interest in linking these forms visually.

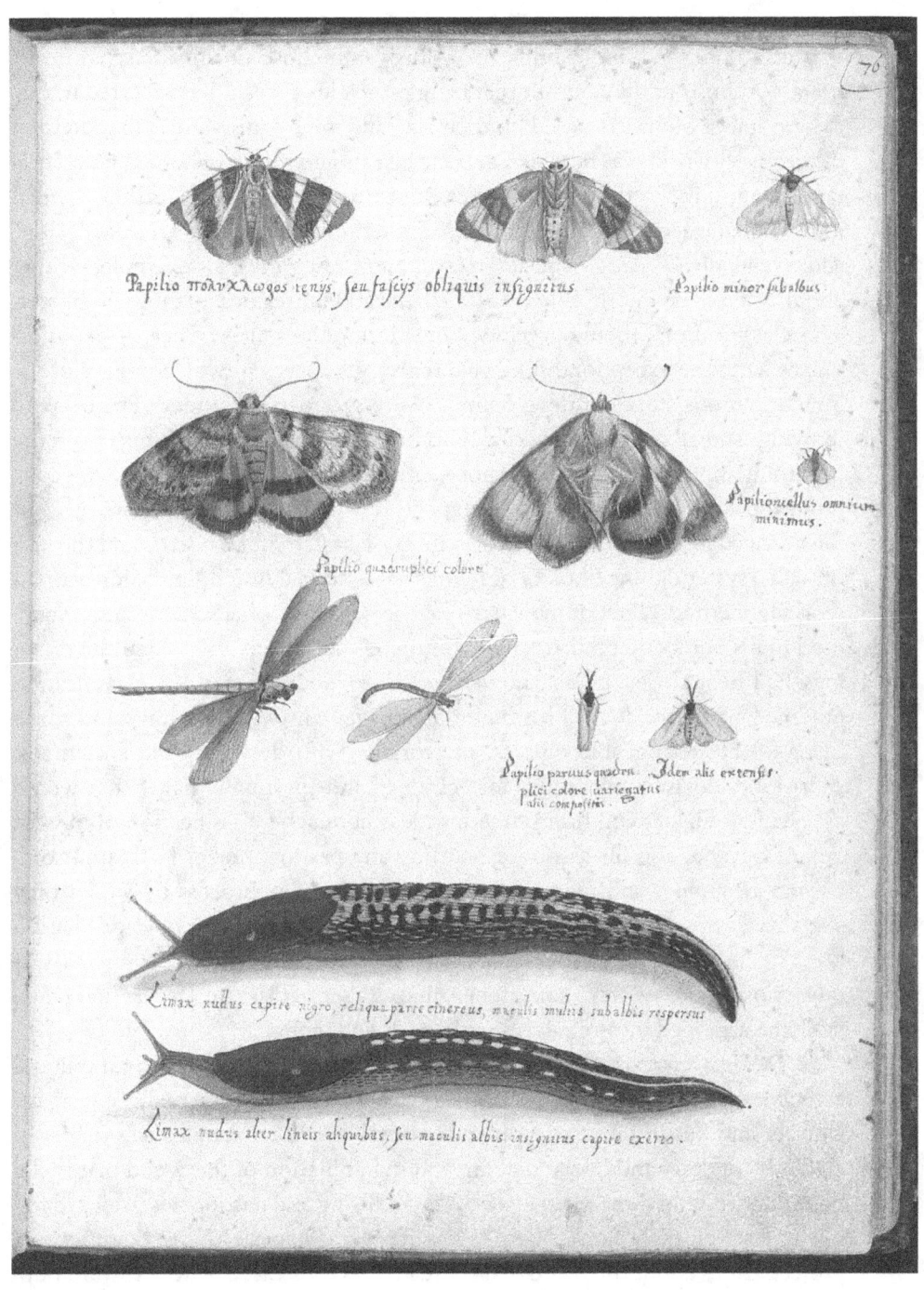

In translating his research from the form of drawings and specimens to that of a published book, however, Aldrovandi chose not to focus on insect metamorphosis as his central organizing principle. Instead, he preferred to create groupings of insects based on their size and type. Aldrovandi's first division of the insect world was between aquatic insects and terrestrial ones. Terrestrial insects outnumbered aquatic insects, and this category was further divided into winged insects (*alata*) and insects that did not have wings (*aptera*).[22] Thus, Aldrovandi divided single species into different categories; for example, writing about caterpillars in one chapter and moths and butterflies in another chapter. This classification scheme precluded illustrating the stages of insect metamorphosis, with the exception of the chapter on silkworms in which images of the insects' various stages of development were grouped together.[23] The book of drawings shows that Aldrovandi did not use this classificatory system from the beginning, since insects of different types and sizes—winged and nonwinged—often appear on the same page, as on page 76 where two slugs appear with various winged insects. The use of a classification system in *De insectis* and the lack of such a system in the book of drawings lends crucial insight into Aldrovandi's working methods. First, it indicates that the organizational scheme Aldrovandi used in his book emerged at some point after the project of making the drawings had begun. Second, it shows that Aldrovandi's system of classification emerged as the product of his material engagement with drawings and specimens as interchangeable objects. In working with drawings and specimens, Aldrovandi devised techniques for "cutting" and "pasting" images that would be essential for helping him refine his ideas about the classification of insects. For Aldrovandi and his associates, cutting and pasting images facilitated comparing, arranging, and rearranging insects, a complex process by which they visualized a world of insects through active engagement with physical materials. The insect world as they envisioned it for the purpose of *De insectis* was one whose contents were comprehensible as a set of discrete objects that could easily be arranged into regular, simple grids and tables.

Cutting and pasting was for the most part a virtual rather than a literal activity; the pages of Tomo VII are intact and were not actually cut apart. Cutting and pasting images from this virtual specimen cabinet was achieved through copying, and in the making and manipulation of the wood plates. To better understand these virtual practices of cutting and pasting it is useful to follow one set of images on its path through several stages of Aldrovandi's insect project. Clues about the route that these images took on their way toward publication are found in one of Aldrovandi's entries in his notebooks for 1593. The entry, titled "Index insectorum in tabulis depictorum Ao. 1593," consists of a list of the woodcuts of insects completed to date by Schwindt.[24] In this list, Aldrovandi noted the name of every insect depicted in each of the woodcuts as well as where the original drawing of each insect could be found in Tomo VII.[25]

The "Index insectorum" is a record of the state of Aldrovandi's progress on his insect publication in 1593, and it provides information about the appearance of the woodcuts and their corresponding drawings at this early stage of the project, at a point in time when only twenty-five of the eighty-one woodcuts had been completed. The "Index insectorum" shows that both the drawings and the woodcuts underwent substantial transformations before the final publication was issued, and that cutting and pasting played a central role in Aldrovandi and Schwindt's working practices.

Several of the woodcuts prepared at this early stage were intended for Aldrovandi's chapter on butterflies and moths and were numbered and organized into tables or *tabulae*. According to the "Index insectorum," a woodcut depicting winged insects had been completed in 1593 and was titled "Secunda tabula." Using the index entry for this woodcut, I have reconstructed its original layout (Figure 2.2). The reconstruction shows that in assembling the images for the "Secunda tabula" woodcut, Aldrovandi and Schwindt collected images from three different pages of Tomo VII: the insects in positions 1, 2, and 5 of the reconstructed woodcut were taken from page 78 (Figure 2.3); those in positions 3 and 4 come from page 77 (Figure 2.4); those in positions 6 and 7 come from page 76 (Figure 2.1). The "Secunda tabula" was therefore a product of the virtual cutting and pasting that took place as part of the task of organizing the insect world and readying it for appearance in print. A discrepancy in numbering between the "Index insectorum" and Tomo VII also shows that drawings continued to be added to the virtual specimen cabinet of Tomo VII after the "Secunda tabula" woodcut had been completed. The butterflies in row 2 of the "Secunda tabula" appear on page 77 of Tomo VII, and in the "Index insectorum" Aldrovandi lists these butterflies as occupying positions 3 and 4 on page 77. However, the butterflies do not in fact occupy positions 3 and 4, and instead are shown in positions 5 and 6. The explanation for this discrepancy is that in 1593, when the "Index insectorum" was written, page 77 of Tomo VII contained drawings of only six insects. In 1593 the butterflies in question did in fact occupy positions 3 and 4, but some time after August 1593 two drawings of butterflies were added—the two that appear in the upper-right corner of the page, occupying the current 2 and 4 positions.

The process of virtually cutting and pasting images did not end with the completion of this woodcut. Sometime after August 1593 and prior to the publication of *De insectis* in 1603, a major change was made to the "Secunda tabula." In the final, published version of *De insectis* Aldrovandi had the image cut apart, and its contents were moved—or "pasted"—into three different places in the book. The four butterflies appear in Table 6 of the chapter on butterflies and insects (Figure 2.5), and the two dragonflies were moved to a woodcut illustrating the chapter on dragonflies (Figure 2.6). The final insect from the "Secunda tabula," a lacewing that Aldrovandi calls "Papilio locustiformis," appears at the

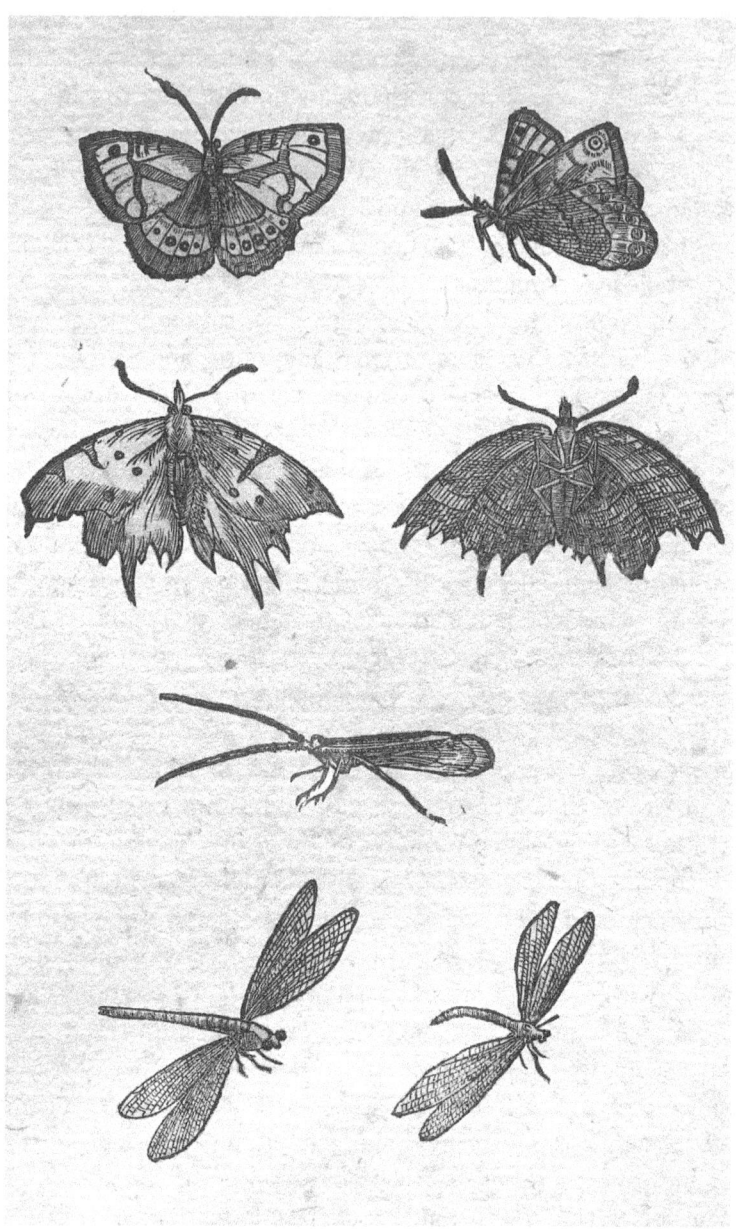

Figure 2.2. Reconstruction of original "secunda tabula" woodcut for Ulisse Aldrovandi, *De animalibus insectis* (Bologna, 1602).

end of the book in the "Paralipomena" (Figure 2.7), a section that served as a catchall for insects not fitting into the other categories Aldrovandi created to organize his book. In this instance, cutting was a literal rather than a virtual activity, since the woodblock upon which the "Secunda tabula" images were carved was actually broken apart in order to adapt to the developing organizational scheme of the book.

The cutting and pasting that occurred with this woodblock is an example of the active engagement with materials that was central to the development of

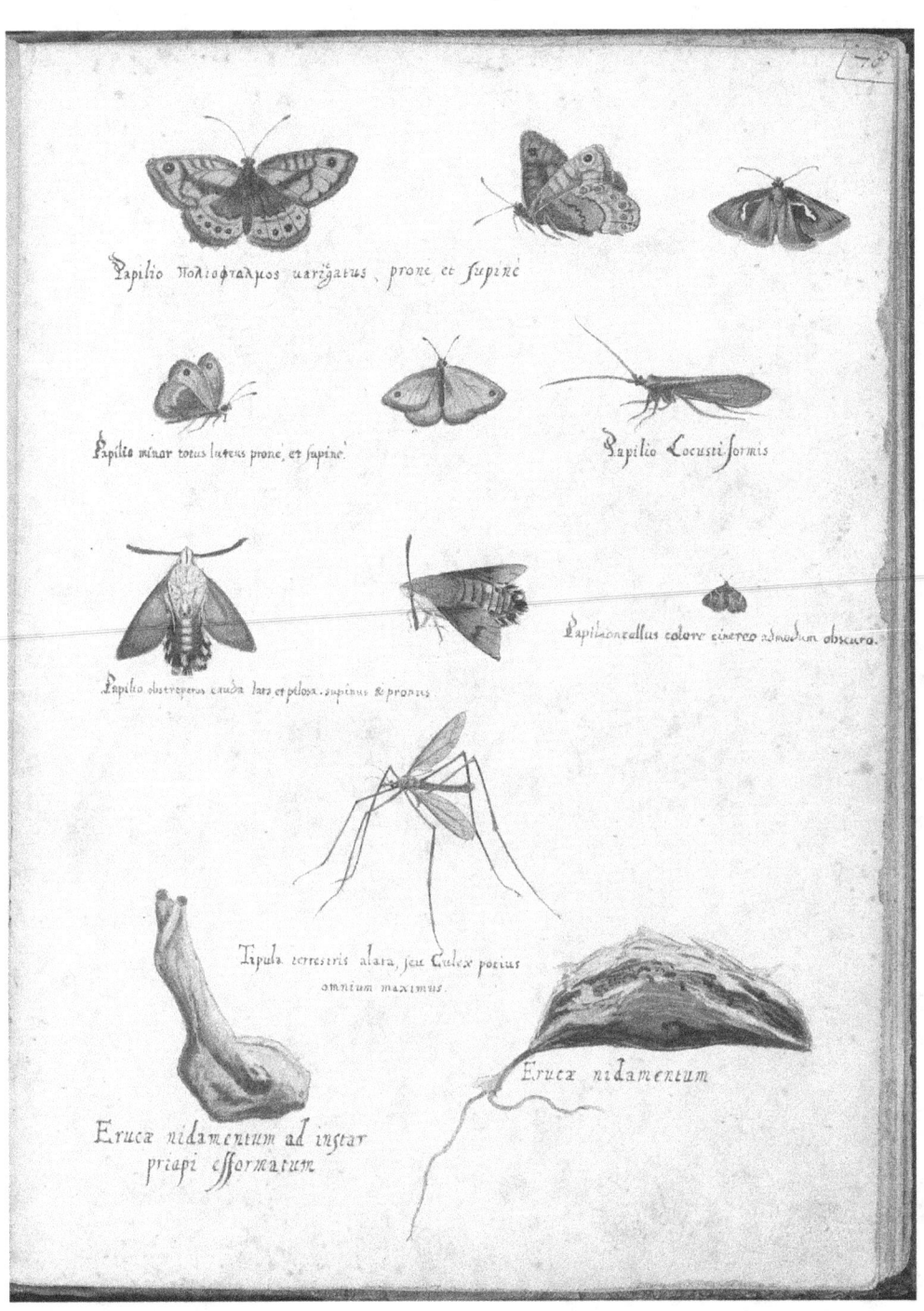

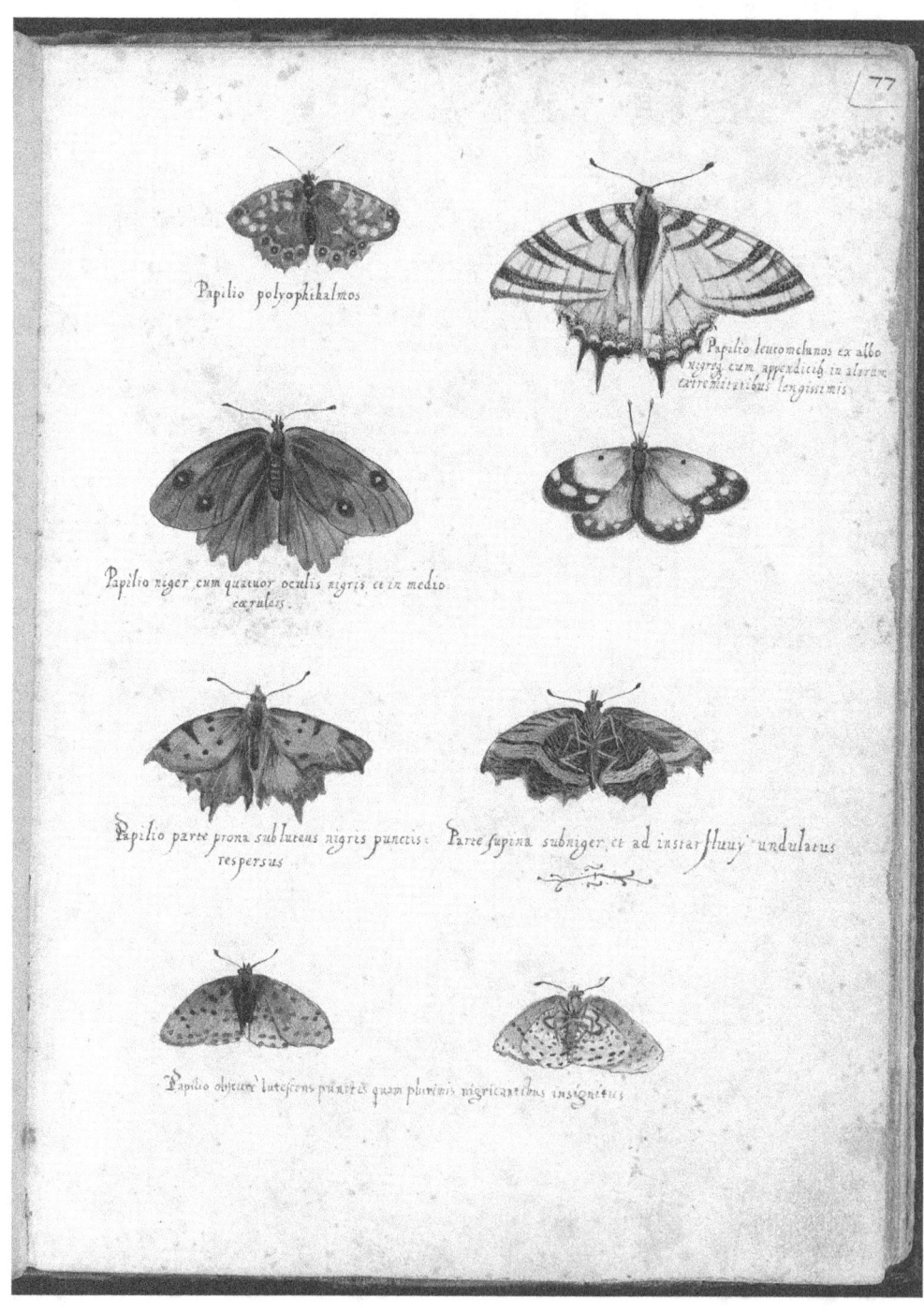

Papilio polyophthalmos

Papilio leucomelanos ex albo
nigroq; cum appendicib; in alarum
extremitatibus longissimis

Papilio niger cum quatuor oculis nigris et in medio
cæruleis.

Papilio parte prona sub luteus nigris punctis
respersus

Parte supina subniger et ad instar fluuij undulatus

Papilio obscure lutescens punctis quam plurimis nigricantibus insignitus

Figure 24. MS Aldrovandi, *Tavole di animali,* volume 7, page 77. Biblioteca Universitaria, Bologna. Courtesy of University Library of Bologna; no reproduction without permission.

ſcunt, habentq́; in medio ferè maculam nigram . Antennæ purpureæ ſunt, in extremo ob- E
tuſæ & albæ . Idem ſecundo loco datur ſupinus, in quo quod in externis alis prope finem
erat, nigticat, hịc virideſcit; ſed macula tamen nigracolorem retinuit nigrum: internæ prope corpus prorſus virides, in fine ad luteum vergunt : cingunturq́; eodem circulo roſeo:
Papilio eſtq́; hoc obſeruatu dignum in internis, quod maculam habeant album cum circulo ferru-
λευκο χλω gineo, quæ extrinſecus non eſt . A potioribus coloribus λευκο χλωρος dici poteſt. quemad-
ρος . modum etiam qui ſequitur pronè, & ſupinè etiam pictus, ſed hunc quia admodum macu-
loſus eſt, λευκο χλωρον variegatum nuncupaui . Prono corpore leucophæo eſt, capite nigri- F
cante, oculis inſtar ſaphyri viridibus, antennis nigris cum appendicibus in ſummitate al-
bicantibus, pectore hirſuto: alis candidis, ſuprà prope caput ad cæruleum vergentibus, maculis magnis nigticantibus inferius reſperſis, dico inferius, nam vbi cæruleæ ſunt, nullam
cernere eſt maculam . Supino corpore magis candicat . Alæ externæ ſunt candidæ, & ni-
Papilio cru gro variæ, internæ ex albo, & viridi eleganter admodum decoratæ . Quintum Papilio-
ciger. nem crucigerum nominaui, quòd externis alis complicatis, albis tęnijs quibus ornatus cernitur, crucem vel potius litteram X exprimat, quanquam id in icone pictor non expreſſe- G
rit . Sunt itaque alæ illæ nigræ, faſcijs, ſeu tænijs albis inſignes, & venis rubicundis deor-
ſum deſcendentibus . Interiores ſunt totæ miniaceæ, maculis nigris, rotundis illuſtres. Sextus idem eſt parte ſupina exſculptus ; in cuius internis alis maculas illas nigras non videas,
& faſciæ, quæ in externis alis prona parte albæ erant, hîc aureæ ſunt . Corpore eſt iſte Papilio pronè luteo toto, ſupino ventre albo nigris guttis triplicis ordinis conſperſò . Oculi
nigri ſunt, & exigui, exiguæ, exilesq́; antennæ . Sextus ac poſtremus alas externas, ac in- H
ternas exteriùs rubicundas ad caſtaneum vergentes habet nigris guttis conſperſas . Supina
parte alæ internæ candicant, exteriores colorem eundem fermè, quem prona earum pars
reſeruant . Corpus pronum atrum eſt, ſupinum cinereum : cinerei quoque pedes ; antennæ nigræ: oculi lutei, prominentes .

Tabula ſexta .

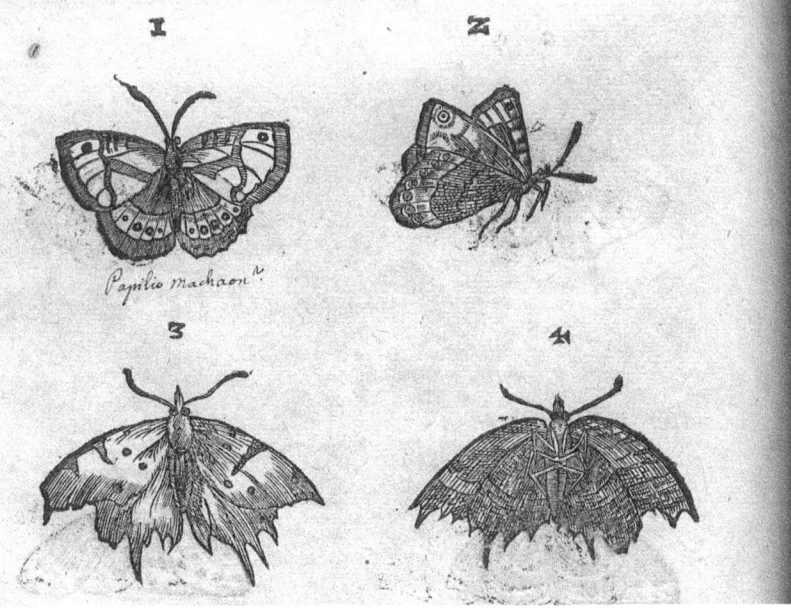

Figure 25. Ulisse Aldrovandi, *De animalibus insectis* (Bologna, 1602), page 244. Special Collections and Archives, University of Idaho Library, Moscow.

Prima earum,quæ à me obſeruatæ ſunt,& binis tabulis depictæ exhibentur,ex mediocri- E
bus eſt,capite,& toto corpore ruffis,alis candidis,extremo macula nigra notatis,pedibus nigris. Secunda capite perlam,ſiue vnionem ſequaquam refert . non enim rotundum eſt , ſed
acutum. Oculos habet enormis magnitudinis, ſubcæruleos. Pectus admodum craſſum ; aluum proceram , tenuem , Tota lutescit , nigroꝗ; vt pictura expreſſit , maculatur . Alæ ſunt
ſubcæruleæ . Tertia alas habet breues,ad ſubcæruleum inclinantes, aluum longiſſimam, tā
ſubtilem,vt mirum ſit,quomodo eam regere poſſit. Oculi quoque vt in proximo , maximi ,
ſed caput orbiculatum . Infra candicat, per dorſum ſuperne ad latera luteſcit , in medio ferrugineo colore eſt. Antēnas habet,admodum breues ſubflauas. Quarta ex mediocrium genere eſt, corpore toto purpureo , aureis zonis vndique interſtincto,alis plane argenteis , nigro maculatis . Quinta alas quoque habet argenteas , ſed non tam reſplendentes. Maculam
in extremo ſingulæ habent nigricantem . Corpus totum aurei fere coloris eſt,nigricantibus
zonis interſtinctum . Pedes ſunt atri . Ex maximis eſt , quæ ſexta in hoc ordine depingitur ,
ſpecioſa ac elegans Perla,capite,pectore,ac dorſo viridis,aluo tota cærulea, per quam linea
fertur à dorſo ad extremum vſque aterrima . Alæ cinereæ ſunt,pedes nigri . Septima oculos
habet virides, tota lutea, nigro maculoſa. Minimarum generis eſt octaua,alas habens argen- F
teas,corpus virideſcens. Nona ex maximis eſt,capite,dorſoꝗ; viridis,aluo dilute flaua, pri-
mùm craſſa,dein tenui,alis ſine maculis. Decima,ac poſtrema toto corpore eſt flauo, nigro
interſtincto,pedibus etiam flauis, alis cinereis ſplendidis .

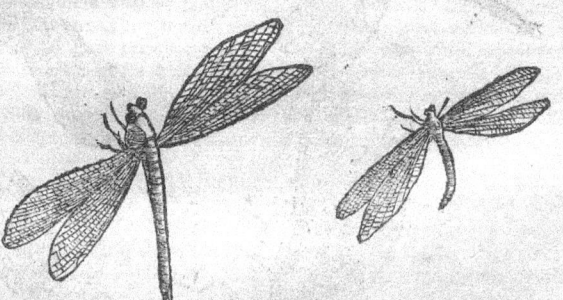

Ex vulgarium genere ſunt,duę iſtæ ſeparatim pictæ , quòd præcedens tabula eas non ca-
peret, prior alas habet ſubflauas, & corpore toto flauo eſt exceptis zonis,quæ ſunt ferrugi-
neæ. Altera tota viridis eſt,maculis quatuor candidis in alarum extremitatibus .
Primo loco picta in ſecunda tabula, ex ijs eſt,quæ circa aquas volitat.Alas habet breues,
& latas,primùm virides, deinde cæruleas , cuius coloris quoque totum eſt corpus à capite
ad anum vſque,præter oculos,& pedes,qui atri ſunt . Oculos aute habet enormiter magnos
ac prominentes. Aluus in exilem acutiem, velut acum deſinit. Secunda etiam aquas proſe- H
quitur,toto corpore viridi, nigris, exiguis zonis diſtincto. Oculi protuberant. Alæ quà cor-
pori hærent argenteo colore,& ſplendore micant, dein cæruleæ ſunt , & argenteis venis re-
fertæ . Tertia minimarum generis pedes habet longiſſimos , Aranearum pedibus non diſſi-
miles, ſed admodum exiles. Corpore, alisque ex luteo ſubalbidis eſt cauda ſimplici.Quarta
ex minimis eſt, tota fere viridis, ſed à lateribus tergoris lutea. Oculi candicant . pedes nigri
ſunt . Quinta cum ſecunda admodum ſimilis,pariter aquas petit,aluo cærulea, pectore viridi,alis primum argenteis,deinde cæruleis.Sexta ex maximis eſt,capite,& pectore viridibus,
dorſi lateribus,vbi alæ enaſcuntur, nigris,aluo longa,cærulea,quam nigra linea lata coloris
aterrimi mediam diſſecat. Pedes atri ſunt . Septima mediocrium generis toto corpore viridi, aurum reſplendente . In dorſo nigricat: alę cinereæ ſunt, & argenteo micant. Octaua aluum habet obtuſam,ac minime bifurcatam, toto corpore ferrugineo, alis candidis, à latere
ferrugineis. Nona, ac poſtrema aluo eſt craſſa, & breui, lutea, nigris circundata Zonis , capite,pedibus,dorſo,& alarum principiis ferrugineis. Alarum cætera candida.

A hi miſſum dixi fol. 279.c. totus eſt candidus.

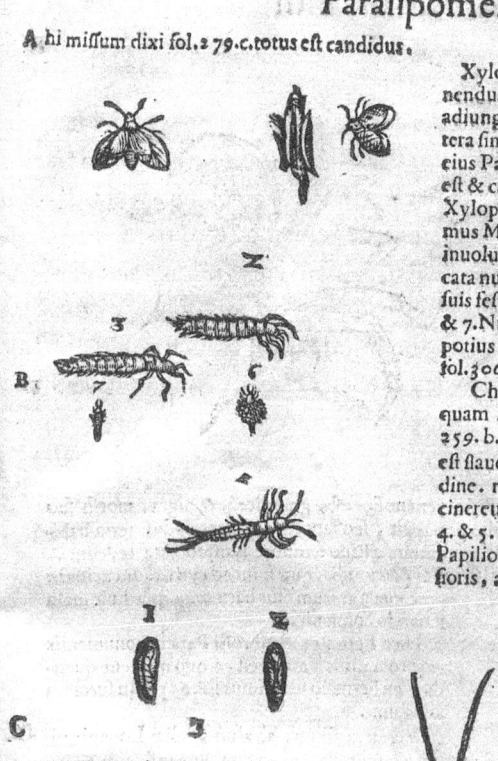

Xylophthori Papilio ſiue Ligniperdę po Ligniperdę
nendus fol. 306.tabula prima . Huic etiam Papilio.
adiungendus erat Xylophthorus minor, cę
tera ſimilis , licet paulo candidior , vnde &
eius Papilio, qui huic appingitur, & minor
eſt & candidior . Habes hic etiam alterius
Xylophthori aquatici (quem ibidem dixi-
mus Malalbergi inuentum) iconem , ſuis
inuolucris & feſtucis ſpoliati , cauda bifur-
cata num. 2. & 3. item alterius minimi cum
ſuis feſtucis , & ſine feſtucis nimirum nu. 6.
& 7. Num. 5. verò iconem Ligniperdæ, aut
potius Phryganij illius , qui ſupradicto
fol. 306. exprimitur nu. 4. cum ſuis feſtucis.

Chryſalis Teredinis illius Erucæ ,
quam Apibus noxiam eſſe ſcripſimus fol.
259. b. pronè & ſupinè exhibita , colore
eſt flaueſcente, ea , qua cernitur, magnitudine. num. 3. depictus eſt eius folliculus
cinereus , quem in aluis texit. Verùm nu.
4. & 5. alia eſt Chryſalis aterrima, è qua
Papilio mihi natus eſt coloris paulò remiſ-
ſioris, aluo procera, pedibus exiguis.

Ex Peſlarū forte ge-
nere fuerit, quod hic de
pingitur Inſectum, licet
antēnis admodum pro-
ceris, & capite exiguo
corpus totum eſt ferru-
gineum, præter oculos,
qui ſunt nigri. De natu-
ra eius nedum explora-
tum quidquam habeo.

Aranea aluo rotun-
da , tota nigra , præter-
quam in principioalui,
vbi macula eſt crocea.
Oua , quæ appingun-
tur, erant alba.

Phalangij ſiue Tarā- Tarantulæ
tulæ vtriuſqᵢ; & maio- hiſtoria.
ris, & minoris à Ferran-
te Imperato Neapolita-
no depicti, hæc eſt icon,
& hiſtoria . Quæ , in-
quit, apud nos dicuntur
Tarantulæ, à Taranto
& circumuicinis locis
nomen inuenerunt
quòd videlicet , ibi &
frequentiores, & notio-
res ſint . Ex Araneo-
rum ſunt genere , ſed

cæteris maiora , ſub terra degunt, & in exordio cauernularum telas texunt albas, denſaſqᵢ
quas tamen ita diſponunt, vt ingreſſum, tranſitumqᵢ non impediant . Prima ſpecies quam-

Figure 27. Ulisse Aldrovandi,
De animalibus insectis (Bologna, 1602),
page 763. Special Collections and Archives,
University of Idaho Library, Moscow.

Aldrovandi's insect categories. The insects that appeared in the "Secunda tabula" traveled a circuitous route through the physical and intellectual structures of Aldrovandi's museum. They began as records of specimens, drawn onto separate pages of Tomo VII by Schwindt, and were then gathered from these pages and deposited into the virtual space of the "Secunda tabula." Later they were separated and moved to new woodcuts according to the new insect categories that Aldrovandi developed. At the time the "Index insectorum" was compiled in 1593 Aldrovandi considered all of the insects pictured in the "Secunda tabula" to be forms of butterflies and moths, since he refers to all of them as "papilio." Later, Aldrovandi seems to have decided to separate dragonflies and damselflies into their own category—"perla"—a term that does not appear in the "Index insectorum." Thus, the process of cutting and pasting was flexible enough to accommodate changes as they unfolded, but the categories that were finalized at the time of publication could not ultimately accommodate every insect in Aldrovandi's collection. Recall the lacewing, or "Papilio locustiformis," of the "Secunda tabula." The name Aldrovandi gave to this insect reflects his uncertainty about how to categorize it. A "Papilio locustiformis," or "butterfly in the form of a locust," was for Aldrovandi a hybrid that did not fit into his chapter on butterflies nor into his chapter on locusts. In addition, its presence in the illustrations for either chapter would have disrupted the visual unity Aldrovandi achieved by grouping insects by type and size. Cutting and pasting enabled Aldrovandi to set limits and boundaries on the insect world as he worked to contain it within the space of the printed page, but it was not a technique that resulted in an infinitely variable and adaptable system of classification. The case of the Papilio locustiformis shows that the cost of establishing these limits was the elimination of creatures that did not fit the categories.

Aldrovandi's book of insect drawings served as both a source for the woodcut illustrations and an active organizational tool. Schwindt's drawings are not only a record of Aldrovandi's insect collection but also served as a set of materials that could be worked with in ways that were not possible using the physical collections alone. Cutting and pasting images between contexts was central to the way that Aldrovandi organized and interacted with these materials, and the volume of insect drawings served as a site for him to work through and develop his ideas about the insect world. Whereas Aldrovandi's insect specimens were dispersed throughout the cabinets in his museum, the virtual specimen cabinet formed in the pages of Tomo VII provided a place for concentrating his attention on insects. Users of the volume of insect drawings could page through it easily to make quick comparisons between images on different pages, thereby permitting a large number of insects to be accessible in a small and relatively portable format. As a virtual specimen cabinet, Tomo VII also served as a mental space in which Aldrovandi could devote his full attention to the new topic he was carving out of the enormous variety of his collections.

In describing the epistemological structures and practices surrounding the concept of *museaum* in early modern Europe, Findlen has noted that "emerging scientific journals often included words such as 'repository,' 'collection,' and 'museum' in their titles to underline the reductive nature of the enterprise, for the pages formed intellectual walls in the same way that the perfect shape of the theatre closed and completed a concept."[26] In much the same way, Aldrovandi's insect imagery helped to form the intellectual walls that were necessary to contain the insect world in the medium of print. The establishment and maintenance of these walls were achieved in part through the dynamic, active processes of cutting and pasting.

Thomas Moffet: Cutting and Pasting a Virtual Community

The illustrations and drawings for Aldrovandi's *De animalibus insectis* are an example of the ways in which one naturalist in early modern Europe managed the transition from specimen to image to print. In order to transform nature into print, Aldrovandi and his artists devised practices that enabled them to translate three-dimensional objects into the two-dimensional form of the printed page, and they used the technique of cutting and pasting images and specimens as a strategy for negotiating this transition. For the English naturalist Thomas Moffet and his circle, cutting and pasting also functioned as a mode of organizing and understanding the insect world. The *Theatrum insectorum* was the product of a community of naturalists that spanned several generations, ranged across wide geographical distances, and encompassed a variety of scholarly backgrounds. The notes, specimens, and images that formed the basis of the book changed hands several times before coming into Moffet's possession, and they were supplemented by materials from Moffet's correspondents and those of his predecessors.

Moffet took over the project in 1588 after the death of his friend and collaborator Thomas Penny (c. 1530–1588). Penny had been collecting, studying, and drawing insects since at least the early 1560s, and he obtained many of the materials on insects that would eventually come to be published as the *Theatrum insectorum* from the great Swiss naturalist Conrad Gessner. Penny continued to add materials to the collections and was working toward publishing the research at the time of his death in 1588, but it was Moffet who completed the initial manuscript for the book in March 1589. Shortly after completing the manuscript and having a title page engraved for its publication, Moffet entered a period of personal and professional upheaval following the death of his patron Robert Devereaux, the second Earl of Essex. Although Moffet continued to work on the insect project, he did not live to see it into publication. After Moffet's death in 1604 the manuscript passed to Moffet's apothecary, known as Mr. Darnell, but a number of other naturalists seem to have had access to the manuscript during Darnell's possession of it. The manuscript eventually came to be owned by Theodore de Mayerne, a Huguenot physician who had settled in London in 1611, and in 1634 Mayerne had the manuscript published. An English translation of the *Theatrum insectorum* was included in Edward Topsell's *Historie of Four-Footed Beasts* of 1658.[27]

Moffet and Penny probably formed their friendship while Moffet was a student at Trinity College, Cambridge. Moffet began his studies there in 1569, the same year that Penny returned to England after four years of traveling in Switzerland, Germany, and France. Although Penny lived in London upon his return, he traveled extensively within England during this period, and he appears to have begun working with Moffet in the areas of natural history and medicine at some point during the 1570s.[28] Moffet left England in 1578 to study medicine in Basel, and afterward he traveled to Italy, Spain, and Germany. By 1584 Moffet had returned to England and was settled in London, where he and Penny worked together on the book on insects in the years preceding Penny's death. The materials on insects that formed the nucleus of their book were obtained by Penny during his travels in Europe from 1565 to 1569, during which time his interest in natural history was stimulated and encouraged. Penny arrived in Zurich in 1565 to meet Conrad Gessner; Penny was one of the many collectors who sent plant specimens for Gessner's *Historia plantarum*, and he is named as the source for 22 of the approximately 375 plants that would appear in the book.[29] Soon after arriving in Zurich, Penny presented Gessner with drawings and specimens of English plants, as well as specimens of plants that he had gathered on his journey through Switzerland. In exchange for these botanical materials, Gessner made a gift to Penny of the pictures and notes on insects that he had been gathering over the years. Charles Raven speculates that because Gessner was primarily occupied with his botanical work at the time, he may have passed these materials on to Penny because he felt he no longer had time to work on them himself.[30]

This gift to Penny reflected the Swiss naturalist's respect for Penny's abilities and skills as a naturalist, but the gift was also an important entrée for Penny into Gessner's community of correspondents in botany and natural history. It was during this period that Penny began exchanging plant specimens and images with a distinguished circle of naturalists that included Carolus Clusius, Mathias de L'Obel (Lobelius), Joachim Camerarius the Younger, and Jean Bauhin. After Gessner's death, Penny embarked on a botanizing expedition that took him west to Geneva where he found many new plants, and he sent specimens and pictures of the plants to his new correspondents.[31] Penny would continue exchanging specimens, images, and information with these naturalists and others as he began work on the project on insects, drawing on the contacts and practices he developed for botany and using them as the basis for his approach to the new subject of insects. Penny's work on the *Theatrum insectorum* began in earnest in the 1570s after he had returned to England, and it is possible that his experiences collecting plants on his journey back to England influenced his shift in interest. After leaving Switzerland, Penny traveled on to France and Germany, arriving in Montpellier at some point in 1566. In contrast to the discoveries he had made in Switzerland that had earned him the respect and friendship of Gessner and his distinguished associates, Penny's botanizing efforts in France yielded less exciting results. Raven notes that "in Montpellier itself Penny's finds naturally become less notable; for the district was the centre of botanical studies and had been thoroughly explored."[32] Though Penny had established a reputation for himself as a skillful collector and painter of plants in Switzerland, his experience in Montpellier may have prompted him to pursue his insect studies with renewed vigor since the novelty of the subject made it an ideal area in which to sustain and build his reputation as a naturalist.

In any case, Penny maintained his correspondence with several of the naturalists he met during his travels, and while working on insects he continued to cultivate contacts across Europe—obtaining information, specimens, and drawings from this widely dispersed network. For Penny and his correspondents, visual images of insects sometimes stood in for specimens. Over the years, Penny received several unusual flies, a moth, a butterfly, four rare beetles, and an African mantis from Clusius. Camerarius sent an unusual fly, several beetles, and a caterpillar. In some cases, the text in the *Theatrum insectorum* specifically notes that Penny received an image, as opposed to a specimen, as in the case of the rare "nose-horn beetle" which "Carolus Clusius sent painted from Vienna."[33] In other cases it is less clear whether Penny received a specimen or a drawing, as in the following description of another nose-horn (or rhinoceros) beetle, this one received from Camerarius (Figure 2.8):

> It produceth its young one from the ground by itself, which Joachim
> Camerarius did elegantly express, when he sent to Pennius the shape of
> this Insect out of the storehouse of natural things of the Duke of Saxony;
> with these Verses: *A Hee begat me not, nor yet did I proceed, From any Female,*
> *but by myself I breed.*[34]

Moffet here refers to the "shape" of the insect, a somewhat ambiguous term that could refer to either a specimen or an image. The rhinoceros beetle illustration that accompanies this passage in the *Theatrum insectorum* was copied from Joris Hoefnagel's *Archetypa* of 1592 (Figure 2.9). A woodcut copy of Hoefnagel's rhinoceros beetle also appears pasted into the margin of the corresponding page of the *Theatrum insectorum* manuscript (Figure 2.10), where many of the original drawings for the book are preserved.[35] Penny died before the publication of Hoefnagel's *Archetypa,* and Moffet states that the "shape" was sent to Penny, so the presence of the Hoefnagel beetle is something of a puzzle. Nonetheless, the

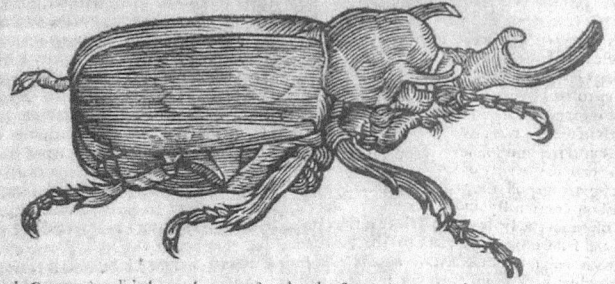

Ioach. Camerarius did elegantly express, when he sent to *Pennius* the shape of this Insect out of the storehouse of natural things of the Duke of Saxony ; with these Verses :

> *A Hee begat me not, nor yet did I proceed*
> *From any Female, but my self I breed.*

For it dies once in a year, and from its own corruption, like a Phœnix, it lives again (as *Moninus* witnesseth) by heat of the Sun.

> *A thousand summers heat and winters cold*
> *When she hath felt, and that she doth grow old,*
> *Her life that seems a burden, in a tomb*
> *Of spices laid, comes younger in her room.*

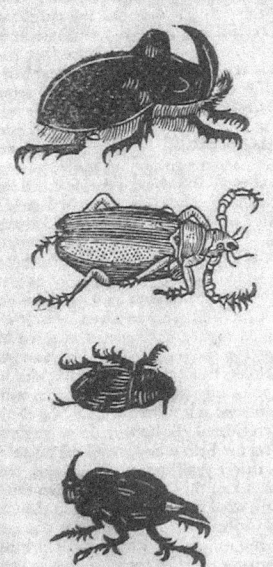

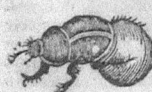

The second kinde of Nose-horn very rare and worthy to be seen, sacred to Mercury, *Carolus Clusius* sent painted from *Vienna*, where it is very frequent, the form is as you see it : it would seem all pitch colour, but that the belly is a full red ; that crooked horn in the nose is so sharp, that (what is said of an Elephant going to battle) you would think it had got an edge by rubbing it against a rock. The third Nose-horn, and fourth seem to be alike, but that the former hath wings growing out longer than the sheath covers, but the others are shorter. You would say they were rub'd with shining ink, they are so perfectly all over black. The Ram or *κείαφος*, hath knotty horns, violet colour, a head greenish from gold colour, the shoulders like vermilion, a purple coloured belly, sheath wings of the colour of the head, it goes forward with legs and feet, of a light red, but the wings shut up in the sheath, do fitly express the small whitish membrane of a Cane.

The greater Beetles without horns are many ; namely, that is called *Pilularius*, and another that is called *Melolanthes* ; another purple, one again that is dark coloured ; one called *Arboreus*, and another *Fullo*. Some call the *Pilularius* the dunghill Beetle, because it breeds from dung and filth, and also willingly dwels there. The Greeks call it *κανθαρ*, and *ηλιοκάνθαρος*, and from its form like a cat, *Αιλυρόμορφ®* ; the Germans, *Roßkafer, Kaat,* or *Mißkafer* ; in English, *Dung-beetle, Sharnbugg* ; in French, *Fouille merde,* as you would say *Dung-digger* ; the Latines call it *Pilularius,* because it turns up round pills from the dung, which it fashions by turning it backwards with its hinder feet. *Porphyrie* doth thus describe the nature of it : All your *Pilularii* have no females, but have their generation from the Sun ; they make great balls with their hinder feet, and drive them the contrary way, like the Sun it observes a circuit of 28. daies. *Ælian* saith almost the same. There is no female Beetle, it puts the seed into a round ball of dung, which it rowls and heats in 28. daies, and so produceth its young. They would say

Rrrr   thus

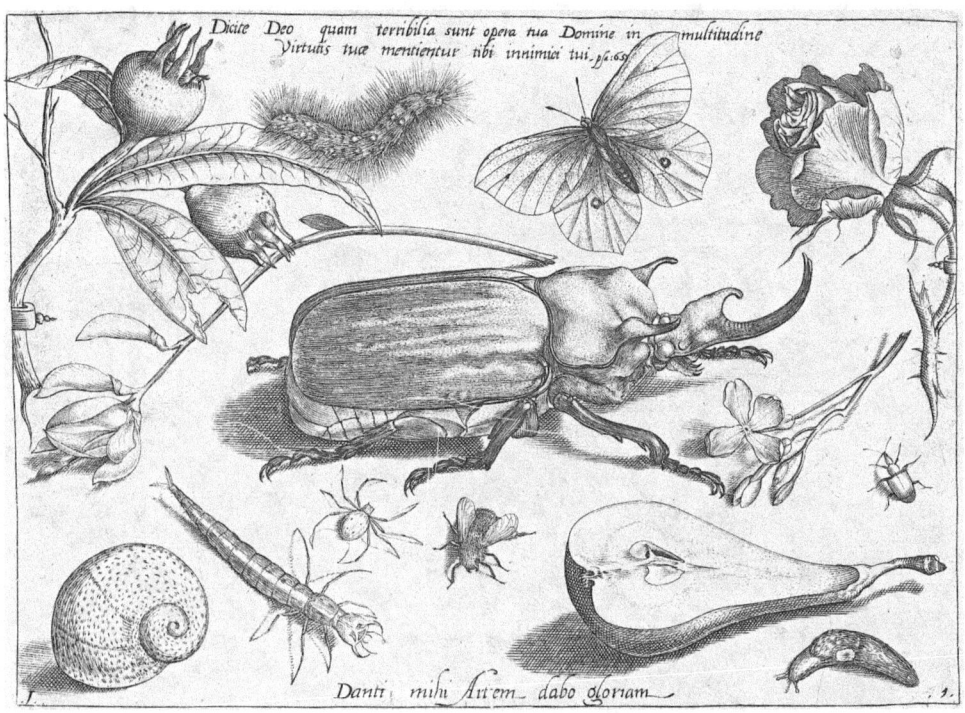

Figure 2.9. Jacob and Joris Hoefnagel,
*Archetypa studiaque patris Georgii
Hoefnagelii* . . . (Frankfurt, 1592), part 1,
plate 1. Research Library, Getty Research
Institute, Los Angeles.

ambiguity of this example points to the fluid boundaries between image and
specimen that existed for this community of practitioners.

Although images could occasionally stand in for specimens for Penny,
Moffet, and their associates—as they did for Ulisse Aldrovandi—images and
specimens were not completely interchangeable. Moffet gives the following
account of another unusual beetle with "shoulders and wings . . . so beautifully
wrought that you would easily swear it were a cloathing of Damask embroidered after the Phrygian manner . . . Pennius first had the picture of it from
Carolus Clusius; but Quickelbergius afterward sent him over the creature
itself."[36] Penny seems to have found it desirable to own both an image and a
specimen of this particularly beautiful and rare insect, but in other cases images
alone were sufficient. In describing a large and brilliantly colored image of a
moth, Moffet indicates that the image captured the appearance of the insect
so perfectly that it precluded the need for textual description: "This did Carolus
Clusius send from Vienna, of so elegant and notable figure, that it is easier to
wonder and admire, than with fit expressions to describe."[37]

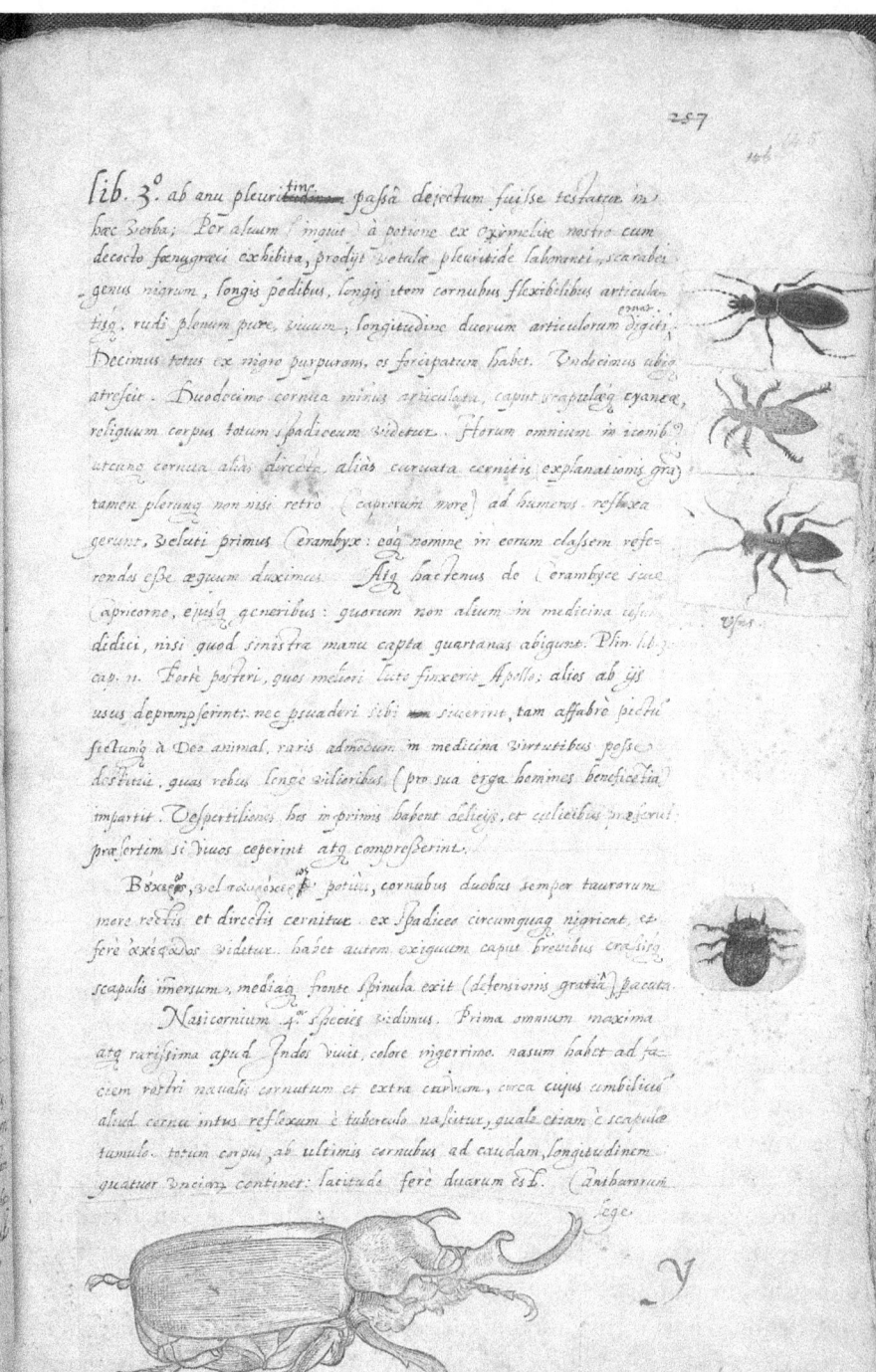

In addition to maintaining Penny's contacts from the Continent, Penny and Moffet also cultivated naturalist contacts in England while pursuing their research on insects. One of Penny and Moffet's chief correspondents in England was Sir Edmund Knyvet, who sent them pictures of insects he drew himself. Moffet seems to have been particularly grateful to Knyvet and his nephew Thomas for their assistance, as indicated in the following note about locusts:

> These did that famous knight Sir Edmund Knivet, freely send in picture
> to Pennius for the enlargement of this work, for he is a Knight that is very
> courteous to learned men, and singularly noble both by descent and
> vertue, and famous for his curious search into the knowledge of natural
> things.[38]

Thomas and Edmund Knyvet were members of a prominent Norfolk family who shared an interest in the study of nature and amassed a collection of books and natural history specimens that they kept in a private museum at their home in Ashwellthorpe. Edmund seems to have spent a fair amount of time collecting and drawing insects in the countryside near his home. He sent Moffet and Penny a "worm of wonderful form . . . exactly deciphered in his own hand . . . very common in Norfolk in England."[39] Images were central to Edmund Knyvet's study of the natural world, and his library contained many illustrated books of natural history. In his library a volume containing pictures of birds was heavily annotated with the English names of the birds added in his hand, and it seems to have been used as a reference or field guide.[40] Knyvet's study of the natural world was embedded in practices involving collecting specimens and creating and using images. Knyvet engaged in these activities both locally and over a distance, in the context of his own museum and through his correspondence with Moffet and Penny. As with Aldrovandi, the interplay between images and specimens was central to the understanding of the insect world by Knyvet, Moffet, and Penny. In at least one instance, that of a feathery silver fly, we can track how the exchange and study of images shaped Penny's perception: "Pennius received one of these painted, from Edmund Knivet: afterwards he often saw them in hedges, and places set with privet."[41] It was not until after Penny had received Knyvet's image of the silver fly that he became aware of its existence, an awareness that then led him to search out its habitat, with which he was already familiar.

When Moffet inherited Penny's collection of insect materials, he also inherited Penny's personal contacts. In addition to maintaining these associations, Moffet continued his predecessor's practice of obtaining information through correspondence. Both Moffet and Penny were trained in medicine, and their contacts reflect their connection to a range of medical practitioners. A surgeon, Edward Elmer, sent Penny a fly with silver wings from Moscow as a "great present," and he later sent Moffet a horned beetle from Russia.[42] Other people who sent specimens and information about the medical aspects of insects included a Dr. Banchurius of Paris who sent Penny a report on scorpions, and a Dr. Barbar who confirmed information on a type of fly that infests the ears of dogs.[43] A cure for deafness using crushed earwigs was sent by Josephus Michaelis, "an Italian and a famous Physician," and a physician from Geneva, Antonius Saracenus, sent Penny drawings of a mantis and a water beetle in 1578.[44] A dark-brown cicada was sent by Ludovicus Armacus, of which the text states: "a very diligent Chirurgeon, [he] brought [it] from Guinea, and gave it to Pennius."[45] This same Ludovicus Armacus later sent Moffet an insect from Africa, apparently subjecting himself to the pain of its bite in pursuit of useful information about it:

> This Scolopender being provoked bites so sharply, that Ludovicus
> Armarus [sic] (who gave me one brought out of Africa) could scarce
> endure him to bite his hand, though he had a good glove on, and a
> double linnen cloth; for he strook his forked mouth deep into the cloth,
> and hung a long time, and would hardly be shaken off.[46]

The intensity of Ludovicus Armacus's dedication to insect research seems to have been matched by that of the scolopender's bite, and is an example of the sometimes unpleasant way that knowledge could be formed through material and bodily engagement.

In addition to these naturalists, physicians, and surgeons, Moffet and Penny called others into the service of their insect research, people who provided information but may have only shared a passing interest in insects. Jacob Garret, an apothecary, reported on an experiment he conducted with the wings of a cricket to determine how they produced sound. Both Penny's brother and William Cave, a nobleman of Leicester, provided information about insects and worms that breed in stone.[47] An Italian merchant provided Penny with a drawing of a tarantula, which Moffet found quite pleasing, though somewhat mistaken in its coloring:

> . . . we give you here a picture of it, that was bestowed upon Pennius by
> an Italian Merchant of happy memory; where if you paint the white
> places with a light brown, and the black with a dark brown, you have
> a true spotted Tarantula; I know no man yet that described it as it
> should be.[48]

A number of unnamed travelers also provided specimens of rare insects to Penny and Moffet, including a "Beetle of a purple color" from Constantinople, two caterpillars from Normandy, and two scolopender worms—one from "new

Hispaniola" and another from St. Augustine's promontory in India.[49] William Brewer, who Moffet refers to as "my friend Brewer,"[50] went to great lengths to supply Moffet and Penny with information about insects:

> I did chance to finde (saith *Bruerus*) in a dirty filthy ditch an Insect with
> very long feet, which for the likenesse of the form, you would say was
> one of the larger sort of Gnats coming forth of a soft leathern purse.[51]

Brewer sent Penny and Moffet several other observations, as well as specimens of a dragonfly and a glow worm. Though not an artist, Brewer communicated several of his observations through visual images. A fragment of a letter from Brewer is preserved in the manuscript of the *Theatrum insectorum,* which includes several rough pen sketches of insects in the margins.[52]

In putting together the materials that would eventually be published as the *Theatrum insectorum,* Penny and Moffet amassed a large collection of information, specimens, and images of insects from contacts and correspondents with widely varying backgrounds. The members of this group ranged from some of the most learned naturalists of sixteenth-century Europe, to gentlemen collectors in England, to friends and travelers to foreign lands whose interest in insects led them to endure ferocious bites and filthy ditches, to unnamed individuals who were paid to make observations of insects. Some of these correspondents, such as William Brewer, may not have had a strong interest in insects before Moffet and Penny made their requests for information but instead became more invested in the subject over time. What is important to note here is not only that this community was formed through the exchange of specimens, images, and information, but also that it was brought into existence as part of Penny and Moffet's efforts to establish insects as a subject matter for a book.

In taking on the project of bringing Penny's unpublished research to publication, Moffet was tasked with organizing and paring down what seems to have been a very large amount of material, something Penny was not able to accomplish during his own lifetime. According to Moffet, Penny left the work

> . . . heaped up together on all sides; Hence it was that his Letters were full
> of blots, and confused with doubtful Characters: and they had perished,
> had not I laid them apart, when they were ready to be cast out of doors;
> and with a great sum of money had redeemed all the torn pieces of it.

In addition to rescuing the insect research from destruction, Moffet also states that he made important contributions of his own of both texts and images:

> . . . I have inserted intire Histories, and above a hundred and fifty
> pictures, which Gessner and Pennius knew not; I have mended the

method and language, and I have put out above a thousand tautologies, trivial matters, and things unseasonably spoken . . .[53]

What exactly did Moffet do with the insect materials when he states that he "laid them apart," and what kind of "mending" of method and language did he undertake? Moffet has been judged harshly by some earlier historians of science for what they saw as his indiscriminate editing of Penny's notebooks. Charles Raven condemned Moffet for possessing "a touch of fancy, a flair for telling a phrase, a literary gift which is wholly unlike the dry scientific precision of Penny's style . . . Penny alone has claim to merit as a scientist."[54] Raven's judgment of Penny's methods as "precise" and "scientific" in comparison to Moffet's fanciful and "unscientific" arrangement of the notebooks was based on an idealization of Penny as a forerunner of a modern scientist, and of Moffet as a dilettante and a dandy who was more concerned with social standing and superstitious medical practices than with science. No doubt, Moffet's prose style is sometimes dramatic, as seen in his heroic account of rescuing of Penny's notes, and he certainly had a tendency to engage in self-promotion, but the *Theatrum insectorum* must be understood as a typical work of early modern natural history. It contains a wide range information culled from numerous classical and contemporary sources, and although many of these sources are cited and some individual contributions can be traced, the book is a product of collaborative efforts.

The *Theatrum insectorum* provides a rich account of the practices of a community of early modern naturalists in its formative stages and of the ways they came to work with one another. Moffet was careful to include detailed information about the people with whom he and Penny corresponded, in part to convince readers of the worthiness and legitimacy of the project through its association with the respected naturalists whose names appear in the book. Moffet's attention to the identities of the people who provided specimens, images, and observations was important for constituting this new community of naturalists whose focus was on insects and for establishing his identity as a trustworthy source. The members of this community were in many cases not known to one another, and would not have considered themselves part of a group of researchers specifically concerned with the study of insects. In gathering and publishing their materials on insects, Penny and Moffet forged a community of like-minded practitioners whose identity would only be crystallized within the pages of the book. In the course of Moffet's professional life he occupied multiple roles within medical, natural history, and courtly circles.[55] His contributions to the text and illustrations of the *Theatrum insectorum* reflect his desire to be accepted as a peer in the networks of elite naturalists in Europe and to be regarded as an important broker of natural knowledge within the English context. In bringing together in one place the dispersed fragments of Penny's research, the text and images of the *Theatrum insectorum* helped to constitute insects as a field of study and as objects of analysis for early modern European naturalists, despite the omission of the names of several key players. Moffet tapped into this emerging community's resources to assemble his virtual specimen cabinet of insects by exchanging written accounts, specimens, and images. For both Ulisse Aldrovandi and Thomas Moffet, images stood in for specimens in a variety of contexts, serving as virtual specimens in a number of different ways. The capacity of images to withstand the rigors of repeated handling and transportation greatly facilitated their use and exchange among early modern investigators, allowing them to circulate locally within individual collections as well as internationally within and among networks of practitioners. This activity of circulating images resulted in the simultaneous construction of insects as a subject of analysis and a community of practitioners organized around that subject.

One of the key ways that Moffet constituted his community of insect practitioners was by writing about it—by naming names, so to speak. Another way that he formed this community was by picturing it. Moffet's textual activity had its visual parallel in the elaborate engraved title page he commissioned shortly after completing the manuscript in 1589. Only one impression of this title page is known to exist, and it is bound with Moffet's manuscript (Figure 2.11); a new title page was created for the 1634 edition published by Mayerne, and it is of a much simpler design.[56] Moffet's original title page was engraved by William Rogers of London and consists of portraits of four naturalists surrounded by insects, flower vases, masks, and other grotesqueries embedded in a triumphal arch. The four naturalists—Moffet, Penny, Edward Wotton, and Conrad Gessner—represent the past and present of research on insects. Edward Wotton (1492–1555) was not personally known to Penny or Moffet but was an important source for them. Much of Wotton's work on insects, found in Book Nine of his *De differentis animalium* (1552), was incorporated into the text of the *Theatrum insectorum.* The title page thus pays homage to Penny and Moffet's predecessors Wotton and Gessner while at the same time elevating the authors— along with their book—to this illustrious heritage. Thus, in the title page as in the text, Moffet brings together persons both living and dead within the virtual space of the book by linking people who were separated by both time and space to create a community around the subject of insects. Although it has previously been assumed that the Rogers title page was completed in 1590, the year Moffet took out a license in the Hague to publish his book, a slightly later date should be ascribed to since many of its elements are copied from Joris and Jacob Hoefnagel's *Archetypa studiaque patriis Georgii Hoefnagelii . . .* of 1592. The vases of flowers appearing in the upper-left corners are based on vases of flowers from the *Archetypa,*[57] which Rogers has cleverly transformed into hives surrounded by flying bees. Several of the insect forms in the title page are also taken from the *Archetypa,* such as the small butterfly appearing in the scrollwork to the left

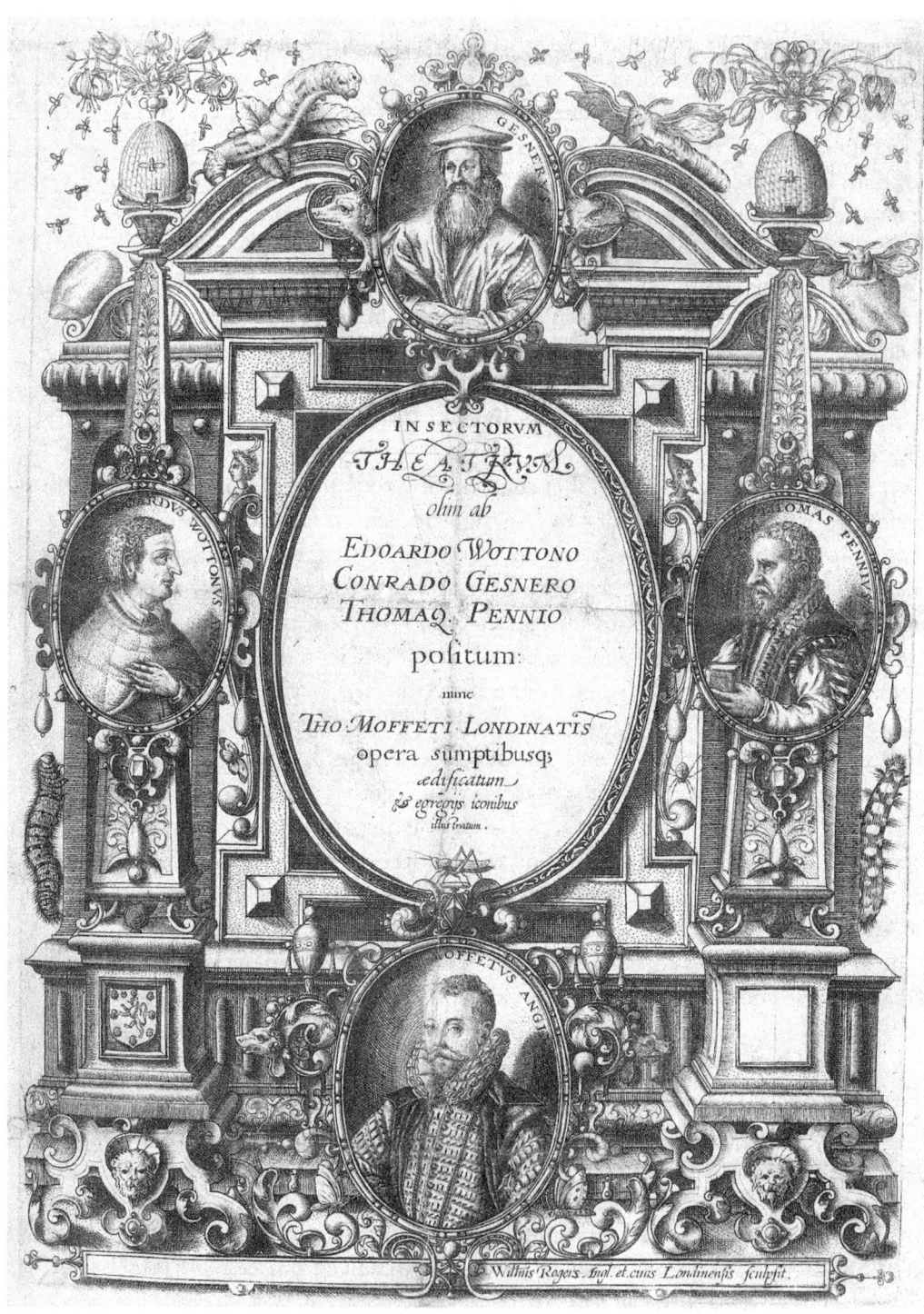

of the central medallion, the spider in the scrollwork opposite this butterfly, and the grasshopper above Moffet's portrait.[58] In these examples of virtual copying and pasting, it seems that Moffet was not eager to have the contributions of the Hoefnagels recognized, perhaps out of fear of losing priority in this new area of knowledge.

Cutting and pasting nevertheless emerged as important strategies for organizing text and visual material, as it did for Aldrovandi, but for Moffet this activity took a different form. Whereas Aldrovandi engaged in a practice of virtually cutting and pasting images, Moffet literally cut and pasted images into his manuscript. Like Aldrovandi, Moffet used cutting and pasting to create groupings of insects by type and size, and both naturalists were able to use the technique to respond to their changing understanding of insect categories. For Moffet, however, visual images sometimes posed a different challenge to his attempts to organize insects and present them in a consistent and regular form. In the manuscript for the *Theatrum insectorum,* images are unruly objects appearing as pasted fragments that do not always conform to the boundaries of the page. At times images even threaten to disrupt the ordering function of the text and its role in framing and containing the variety of insect forms. One image in particular, a watercolor drawing of a swallowtail butterfly (*Papilio glaucus* L.) by John White, seems to have prompted Moffet to come up with a new way of arranging text and image in his manuscript.[59] White's drawing of the large black-and-yellow butterfly is the largest image in the manuscript, and it is pasted into the upper half of folio 96 in the chapter on butterflies and moths (Figure 2.12).[60] Earlier in the manuscript, drawings of moths and butterflies are pasted into the margins adjacent to the text that describes them, until we arrive at the page containing White's swallowtail. Too large to fit into the margin, the life-size image occupies a full third of the page. The presence of such a large insect forced Moffet to make room for it in the center of the text rather than employing his usual procedure of pasting the image into the margin adjacent to the text. Moffet seems to have been temporarily disoriented by this intrusion because the handwriting immediately following the drawing is small and compressed, but he quickly recovers and his handwriting returns to its previous elegant form. But for the rest of the chapter, Moffet adopts this new mode of presenting images, and he pastes all of the remaining drawings into the center of the page rather than the margins, even with images that are small enough to fit in the margins. As a result of this new arrangement the text is fragmented into short, numbered passages interspersed with the drawings (Figure 2.13). By parsing the text into discrete, isolated passages Moffet was able to better associate the drawings with the words that described them; what initially began as a response to the intrusion of an unusually large image culminated in a new way of ordering text and image. By creating a visual link between the information and the illustration, these arrangements encourage the reader to take in text

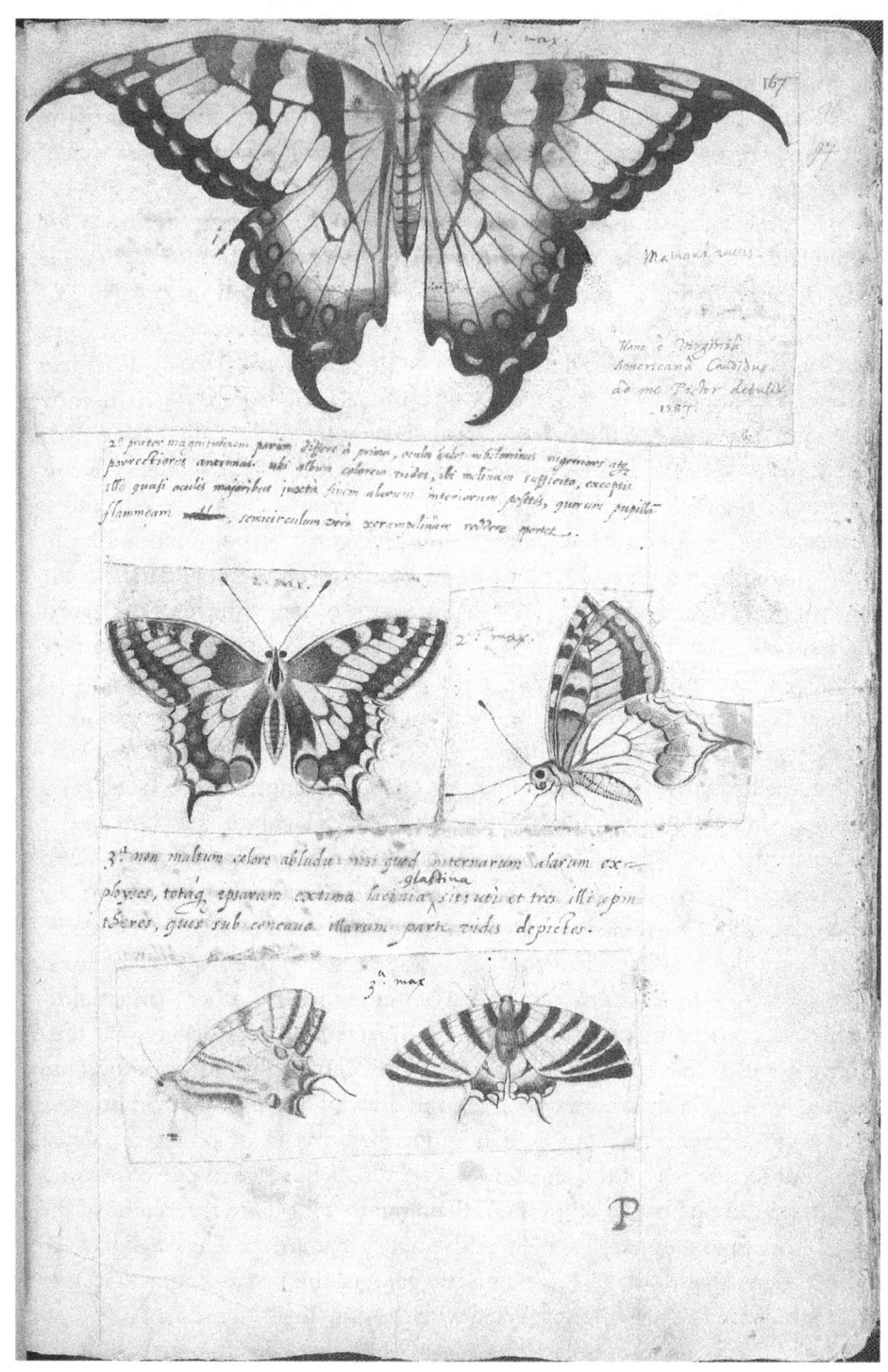

98. 97

6.ᵃ ala superiores extra nigrescunt per mediam partem limbo quodam
obsoletius rubido currente. extremitate ipsorum paruo, guttisq niueis
micant, obscuris per ambitum crenulis asperatæ. intus autem limbus
ille priorem atq saturiorem exprimit colorem, et juxta radicem
cærulæ euidentiæ. Inferiores alæ alteram intus, alteram foris faciē
ostendunt. foris fuscæ sunt totæ, excepto spinosa uestita subrubente limbo
perpusillis nigris 4 soricülis, ex Oualis duobus polychrois simul
positis notatæ. intus autem nigel talæ monstrant, sed ex nigro
paruo cærco uermiculato in exemplo substilis uniquncæ des deſinit.
corpus illi nigrum, oculi ateßuæ, pedes concolores, fuſci.

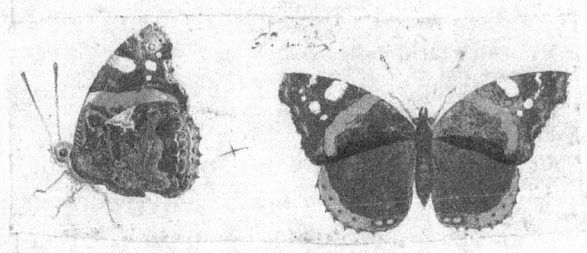

7.ᵃ corpore toto picea, in singulis dorsi tamen meiſuris punctulos
duos gerit albiſsimos. alas ex flauo rubeſcentes maculæ nigræ candi-
de�q albæ adornant. Oerum munifica rerum parens natura extre-
mam alarum oram potiſsimum decorauit, quæ nonuullis denticellis
serratim æquē interuallis distantibus donatæ, in quarum fimbria
uiginti clauæ cærulei filo nigre transfixi mirificum edunt splen-
dorem. /

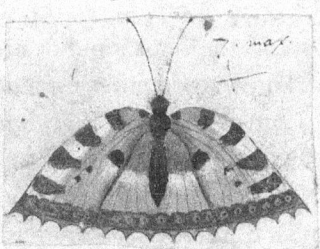

and image together. Moffet thus used the technique of cutting and pasting to bring the natural world to order within the medium of print, and in so doing developed an arrangement of text and image that could conform to the format of the printed page while preserving the variety of shapes and sizes found in the insect world.

The *Theatrum insectorum* is not only a record of the material and visual practices Moffet developed for constructing his version of the insect world but also a record of the intellectual exchange and collaboration that characterized the study of nature in late sixteenth-century Europe. Although Moffet was part of a far-flung network of domestic and international correspondents, for Moffet these collaborations were anchored in the day-to-day experiences of his specific milieu. Moffet was a member of a community of naturalists centered around the neighborhood surrounding London's Lime Street, which also included Thomas Penny, James Cole, Mathias de L'Obel (Lobelius), Emmanuel van Meteren, and James Garret. As Deborah Harkness has shown, the Lime Street community was international in its makeup, consisting of native English people as well as immigrants who corresponded with other botanists, naturalists, and students of nature throughout Europe. [61] On the whole, members of the Lime Street community were financially self-sufficient and did not rely on patrons for their livelihood, and they engaged in intense collaborations based on their shared interests in exploring nature and solving common problems. One drawback to this highly personalized and economically independent community was that its members did not feel compelled to publish their findings, with the result being that their activities and contributions quickly faded from memory after their deaths.[62] Moffet was a key member of the Lime Street community, but his experiences with regard to publishing his work were slightly different from those of other members; although the *Theatrum insectorum* was not published during his lifetime, Moffet did make great efforts to see the work to publication. As I will discuss later in this chapter, the book's publication in 1634 insured that Moffet's work on insects *was* known to later seventeenth-century practitioners, and it played an important part in continuing the community-building activities centered on insects that Moffet helped to establish. In other ways, too, Moffet did not completely fit the Lime Street model, notably because he had courtly connections and was a member of the College of Physicians. His dealings with that institution, his patron the Earl of Essex, and his support for several un-trained medical practitioners sometimes put him into awkward situations with regard to patronage—as Frances Dawbarn notes, his "loyalties were compromised by his triple role as client, broker, and Fellow of the College."[63] Nevertheless, it was Moffet's position as a broker of knowledge that allowed him to create a book in which a new subject and a community devoted to that subject could be formed. The unique nature of the book as a virtual collection and community enabled it to function in this way long after it was set in print.

Studying Insects in the Later Seventeenth Century: Engaging with the Sources

Moffet's insect research would not have become known to the wider public in printed form without the intervention of the French physician Theodore de Mayerne (1573–1655), who arranged for the publication of Moffet's manuscript in 1634. Mayerne's interest in Moffet's manuscript came about as the result of a number of converging interests shared by both men, among them medicine, Paracelsianism, the investigation of nature, and the materials and techniques of painting. Prior to his move to London in 1611 Mayerne served as court physician to Henri IV, and he was involved in the debates about Paracelsianism that divided the French medical community there during the first decade of the seventeenth century. In England, Mayerne continued to serve royal patrons as the court physician to Elizabeth I and James I, but he was also active in a range of intellectual pursuits that brought him into contact with practitioners from varying backgrounds, in both face-to-face contexts and through correspondence.[64] Moffet and Mayerne's interests in chemical medicine meant that they both would have been very familiar with distillation manuals and medical recipe books. As discussed earlier in this chapter, for English practitioners one of the most important of these books was Conrad Gessner's *The newe jewelle of health* (1576). This book came to be published in England under circumstances similar to those that gave rise to the *Theatrum insectorum*. *The newe jewelle of health* was published through the collaboration of a surgeon named George Baker and a book and manuscript collector named Thomas Hill, who may also have been responsible for translating the book into English. Hill died before the book was published, leaving Baker to finish the manuscript and see it to publication, a situation that had its parallel in the passing of Gessner's insect researches to Penny, Moffet, and ultimately to Mayerne.

It has been noted that the *Theatrum insectorum* is in many ways a medical treatise, containing as it does many accounts of the effects of ingesting or physically engaging with insects.[65] In its overall structure as a compilation of information gathered from numerous sources over the course of decades, the *Theatrum insectorum* also resembles the books of medical recipes that gathered together information from similarly diverse sources. Moffet was involved in several other publications in which such methods were used, and Mayerne was also closely tied to this culture of compiling and collecting information. Mayerne is best known today for an extraordinary unpublished manuscript he produced over the course of thirty years that treated art, alchemy, and numerous other material processes. Mayerne's manuscript is an important source for present-day art historians because it is a rich source of information on early modern painter's pigments and processes. It is also an intriguing working document that is part recipe book and part painting manual, as well as a record of Mayerne's voracious appetite for knowledge about material, visual, and natural processes.[66] Mayerne's lengthy entry on lapis lazuli is characteristically broad ranging in the applications and effects it encompasses. Mayerne uses the language of both alchemy and medicine to instruct artists in how they can make paint using the "tincture" of lapis lazuli: "Take lapis Lazuli which is blew and full of yellow veines what quantitie you will, breake it in frustuls as bigg as a beane, in an iron mortar: Then putt it in a crucible, uppon a grate fier till it bee red hoate." He later describes the resulting substance as an "ointment," and relates some of its virtues: "It also comforteth the brayne, and theref[ore] is very proffitable agaynst frensies, vertigo, palpitatio cordis, melancholia and other sicknesses of the spirits."[67] For Mayerne, the concepts of alchemy were central to his under-standing of the transformative processes of nature, and it was an important aspect of his interest in insects. Mayerne contributed his own preface to the *Theatrum insectorum,* and he makes clear his view of the connection between insects and alchemy. He inveighs the reader to "consider the generation and beginning of insects, and . . . weigh the various transmutations which they undergoe . . . And if animals and plants be transmuted, why should that be denied to met-als?"[68] Unlike Mayerne, Moffet's brand of Paracelsianism seems to have been limited to chemical medicine and did not include a belief in the transmutation of base metals into gold.[69]

Despite their differences of opinion regarding transmutation, Mayerne and Moffet's interests and approaches were compatible on many levels, and their collaboration on the *Theatrum insectorum* was a success, although they never met one another. Through their combined efforts the *Theatrum insectorum* was finally published, and its collaborative and communal character was maintained throughout the seventeenth century. Aldrovandi's *De animalibus insectis* and Moffet's *Theatrum insectorum* continued to be used as reference works on insects well into the seventeenth century, along with later illustrated publications on insects by authors such as Jon Jonston, Jan Goedaert, and Martin Lister. Later practitioners used this growing corpus of insect publications as guides for their own study of the natural world, and illustrated natural history books continued to form communities long after the initial phases of their research and publica-tion had been completed. The *Theatrum insectorum* in particular served as a site around which practitioners continued to come together to study insects, but the work of these later practitioners shows that the parameters established by books such as the *Theatrum insectorum* and *De animalibus insectis* were by no means fixed. The two examples offered below demonstrate how later natural-ists both replicated the practices of their predecessors while at the same time actively reworked these publications for their own uses and purposes.

The first example is a copy of the *Theatrum insectorum* containing annota-tions dating from the 1690s.[70] The notes record details of the unknown owner's observations of insects and collecting activities and contain references to other books on insects. A possible candidate for the author of these annotations is Charles Dubois, an "amateur" naturalist who was active in London during the 1690s. Dubois compiled a notebook of drawings and notes about butterflies and other insects he collected between 1692 and 1695, and given his keen interest in insects he is very likely to have owned a copy of the *Theatrum insectorum*.[71] Although this attribution cannot be confirmed, the annotations nevertheless contain useful information about how the book was used and provide a glimpse into the late seventeenth-century community of naturalists whose interests included insects. The book seems to have been used as a sort of field guide by its owner, who recorded observations of insects directly into its pages, usually in the margins next to the printed images. Unlike Aldrovandi's massive tome, Moffet's publication was somewhat portable, measuring approximately 12½ inches by 8¼ inches by 1½ inches. As such, Moffet's book could conceivably have been carried in a traveling bag, though it does not seem to have been used in this way by the person who annotated it. Instead, observations were more likely recorded in the book shortly after they were made, perhaps transferred there from field notes. The annotations show that the author's investigation of insects proceeded in tandem with study of the illustrated book, field trips, and the study of specimens in his own collection and those of others. On page 106, the author wrote the following next to an image of a butterfly: "This Figure very much resembles the Fritillary Butterfly which I find in Hampstead wood about the end of June." Another note on the same page mentions two other images of butterflies, referring to them by number ("1st" and "3d"):

> 1. I found the 1st sort Jul. 11 1693 in the Chalk-pitt close near Charlton sucking the flower of the Origanum Anglicum . . . this is very beautifull and if curiously examined very different from the 3d kind which cometh from a Catterpillar and feeds on Ragwort.[72]

The owner of the book studied the images in the butterfly chapter very closely, correctly noting that two of the woodcuts were duplicates.[73] On several pages the owner has drawn lines between the images of butterflies and the text that describes them in order to correct a misalignment of text and image that occurred in printing. The technique used by Moffet of linking text and image was adopted by this later reader of the book to the extent that he or she was able to correct printing mistakes, thereby rendering the book more useful as a guide for identifying specimens.

Several notes show that the owner of the book may have had the opportunity to consult Moffet's original manuscript and its drawings in order to study the images in more detail. Above the woodcut image of John White's swallowtail butterfly, the owner has copied out the inscription that appears with the original drawing and includes the citation "Hanc a Virginia Americana candidus ad me Pictor detulit 1587. C Mss Moffet in Mus. D. Charlton." Since Moffet did not include this Latin inscription from the White drawing in his text, the owner of the book may very well have copied the inscription from the original drawing after studying Moffet's manuscript, or the information could have been passed on from someone else who had seen the manuscript. The notation "Mus. D. Charlton" is most likely a reference to the museum of William Courteen, an English merchant and collector of the late seventeenth century who took on the name of Charlton in order to divert his creditors after a series of financial debacles. Courteen spent great sums of money on his collection of antiquities and curiosities, and the inscription above indicates that he could have owned the Moffet manuscript at some point.[74] The owner of the annotated book may have consulted the swallowtail drawing in the manuscript in order to confirm his suspicion that the butterfly was not native to England. The question seems to have arisen with regard to a specimen collected by John Ray, as indicated in the following note from the book, dated July 24, 1696:

> Mr. Ray I understand caught this Butterfly in the north of England from whence Mr. Dale had one that he showed me which I find very much to agree with the Figure & description of this 2 & not of the last or no. 1 [White's swallowtail] as he supposed.[75]

The annotator believed that the specimen shown to him by Mr. Dale did not correspond to White's swallowtail, as Dale and Ray supposed; instead, the owner of the book believed it corresponded to a butterfly pictured on the following page. This question may have been resolved by consulting the inscription on the original drawing, which clearly states that the butterfly came from America. This series of annotations shows the continued importance of images for the study of insects in the seventeenth century, and that communities of practitioners continued to coalesce around networks of specimens, books, and drawings. The collaborative relationships between Moffet, Penny, and their associates that took shape in the printed pages of the *Theatrum insectorum* were replicated many years after its publication, and the book continued to serve as a catalyst for creating the types of relationships and interactions that originally brought it into existence.

Users of the *Theatrum insectorum* did not consider it a static body of information but rather a source with which they could actively engage. As the example above shows, the book shaped how one anonymous annotator organized his or her investigations of the insect world by providing a virtual space for consulting specimens and images, as well as a place for working through questions raised in conversations with other naturalists. Another late seventeenth-century practitioner, the English artist-naturalist Alexander Marshal (c. 1620–1682), also used the *Theatrum insectorum* as a tool for organizing his research on insects, and the technique of cutting and pasting was an important part of Marshal's work.

Marshal is known today primarily for his extremely skilled botanical paintings, of which most surviving examples are collected in an album of drawings held at Windsor Castle.[76] Marshal was also an expert gardener and naturalist with a particular interest in insects, and his collection of insects was known and admired by the English community of naturalists, collectors, and others interested in "curious" matters. During the 1660s and 1670s Marshal made many drawings of insects and recorded his observations of the insects that he raised. Several of Marshal's friends encouraged him to publish his research, and it seems that he may have intended to publish a revised edition of the *Theatrum insectorum*.[77] Unfortunately, like his predecessors Moffet and Penny, Marshal died without ever seeing his research on insects in print.

The majority of Marshal's drawings and notes on insects survive as a series of fragments in an album now owned by the Academy of Natural Sciences of Philadelphia.[78] An eighteenth-century catalogue of Marshal's drawings compiled by William Freind shows that the drawings and notes in the Philadelphia album were originally kept as two separate volumes.[79] One of these two volumes was Marshal's copy of the 1658 edition of the *Theater of Insects*, into which he pasted his own insect drawings and notes. The other volume of Marshal's insect materials was listed by Freind as the "Supplemental Volume of Painted Insects with M.S.S. Annotations."[80] The drawings and text fragments in the Philadelphia album show that Marshal obtained insect specimens from a network of acquaintances and associates, as well as through his own collecting efforts, and that his collection included insect specimens from India, Africa, and North America.[81] Marshal annotated and pasted drawings into thirteen pages of his copy of the *Theater of Insects*; several of the drawings in the Philadelphia album retain fragments of the pages on which they were pasted. These fragments, combined with the Freind catalogue entries, make it possible to understand something of Marshal's working methods. Freind notes that Marshal added drawings of locusts in three different colors to page 983 of the *Theater of Insects*; one of these can be matched to figure 2 on folio 25 of the Philadelphia album (Figure 2.14).[82] In this image, Marshal has drawn a purple locust of the same dimensions as the African locust pictured on page 983 (Figure 2.15), and he has pasted it directly on top of the woodcut image. Marshal surely must have made his drawing in consultation with a specimen, although his note accompanying the drawing indicates that it came from Hispaniola rather than Africa, as Moffet's did.[83] Regardless of the origin of Marshal's purple locust, his observations and visual representation of the specimen were filtered through his experience of reading and viewing Moffet's text and image of the same type of insect. Moffet's entry on the insect is as follows:

> I procured one from Barbary that was brought out of Affrick with some
> cost to us, slender, five inches long, hooded, the head pyramidal, very

long, out of which almost at the top came forth two little broad cornicles about an inch long, much like that Turbant, which the Turkish Janizaries use with two feathers in it: a little below the root of it come forth two eyes standing out, great, dark red, the body long, of a bloud red purple; the tail like a Swallow two-forked, four wings of somewhat an ash-colour, deckt with certain dunnish spots; the four former feet and shanks very slender; the hinder strong, brawny, and long, and by reason of the spots drawn athwart all along the thighs blackish.[84]

Marshal's drawing of the purple locust corresponds very closely to Moffet's written description from the "bloud red purple" coloring of its body to its "ash-colour" wings with "dunnish spots," and it adopts the same stance as the insect pictured in the woodcut, differing from it only slightly in the omission of one of the rear legs.

Although Marshal followed Moffet very closely in his rendering of the purple locust, not all of his revisions and additions to Moffet were pasted into his copy of the *Theater of Insects*. The larger part of his insect material was kept in the sixty-three-page "Supplemental Volume." However, the connection to

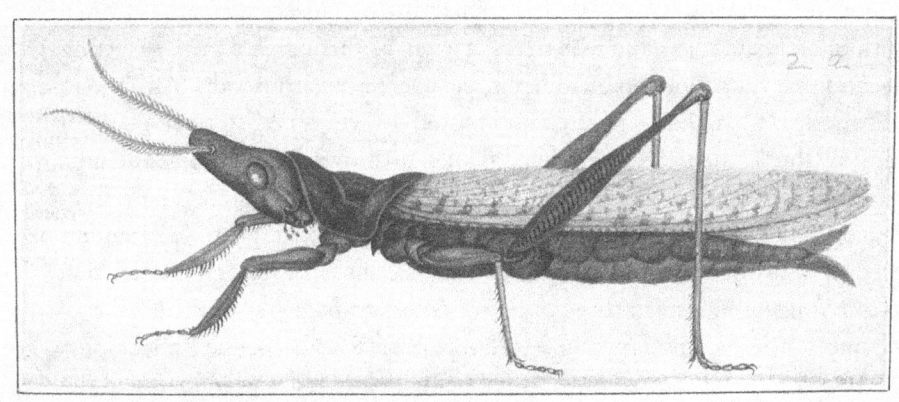

Figure 2.14. Alexander Marshal, *Purple Locust*, Marshal Album, Coll. 941, page 25, figure 2. Courtesy of The Academy of Natural Sciences, Ewell Sale Stewart Library, and the Albert M. Greenfield Digital Imaging Center for Collections, Philadelphia.

their Gods. Of this *Italian Mantis* (whoſe figure we do here repreſent) *Rondeletius* makes mention in his book *de Piſcibus,* in theſe words : *It hath a long breaſt, ſlender, covered with a hood, the head plain, the eyes bloudy, of a ſufficient bigneſſe, the cornicle ſhort, it hath ſix feet like the Locuſt, but the foremoſt thicker and longer than the other, the which becauſe for the moſt part ſhe holds up together* (*praying-wiſe*) *it is commonly called with us* Preque Dieu, *the whole body is lean.*

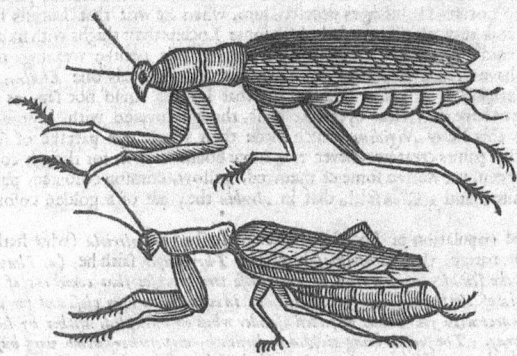

So divine a creature is this eſteemed, that if a childe aske the way to ſuch a place, ſhe will ſtretch out one of her feet, and ſhew him the right way, and ſeldome or never miſſe. Her tail is two forked, armed with two briſtly prickles : and as ſhe reſembleth thoſe Diviners in the elevation of her hands, ſo alſo in likeneſſ of motion ; for they do not ſport themſelves as others do, nor leap, nor play ; but walking ſoftly, ſhe retains her modeſty, and ſhewes forth a kinde of mature gravity. Though *Pennius* affirms that he often ſaw this kinde at *Montpellier,* yet in his papers he ſaith that he received the figure of it from the worthy *Antonius Saracenus,* a Phyſician of *Geneva.*

Another ſpecies of this *Mantis,* *Carolus Cluſius* ſent from *Vienna* exactly deſcribed, being brought thither out of *Greece,* which is like unto the former in ſhape and magnitude, but of another colour beſtowed on it either by nature or the place where it lives ; for it hath cornicles of a full yellow, the eye of hyacinth colour, the wings of a faint yellow, the reſt of the body of Amethyſt, only that the feet ſhanks, as alſo the joynts of them were more hairy and white, and the clawes of the fingers bended backward were black.

I procured one from *Barbary* that was brought out of *Affrick* with ſome coſt to us, ſlender, five inches long, hooded, the head pyramidal, very long, out of which almoſt at the top came forth two little broad cornicles about an inch long, much like that Turbant, which the Tur-

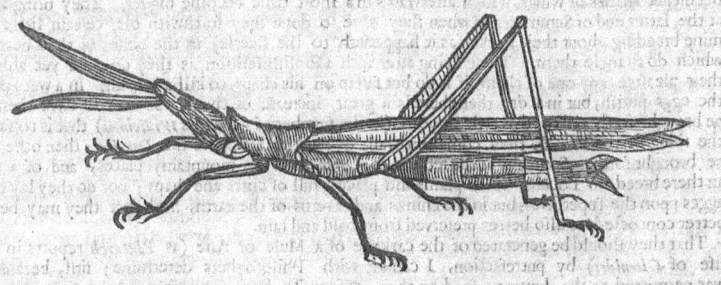

kiſh Janizaries uſe with two feathers in it : a little below the root of it come forth two eyes ſtanding out, great, and of a dark red, the body long, of a bloud red purple ; the tail like a Swallow two forked, four wings of ſomewhat an aſh-colour, deckt with certain dunniſh ſpots ; the four former feet and ſhanks very ſlender ; the hinder ſtrong, brawny, and long, and by reaſon of the ſpots drawn arhwart all along the thighs blackiſh. And this of the common or ordinary and winged Locuſts, and of the rarer ſorts ſhall ſuffice to have been ſaid ; unleſs the Reader ſhall think fit with me to add more differences of them. The face of the ordinary Locuſts is fierce, long, wrinkled, fenced as it were with ſcales, which even cover the mouth : in the upper part they have teeth faſtned that are broad, black, and very hard, with which they eaſily eat ears of corn, and ſcranch them with a great noiſe. The *Greek* and *African* Locuſt appears with a ſhor-
ter

Moffet was present in the Supplement as well. Freind notes that Marshal pasted a drawing and a written description of a beetle into page 978 of the *Theater of Insects,* but he goes on to say that several pages of the Supplemental Volume contained "Additional Observations to be inserted at the End of Page 978, in Mouffet, with Two paintings of The Lantern Fly, and One of a Small Night-fly, with Hist[orie]s of Each."[85] These two paintings of the lantern fly correspond to the drawings on page 23 of the Philadelphia album (Figure 2.16). Here, Marshal presents two views of this monstrous and impressive creature—from above, showing the insect with its wings extended, and in profile, facing to the left. Marshal's rendering of the lantern fly recalls elements of Hoefnagel's open-wing stag beetle discussed in chapter 1— in particular, the sinewy veins of the translucent wings. Marshal's lantern fly also brings to mind Aldrovandi's virtual specimen cabinet, in which the dual view is used in order to illustrate multiple facets of the insect's appearance. Though his presentation of the lantern fly draws on these older conventions for illustrating insects, Marshal has made this insect uniquely his own, in both the spectacular image he created and in his written notes. Marshal writes that he obtained the specimen from an acquaintance whose brother lived in India for many years, and he describes how such insects were kept and fed in India. Marshal seems to have kept one or two specimens of the lantern fly alive, and was thus able to report his own observations of its physical appearance and feeding habits. He lays claim, moreover, to being the first to write about and draw this particular type of light-emitting insect: "This fly having never before been spoken of, nor represented." To make his association with the lantern fly complete, Marshal included an image of himself with it. Freind notes that "on the Back of this Leaf is pasted a Head drawn upon blue Paper by Mr Marshall [sic]; supposed to be the Portrait of Himself done by his own hand." This self-portrait is now lost, and no other likenesses of Alexander Marshal are known to exist, but it is an important reminder of how an artist-naturalist such as Marshal constructed his professional identity through creating and cutting and pasting images. Although the lantern fly image and notes were very likely intended for a revised edition of Moffet, Marshal retained ownership of the insect by cutting and pasting himself into the image.

Bringing Forth "Nocturnal Studies" into Light

Great amounts of time, money, and effort were devoted to the publication of Ulisse Aldrovandi's and Thomas Moffet's books on insects. Both publications required the organization of vast amounts of information collected over many years that took the form of drawings and other types of images, written observations, texts by classical and contemporary authors, and specimens. Such a varied and unruly body of information could not be easily accommodated to the medium of print. Both Moffet and Aldrovandi had to devise strategies to

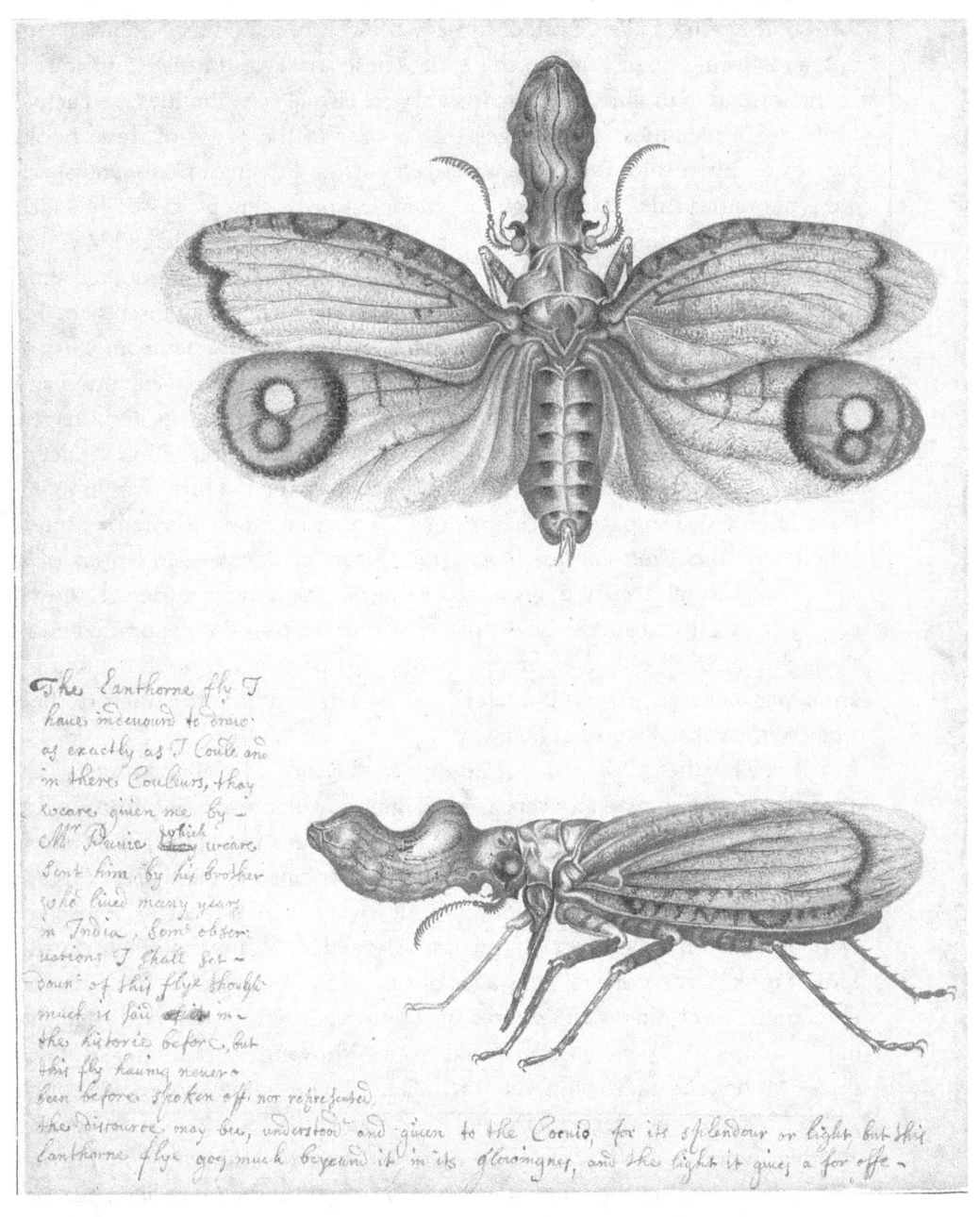

The Lanthorne fly I
haue indeuourd to draw
as exactly as I Coule and
in there Couleurs, thay
weare giuen me by —
Mr Prince which weare
sent him by his brother
who liued many years
in India, some obser-
uations I shall set —
down of this flye though
much is said of it in —
the histories before, but
this fly hauing neuer-
been before spoken off, nor represented,

the discourse may bee, understood and giuen to the Cocuio, for its splendour or light but this
lanthorne flye goes much beyound it in its glowingnes, and the light it giues a for offe —

Figure 2.16. Alexander Marshal,
Two Views of a Lanthorne Fly, Marshal
Album, Coll. 941, page 23. Courtesy of
The Academy of Natural Sciences, Ewell
Sale Stewart Library, and the Albert M.
Greenfield Digital Imaging Center for
Collections, Philadelphia.

manipulate, edit, and trim the insect world to fit it into the two-dimensional form of the printed page. Both of these naturalists relied on techniques of cutting and pasting to accomplish this task. These strategies in turn contributed to the formation of virtual specimen cabinets, virtual communities, and actual professional identities. The images that appear in the pages of these books did not simply record preexisting arrangements of drawings, notes, and specimens but rather emerged through the multiple, convergent processes by which these naturalists came to organize their vision of the insect world and in which images and image making played a central role. Later naturalists adopted similar strategies for managing the information they gathered about insects and for guiding them in their own investigations. The *Theatrum insectorum* in particular served as a site for later naturalists to continue the process of exploring and ordering the insect world through active engagement with texts and images. While naturalists such as Aldrovandi and Moffet expended great effort on defining the parameters of the insect world, the scope of their control was in many ways limited to the pages of their books. The anonymous annotator's interaction with the *Theatrum insectorum* and Alexander Marshal's revisions of it show that natural history publications were not static repositories of knowledge. Just as nature came to be contained within the pages of the printed book through the continually evolving processes of reworking, regrouping, and rearranging materials, so too did later practitioners continue to structure their study of the natural world in this way.

For all of the collaborative relationships that formed around Moffet's *Theatrum insectorum,* however, there is one significant figure missing from them: Ulisse Aldrovandi. Or rather, Aldrovandi's *name* does not appear in Moffet's book, but his images are present. Moffet's illustrations of silkworms (Figure 2.17) are direct copies of those found in Aldrovandi's book, which themselves were based upon original drawings commissioned by Aldrovandi (Figure 2.18). Moffet took illustrations of the stages of silkworm development from several different pages of Aldrovandi's book. In addition, several are reversed—indicating that copies may have been traced from Aldrovandi's book after Moffet's death. Moffet and Aldrovandi did not collaborate with one another, although Aldrovandi may have been aware that Moffet and his associates were preparing a similar work. In his preface to *De animalibus insectis,* Aldrovandi states that he has heard that another person was making efforts in the same field, but that he himself had not seen the results of this other person's research. He goes on to urge this unknown author to publish the work as soon as possible, and he chides him for refusing "to bring forth his nocturnal studies into light."[86] Aldrovandi does not mention Moffet by name, and he may not have known the identity of his fellow insect researcher, possibly due to the complicated nature of the authorship of the *Theatrum insectorum.* The anxiety expressed by Aldrovandi in his preface about the work of another insect researcher seems to have been

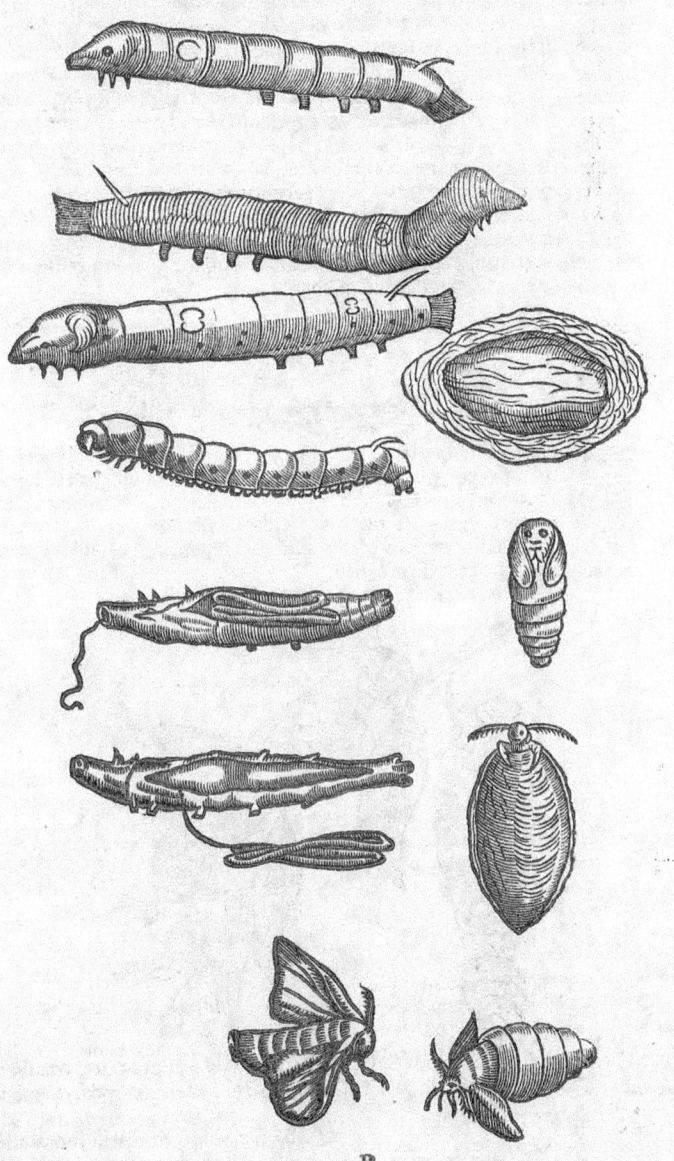

R

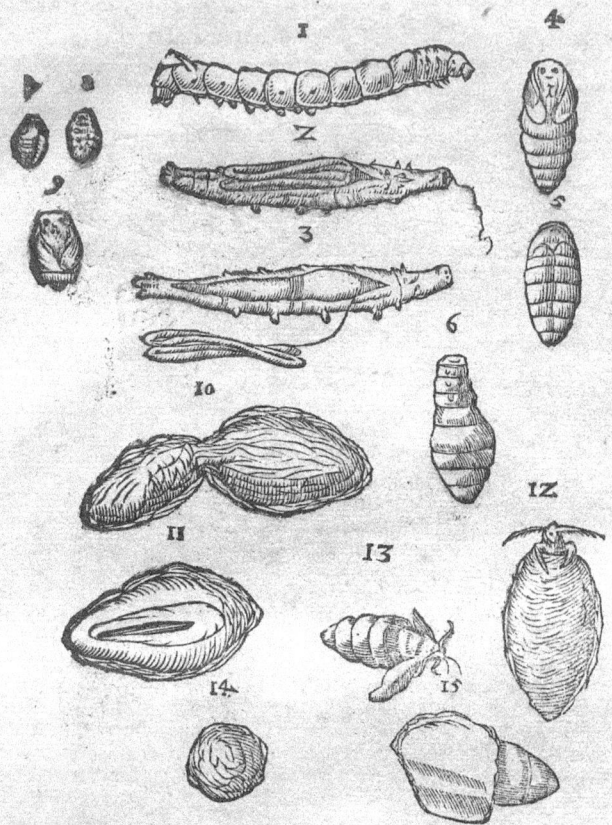

BOMBYCES VETERIBVS ROMANIS ET GRÆCIS FVISSE IGNOTAS.

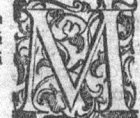

Sericũ non eſt diuersũ à bombyci no.

ᵃ Li.5.Hiſt. cap.19. Texere vbi prius inuē tum.

VLTI, quòd Erucas haſce ſiue Bombyces cùm Græcis, tùm Romanis, incognitas fuiſſe cernerent, Sericumǭ; Plinius & Virgilius à Seribus Scythiǫ populis ex arborum frondibus depecti ſcriberent, Sericum à bombycino diſtinguendum eſſe exiſtimarunt; ſed meo iudicio falluntur. Bombyces ve rò non agnouiſſe, nec Grǫcos, nec Romanos veteres ex ijs facilè colligitur, quæ de eis memoriæ prodiderunt. Tradit ᵃ Ariſtoteles fieri ex quodam Verme grandiore (citabo autem locum prout priùs reſtitui) qui veluti cornua protendit, ſuiǭ; generis eſt, primò toto immutato Erucam, deinde qui Necydalus appellatur (inua- ,, lidam dixeris) ex hoc Bombylius, hanc variam formarum ſucceſſionem, in ſemeſtri tem- ,, poris ſpatio compleri. Ex hoc animalis genere bombycia illa mulieres nonnullas retor- ,,

Figure 2.18. Ulisse Aldrovandi, *De animalibus insectis* (Bologna, 1602), page 282. Special Collections and Archives, University of Idaho Library, Moscow.

well founded. Not only did Moffet's book become more well known to certain circles of naturalists, but it also contained original research by Aldrovandi that was not acknowledged. Given the many hands that contributed to the authorship of the *Theatrum insectorum,* it is not possible to determine whether Moffet himself was responsible for using Aldrovandi's work without properly citing its author; doing so would have constituted a serious breach of etiquette within the "economy of obligation" in which Moffet and his collaborators operated. Strict codes of conduct among these naturalists were meant to keep disputes in check. As Harkness notes: "Adopting proper forms of address in letters, acknowledging learned assistance, and promptly repaying obligations were ways of expressing respect and collegiality."[87] Whether the failure to acknowledge Aldrovandi's illustrations was deliberate or unintended, it is an indication of the uncertain status of collaboration and competition in the study of natural history in sixteenth-century Europe and of the challenges facing those who ventured into new territory where subject matter and proper modes of social interaction were still under construction.

As I described in chapters 1 and 2, insects emerged as new subject matter during the 1580s and 1590s for European artists and naturalists in courtly, academic, and medical contexts. In this chapter I examine representations of insects in European still life painting from circa 1590 to 1620 and the establishment of insects as subject matter during the formative period of this new genre of painting. As artists developed conventions for representing objects in this new genre, they drew heavily from cultural practices associated with collecting and displaying the natural world, in particular those associated with kunstkammers, cabinets, and natural history. Still life artists chose to include the relatively new subject of insects in this new genre of painting, and this chapter addresses the convergence of these two new areas of artistic, scientific, and visual activity. The significance of the decision to include insects in early still life paintings is best understood by considering the context of these paintings within early modern European cultures of collecting. Artists such as Balthasar van der Ast, Ambrosius Bosschaert the Elder, and others in their circle actively participated through their still life paintings in defining new roles for material objects in connection with the new ways of studying, observing, and understanding the natural world that were developing in early modern Europe. In carving out new artistic territory for themselves, these artists used insects to appeal to the taste for exotic products of nature that dominated in courtly and noble circles. Because insects were newly collectable objects well suited to techniques of visual representation that emphasized small size, delicate structures, and intricate forms, they served the needs of this emerging artistic community and the developing persona of the master still life painter. However, still life painters offered their audiences carefully constructed worlds and were therefore attentive to the histories and associations of the other objects they presented. By picturing insects along with more established collectable objects such as shells and coins, still life artists generated an interest in insects that was connected to the new genre of painting, and ultimately to the artists themselves. Still life painters were purveyors of exotic goods that took both two-dimensional and three-dimensional form, and in this role they helped insects to attain the dual status of collectible image and collectible object.

Objects in Early Floral Still Life Paintings

When artists at the turn of the seventeenth century began to explore and define the new genre of still life painting, they had to decide which types of objects to include in their compositions. In these early still lifes, colorful and exuberant bouquets of flowers in rare or costly vases were the usual focus, and it has often been noted that artists mixed imagination with reality by combining plants that flowered at different times of year. Accompanying the bouquets are small items such as insects, shells, coins, medals, and jewelry that are strewn on the surfaces of tables and ledges or appear in the corners of niches. Why did early still life painters choose to include these particular objects, and why were insects among them? The answer to the question may at first seem obvious, since insects, flowers, and shells are part of the natural world, while coins, medals, and jewelry all have their material origins in the earth. What explains the shared presence of these items in early still life paintings, however, is not solely their connection to nature and the earth but also their status as objects suitable for display in cabinets and collections. Taken as a whole, flowers, coins, shells, insects, and other items can be understood as objects on display that are similar in form and arrangement to the rare and exotic objects found in a kunstkammer or natural history collection. In choosing to include these particular objects, early still life painters relied on principles similar to those followed by owners and arrangers of such collections and made their selections based on the criteria of rarity, small size, durability, and visual appeal. Objects possessing such qualities made them suitable for display in both the three-dimensional space of the kunstkammer and the two-dimensional setting of a still life painting.

Although still life painting as an independent artistic genre emerged at the turn of the seventeenth century, a variety of precedents existed for the visual representation of small objects. Natural history illustration and manuscript illumination form the two major visual predecessors of still life painting by providing artists working in the new genre with a conceptual framework for thinking about the relationships between the natural world, visual images, and collectable objects. In the sixteenth century, artists who produced illustrations for both printed books and hand-painted manuscripts presented the natural world as a series of precious objects to be examined, contemplated, and admired for the complexity and rarity of their physical forms. Unlike shells and coins, insects were relative newcomers to the economies of exchange upon which early modern European collecting was founded. Early still life painters contributed to the construction of insects as collectable objects by making them visible as appropriate items for a collection. By choosing to include insects alongside items whose value as collectable objects were already long established, still life painters were in effect providing a pedigree of sorts for insects. At the same time that they were making the case that insects should be included in the class of rare, collectable, and valued objects that went into a collection, artists were also using the cultural status of these objects to establish the cultural worth of still life painting itself. Like the drawings and illustrations of insects in the collections of Ulisse Aldrovandi and Thomas Moffet, still life paintings could serve as substitutes for physical specimens and could draw attention to aspects of a specimen that could not be viewed in the physical space of a collection. And, like the images in natural history collections and publications, still life paintings could be composed through a process similar to cutting and pasting, where flowers from different seasons and objects owned by different people could be brought together in a single (virtual) space. Still life paintings functioned as both two-dimensional images and three-dimensional objects worthy of display in the very spaces occupied by the objects shown in the paintings. As visual records of particular objects they could be understood as virtual collections, and in this respect the painting itself was transformed into a collectable object. Early still life paintings are therefore important examples of the ways that early modern European knowledge of the natural world was formed through possessing, displaying, and studying material objects, and of a particular type of knowledge that emerged at this intersection of material and visual culture. In developing techniques for translating three-dimensional objects into two-dimensional images, early still life artists helped to establish the status and value of objects within shifting networks of material and cultural exchange while also laying the groundwork for questioning the relationship between image and object.[1]

Insects and Early Floral Still Life Painting

The earliest European still lifes in oil date to around 1600 to 1603, but earlier examples on vellum appear in the last quarter of the sixteenth century. Insects appear in many of these early still life images as elements in floral bouquet pieces. Early floral still life paintings differ from those of the middle and later seventeenth century in several key ways. In contrast to the predominance of Dutch artists after around 1630, Flemish artists dominated in the early development of the genre. In addition, still life paintings after 1630 were closely associated with the golden age of the Dutch Republic, notably through its mercantile culture and growing market for art within the upper and middle classes. Although commercial interests and urban settings were also important factors in the development of early still life paintings, these earlier paintings had a closer connection to courtly and elite culture. Thus, the concerns and interests of these patrons, especially their enthusiasm for collecting wonders and rarities from the natural world, were key factors in the development of early still life. Their cabinets and collections provide the context for understanding the categories of objects that were deemed worthy of inclusion in early still life paintings.[2]

Insects were among the types of objects that appear within the rarefied spaces of both still life paintings and the cabinets of collectors. In still life paintings, the most frequently depicted insects were butterflies, moths, beetles, grasshoppers, dragonflies, and the common housefly. Apart from the housefly, artists generally favored insects with colorful and showy patterning, intricate structures, or both. In Roelant Savery's *Bouquet of Flowers with Two Lizards* from 1603, for example, a bouquet of flowers sits in a shallow niche surrounded by a variety of objects, insects, and other creatures (Figure 3.1). A large beetle, a large moth, and a housefly occupy the upper corners of the niche, while another large beetle, two lizards, and several shells occupy the ledge below. An easily overlooked large tiger moth rests upon the glass vase, and its size and positioning makes it difficult to distinguish it from the flowers.[3] Like other insects that appear within bouquets in early still life paintings, Savery's tiger moth offers surprise and delight to the attentive viewer who looks carefully enough to discover its presence. Chosen for their visual complexity and rarity, the insects Savery selected for this painting are similar to the other rare and exotic objects in early still life paintings.

Insects served several purposes in still life paintings. Symbolic interpretations of still life usually associate butterflies and moths with transience, or the stages of Christian life moving from birth and death to resurrection.[4] Flitting among the petals of the rare blooms, or striking an inquisitive, attentive pose in the foreground, insects also serve to evoke life and movement within otherwise static images. A parallel function is their role in signaling decay and death, as in Ambrosius Bosschaert the Younger's painting of a dead frog in which flies buzz around the corpse of a frog, thereby hastening its decay while at the same time heightening the contrast between its stiffening form and their lively movements.[5] Like other artists working in the genre, Bosschaert uses insects to invite the viewer to simultaneously marvel at the natural processes on display and at the artist's ability to create a lifelike image. This jarring contrast between life

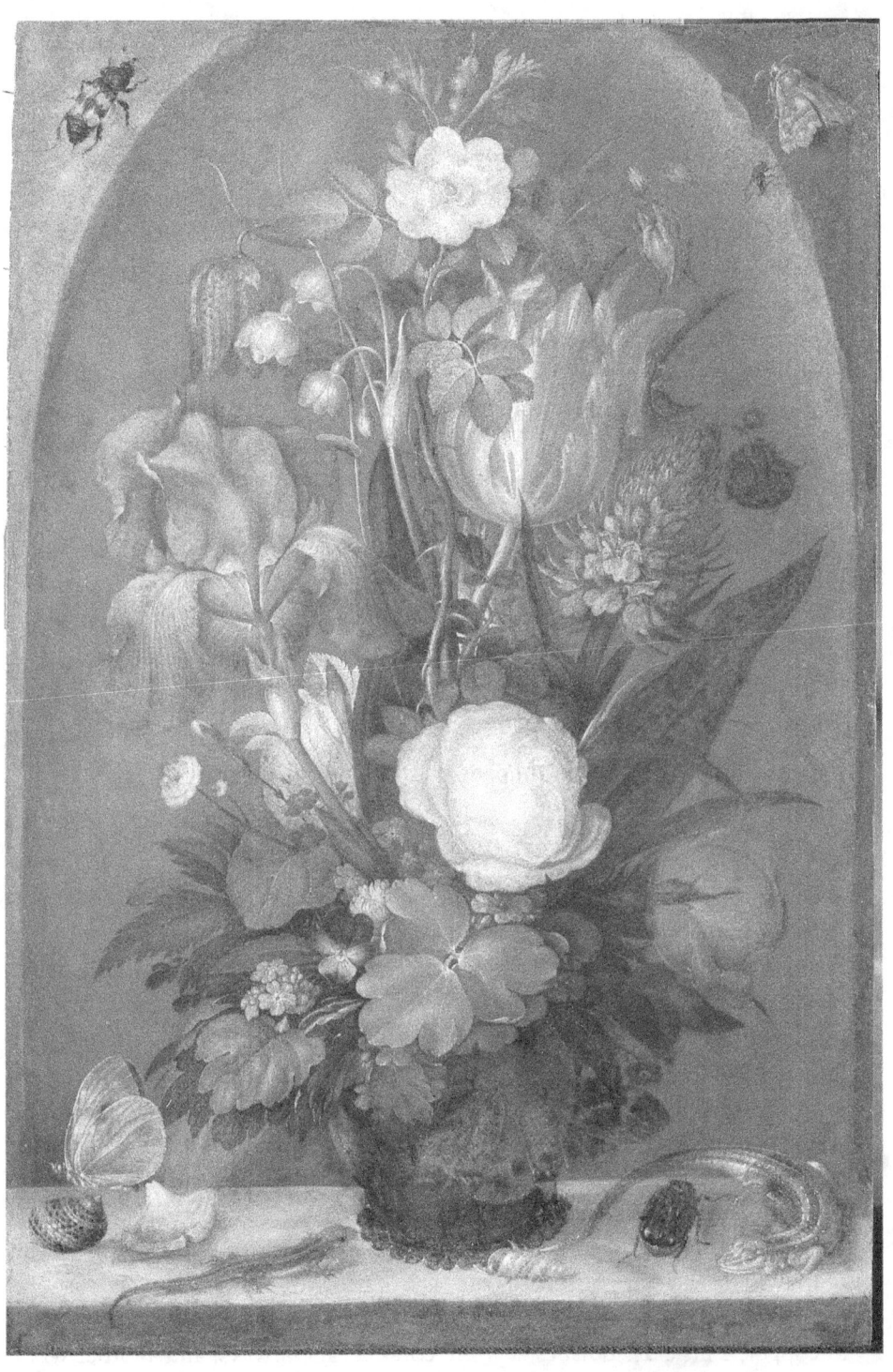

Figure 3.1. Roelant Savery, *Bouquet of Flowers with Two Lizards*, 1603. Collection Centraal Museum, Utrecht.

and death lies at the heart of the paradoxical appeal of still life. Insects, like the paintings themselves, occupy a position somewhere between living and not living that is in many ways similar to the shifting status of still life as both object and image.

The abilities of still life artists to create naturalistic images can sometimes lull us into overlooking the fact that we are in the presence of an intricately constructed environment with many artificial elements. It is easy to forget that insects such as the ones in Savery's painting generally live outdoors and not in dark interior niches. In nature, most insects can only be glimpsed fleetingly as they fly past our eyes or skitter across a surface. Visual images thus provided a means of controlling an otherwise chaotic natural world by fixing features and movements, and in the case of insects by fixing them in aesthetically pleasing poses where they are visible for the viewer's inspection and delectation. Insects in still life paintings would have been modeled after specimens or copied from existing illustrations. Although no insect collections from the late sixteenth century are known to have survived, collecting insects certainly occurred during this period.[6] Collecting and representing insects, already a serious pursuit in natural history circles by the 1590s, would have been one of the factors contributing to an artist's decision to include an insect in a still life painting. Natural history illustration was another important visual precursor.

Natural History Illustration: Precedents for Picturing Insects

While insects were new subjects for artists toward the end of the sixteenth century, other objects and creatures from the natural world had attracted interest from artists prior to this time and were already established as suitable objects for studying, collecting, and representing. Much of this activity took place in the field of natural history. Although Thomas Moffet's and Ulisse Aldrovandi's illustrated books on insects were not published until the seventeenth century, still life artists interested in insects had other visual precedents to draw from for constructing a framework for the insects. Shells in particular could have served as models of collectable objects from the natural world, since they possessed many of the same characteristics as insects. The appeal of shells and other marine objects to artists is evident from at least the early sixteenth century. Albrecht Dürer on his journey to Antwerp of 1520–21, for example, visited shell collectors and exchanged his books and prints with them for shells and other objects.[7] Dürer's journal records his exchanges with Lazarus von Rafensburg, from whom he received "a great fish-scale, 5 snail-shells, 4 medals of silver, 5 of copper, 2 little dried fishes, a white coral, 4 cane-arrows, and another white coral."[8] He traded his books with another collector, Rudiger von Gelern, for a snail shell and silver and gold coins, and he notes receiving pieces of coral from two other collectors.[9] Dürer's interest in fish and shells was part of his broader exploration of the visual and material forms of the natural world, but it was also connected to the emerging field of European ichthyology.

In the middle of the sixteenth century, European research and activity in the area of fish and shells expanded rapidly. A number of early works in the history of ichthyology were published in the 1550s, and shells were included in all of them. Like the later books on insects, illustrations were key features of mid-sixteenth-century books on fish, and several of the major sixteenth-century publications on fish featured numerous illustrations of intriguing shell forms.[10] The three leading authors on shells and fish—Pierre Belon, Guillaume Rondelet, and Ulisse Aldrovandi—focused much of their attention on specimens from exotic places and those that possessed unusual forms. Early still life painters who included shells in their compositions chose specimens with similarly unusual and intriguing characteristics. The illustrated books by Belon, Rondelet, and Aldrovandi could have provided these artists with many examples from which to model their forms. For example, the spiny shell shown in two views in Belon's book (Figure 3.2) resembles the intricate devil's claw shell in an undated painting by Bartolomeus van der Ast.[11]

Shells were also connected to artists' material practices by serving as both a source of pigment and painter's tool. Rondelet makes reference to the "painter's shell" *(coquille des peintres)* and describes how painters scrape the shell

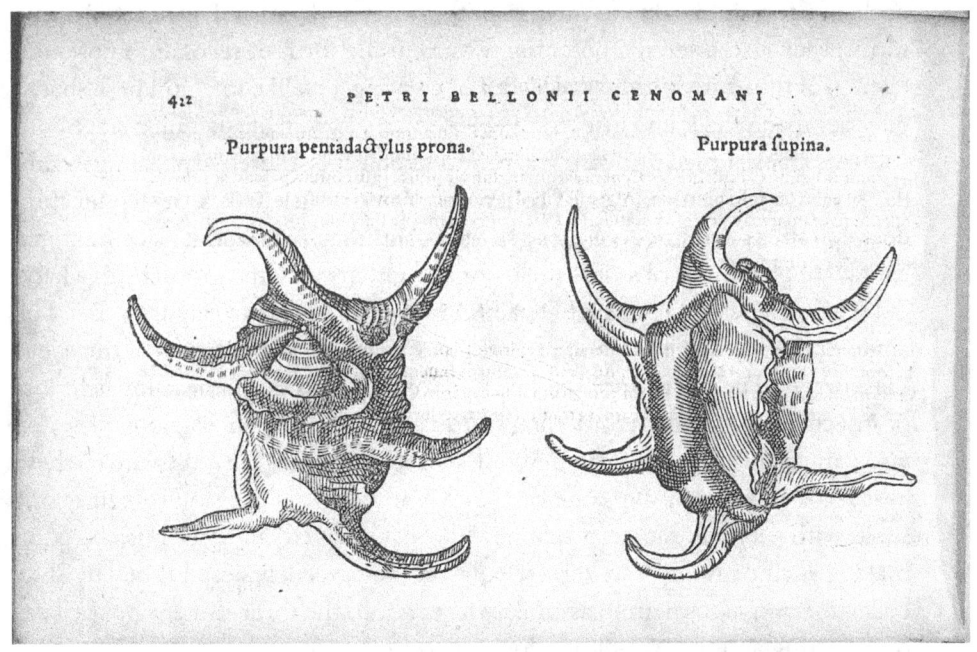

Figure 3.2. Pierre Belon, *De aquatilibus* (Paris, 1553), page 422. Photograph courtesy of Brown University Library.

in order to make pigment. Rondelet also mentions the use of mussel shells to mix and store paint.[12] This use also appears on the last folio of Joris Hoefnagel's *Mira calligraphiae monumenta* manuscript as part of a vignette showing the tools of the illuminator's craft (Figure 3.3).[13] Two halves of a mussel shell, one with gold paint visible in its interior, are pictured along with a bowl and a pitcher, all of which were vessels for mixing and storing pigments.[14] For artists such as Dürer, Hoefnagel, and Van der Ast, shells were collectable objects, tools, and artistic subject matter. Shells were also part of a tradition of humanist learning that promoted the leisurely contemplation of objects from the natural world. As natural objects displaying ingenious forms, shells were seen as a legitimate focus for the erudite scholar or courtier seeking worthy leisure time diversions. Shell collecting as a form of mental relaxation was encouraged by humanist intellectuals as part of the larger context of hexameral literature, in which the biblical account of Creation was amplified and extended partly through careful study and eloquent description of the natural world.[15]

Insects were thus in good company when they were paired with shells in early still life paintings, and the material and visual practices of studying, collecting, and displaying shells supplied a cultural context for still life painters when considering insects as subject matter. The rich intellectual tradition of shell collecting provided a framework for the contemplation of natural objects into which insects could be incorporated. Like shells, insects posed a technical challenge to painters due to their smooth, glossy surfaces and intricate, delicate forms; they also offered a potential reward in the form of recognition of one's talents for those artists who succeeded in creating a lifelike image. These shared qualities made shells and insects suitable subjects for still life painting, since both could be transformed into rare and beautiful objects—thereby appealing in similar ways to the sensibilities of collectors. Van der Ast makes this connection between shells and insects clear in several paintings, such as the placement of a katydid opposite a devil's claw shell,[16] or in another work the grouping of a katydid with other shells in a still life painting from around 1630 (Figure 3.4).[17] The angle of the insect's body and its triangular shape echo the forms of the shells that surround it, and the polished surfaces of the shells recall the shiny exteriors of insects. The parallel positioning of the insect and the shells along with the visual mirroring that occurs in the composition show that the artist viewed these as items as objects in the same category. Van der Ast's decision to pair exotic shells with a katydid suggests that he viewed it as an equally precious object. In making such pairings, Van der Ast built on the precedents established by shell collectors and used them to assimilate insects into the three-dimensional spaces of collecting and the two-dimensional spaces of representation.

Natural history, notably through the specimen drawing, provided another important visual precedent for still life painters. Produced in large numbers in the sixteenth century in natural history settings, specimen drawings offered

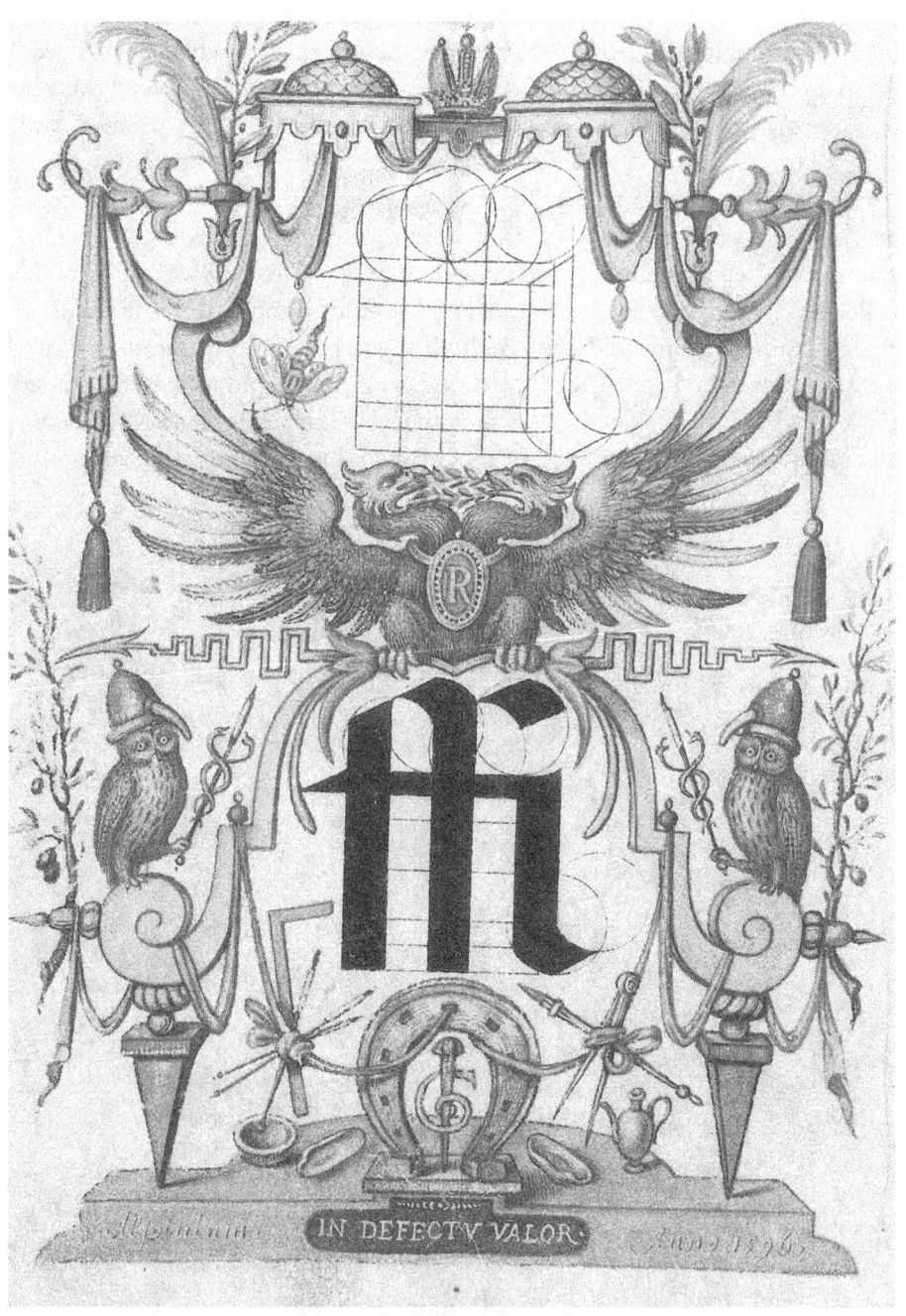

IN DEFECTV VALOR

Figure 3.3. Joris Hoefnagel, *Guide for Constructing the Ligature ffi*, c. 1591–96. The J. Paul Getty Museum, Los Angeles, Ms. 20, fol. 151v.

new ways of thinking about the visual representation of objects from nature in the context of collecting. A specimen drawing typically presents a single specimen, or multiple views of the same specimen, against a white background. Apart from shadows cast by the specimen, the background is usually devoid of contextual elements. The objects depicted are usually highly detailed, with clear edges and borders. Because specimen drawings from the early modern period emphasize external features, they were well suited to species that could be visualized as discrete objects with complex surfaces and textures. Take, for example, a shell presented in two views in a drawing from Ulisse Aldrovandi's collection (Figure 3.5). Because of its small size, it was possible for the artist to represent the shell life-sized, with multiple views presented within the space of a single drawing. The shell's edges are clearly defined, and its surface appears thick and dense. Painted in thin layers of gouache or watercolor, the shell serves as a suitable vehicle for exercising the artist's skills at creating the illusion of

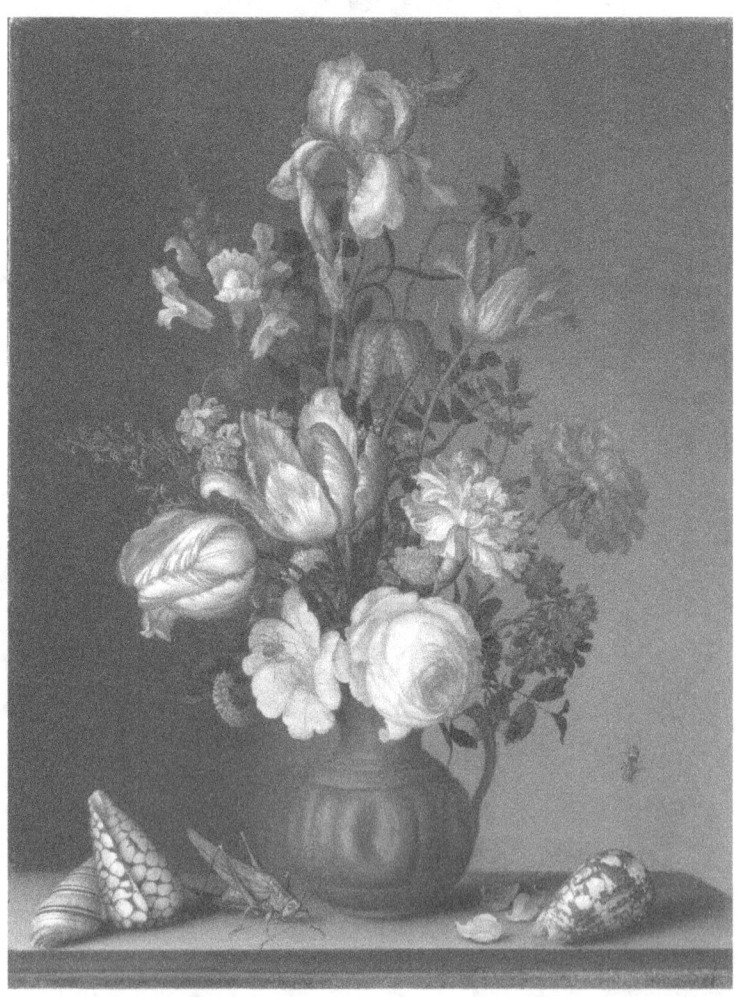

Figure 3.4. Balthasar van der Ast, *Flowers in a Vase with Shells and Insects,* c. 1630. Copyright National Gallery, London/Art Resource, NY.

texture, and the overall effect is that of a heavy object resting upon a flat surface. This shell drawing is part of Aldrovandi's enormous collection of natural history drawings, and it appears in the same volume that contains drawings of insects.[18] Taken together with the insect drawings in the same volume, the shells begin to approximate the subject matter and visual style of still life painting. For Aldrovandi, the drawings provided a space in which individual objects could be taken out of the context of the collection for prolonged and concentrated examination. The results of these investigations were visual images that stress the physical presence of the specimens and present them as highly individualized objects. The drawings are suffused with a mood of serious contemplation, not unlike the quiet spaces occupied by bouquets, insects, shells, and coins in early still life paintings. It is the logic of the specimen that dominates

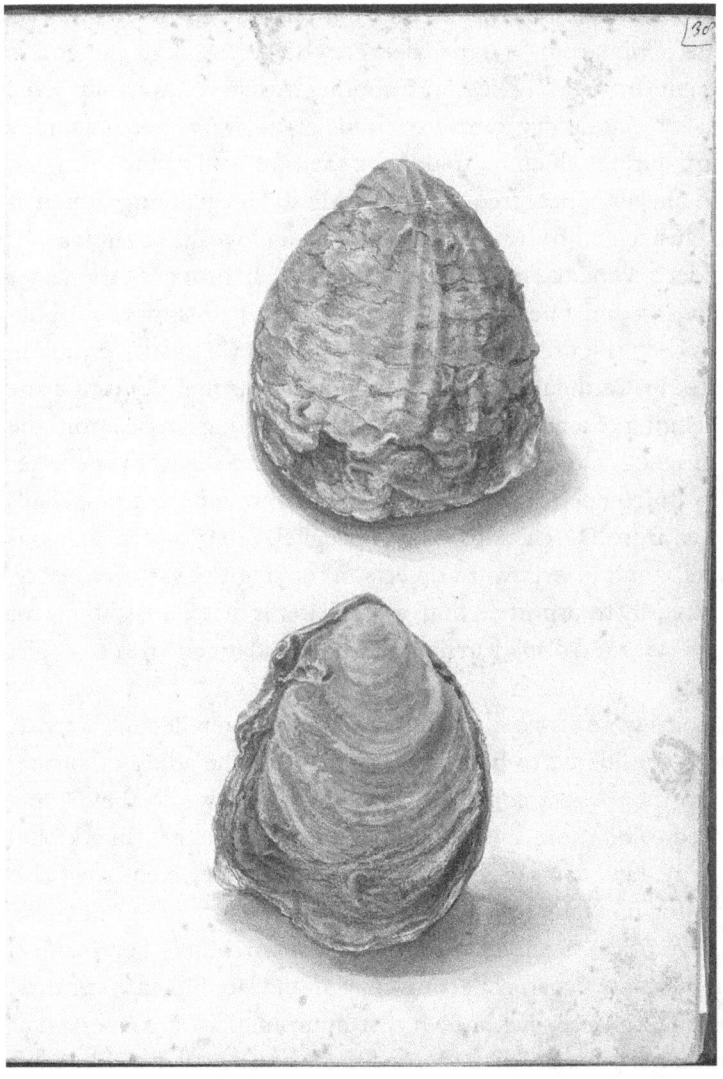

Figure 3.5. MS Aldrovandi, *Tavole di animali,* volume 7, page 38. Biblioteca Universitaria, Bologna. Courtesy of University Library of Bologna; no reproduction without permission.

the visual world of still life, a world in which nature is laid out for the viewer's prolonged examination by being removed from time and space to exist solely as "object."

The Realm of Small Things: "Not Too Plentiful like Stones"

Some of the most popular items for elite and wealthy collectors of this period were coins and medals. These objects had intellectual and artistic pedigrees comparable to those of shells and insects, and as such they appear frequently in early still life paintings. Coins and medals share the physical characteristics of being small, durable, and relatively easy to preserve—criteria that are practical but also essential for early modern European collectors. Some collectors amassed very large quantities of ancient coins and medals, which they displayed in cabinets and in rooms set aside especially for this purpose. Like other objects in collections of this period, coins and medals were also objects of exchange and were often sent as gifts from one collector to another. Artists too were attracted to coins as material objects, as demonstrated in the discussion above of Dürer's travels to Antwerp, during which he traded for shells as well as for coins and medals. Coins and medals appear frequently in early still life paintings, often as references to the elusive and fleeting nature of fame and wealth, as in Jacques de Gheyn the Elder's *Vanitas Still Life* from 1603.[19] Jan Brueghel the Elder included coins, jewelry, and a beetle as accompaniments to a sumptuous bouquet he painted in 1607 (Figure 3.6).[20] In describing a very similar painting in a well-known letter to Cardinal Federico Borromeo, Brueghel wrote that he was preparing a painting of a bouquet of flowers representing species from the royal gardens in Brussels, along with jewelry, medals, and shells. In the letter Brueghel also tells Borromeo that he will be sending him a box of rare shells from India obtained from Dutch ships.[21] For Brueghel, rare flowers, precious coins, jewelry, shells, and insects were objects occupying the same category. Small, durable, and easily transported, they were all items that were suitable for an elite patron to possess and display in his collection, and for an artist to depict in a still life painting.

Like shells, coins were also a site of intense humanist intellectual activity. Ancient coins were considered to be reliable sources for the study of ancient history, and due to their direct connection to the ancient world they sometimes were believed to be more reliable than written documents. In addition to providing the names and dates of rulers and historical events, coins were also important sources for portraits of ancient rulers. This aspect of coin collecting was particularly critical to the Habsburg collector Archduke Ferdinand II of the Tyrol, whose family began collecting coins in the late fifteenth century. The historian Martha McCrory has shown that numismatic sources served as models for the ancestral portraits that adorned the Spanish Hall at Ferdinand's

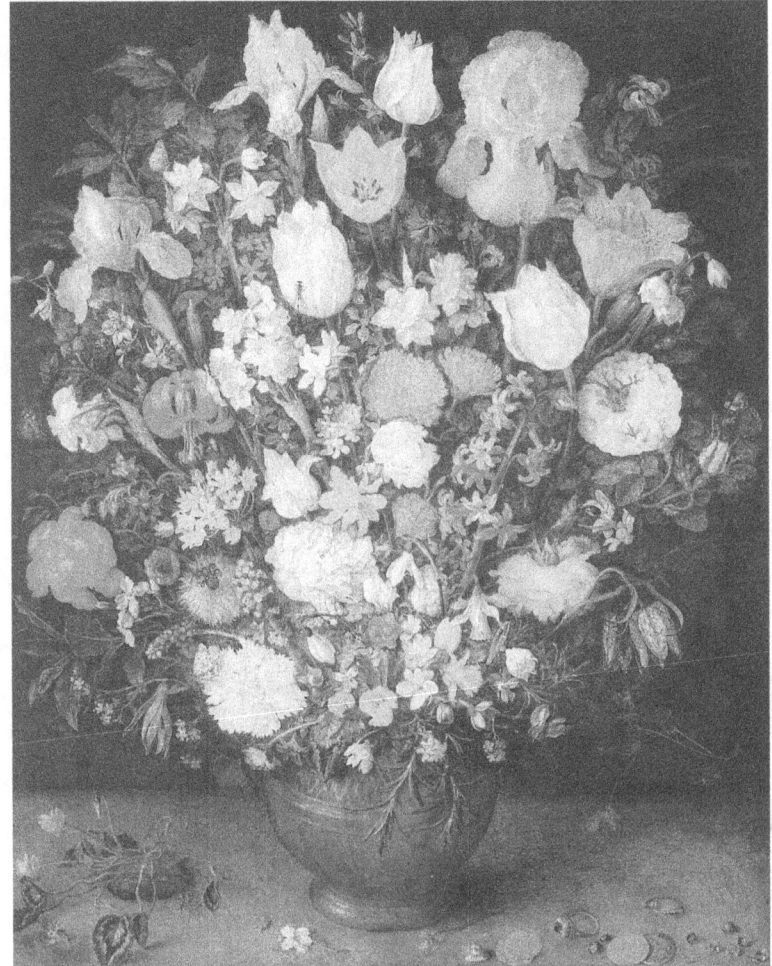

residence at Schloss Ambras. Lacking visual sources for some of his ancestors, Ferdinand contacted Francesco I de Medici to ask if he possessed medallic portraits that could be consulted for the project.[22] In making this request Ferdinand was participating in an established social ritual involving the exchange of coins between collectors. These small, portable objects were often sent in letters as gifts or tokens of friendship. As the historian John Cunnally writes: "The evidence indicates that antique coins circulated easily and rapidly throughout Renaissance Europe, passing from hand to hand among the humanists of every land as a kind of *koine,* a common bond or network of communication, which enabled lovers of antiquity to recognize and acknowledge one another."[23] Like the traffic in natural history specimens, coins circulated within dense networks of exchange among sixteenth-century collectors.

Coins, shells, and insects not only possess physical characteristics that make them suitable objects for exchange and display but also they are connected to the natural processes of the earth. Ancient coins were made of metal, and they were often obtained from hoards buried deep in the ground—the unearthing of which would provide another reminder of the connections between coins and the natural world. Cunnally notes that "Renaissance antiquarians regarded ancient coins as ubiquitous objects, constantly emerging from the earth."[24] Another important factor in maintaining the status of coins as collectable objects was that they, like shells and insects, were both rare *and* abundant. There were sufficient quantities available of the more common insects, shells, and coins to satisfy the desires of many collectors, but at the same time wealthier collectors could pursue rare, hard-to-find examples. In addition to being available in relative abundance, insects, shells, and coins also offered a great deal of variety to collectors. Cunnally points out that even though coins were abundant and easily available, they were still in high demand among sixteenth-century collectors.[25] Because many different types existed, large-scale collecting of these items was justified, as were methods of display that emphasized variety and abundance. A later seventeenth-century author writing on the history of coins and coin collecting noted this delicate balance between scarcity and abundance as one of the reasons why ancient Romans chose metal as the material for making coins. Metal was "frameable into any weight or figure, endures for longer and is not easily broken like wood, more plentiful than Jewels, not too plentiful like stones."[26] Coins, like insects and shells, were well suited to collecting because they were a category of object for which it was conceivable to amass an encyclopedic and comprehensive collection but were not so plentiful as to render them valueless.

As with natural history specimens, illustrated books and manuscripts were an important site for the construction of knowledge about coins and medals in early modern Europe. Numerous books on ancient coins were published in the sixteenth century, and these publications were also of great interest to coin collectors of the period. Books provided information to collectors about the coins and medals they owned as well as those they hoped to own, thereby functioning in some ways as virtual collections.[27] In much the same way that illustrated books on shells and insects could provide collectors with information on specimens they did not own or provide multiple views of specimens, illustrated books about coins could serve as placeholders or replacements for coins that were missing from one's collection. In some instances the connection between books, images, and coins took material form. Parts of Archduke Ferdinand II's coin collection were displayed in specially made books bound in black velvet and adorned with silver and gold clasps.[28] The use of these booklike display cases for coins blurs the boundary between three-dimensional objects and two-dimensional images, and it is this blurring of boundaries between images and objects that also lies at the heart of still life painting. This interchange between three-dimensional object and two-dimensional image also paralleled developments in the realm of commerce, in particular the changing form of money at the turn of the seventeenth century. Although coinage was the medium by which most everyday transactions were conducted in early modern Europe, by the early seventeenth century paper substitutes for coins were beginning to appear. In England in the early seventeenth century goldsmiths began taking on functions that would later form the basis of banking practices by using safes to store their materials and offering these safes to others for storing coins and other valuables. Goldsmiths issued paper receipts for the goods they stored, and these receipts began to circulate as payments for debts.[29] In the Low Countries, the use of paper substitutes for coins occurred in the late sixteenth century during the war with Spain.[30]

An early floral still life by Georg Flegel shows a cockchafer beetle on a table strewn with coins and other small creatures, an indication that still life painters of this period viewed insects as appropriate objects to include in an environment where worth resided in both physical properties and exchange value (Figure 3.7).[31] As with coins, insects were also objects for which paper substitutes in the form of drawings and prints were not only acceptable but also becoming increasingly valued by collectors. With insects, however, an additional element contributed to the value of the visual image beyond its function as a substitute for a specimen: namely, the artist's skill in depicting the insect's lifelike appearance. Still life painters understood this, and by including skillfully painted insects in their compositions they were able to exploit this quality to impart value to the new genre of still life painting. At the same time, still life painters helped to establish the value of insects by including them among other objects such as shells and coins that had established pedigrees as highly collectible items. In this way, new ways of thinking about objects merged with new ideas about the role, function, and value of two-dimensional images. Like other collectible objects from this period, the value of insects lay in their capacity to be, according to Thomas DaCosta Kaufmann, "temporarily or permanently removed from economic circulation."[32] Insects could be removed from nature and transformed into collectible objects, and images of insects presented specimens in a similarly decontextualized but elevated fashion.

Cabinets, Curiosity, and Still Life

Natural history collections such as that of Ulisse Aldrovandi were formed as part of the broader interest in collecting in sixteenth-century Europe. Princely collections formed in the same period included items of *naturalia* and *artificialia,* and objects that mixed natural and human artifice were especially prized. Collectors of the sixteenth and early seventeenth centuries can be seen as possessing universal, encyclopedic ambitions because of their passion for amassing immense numbers of material objects and visual images. But as Lorraine Daston

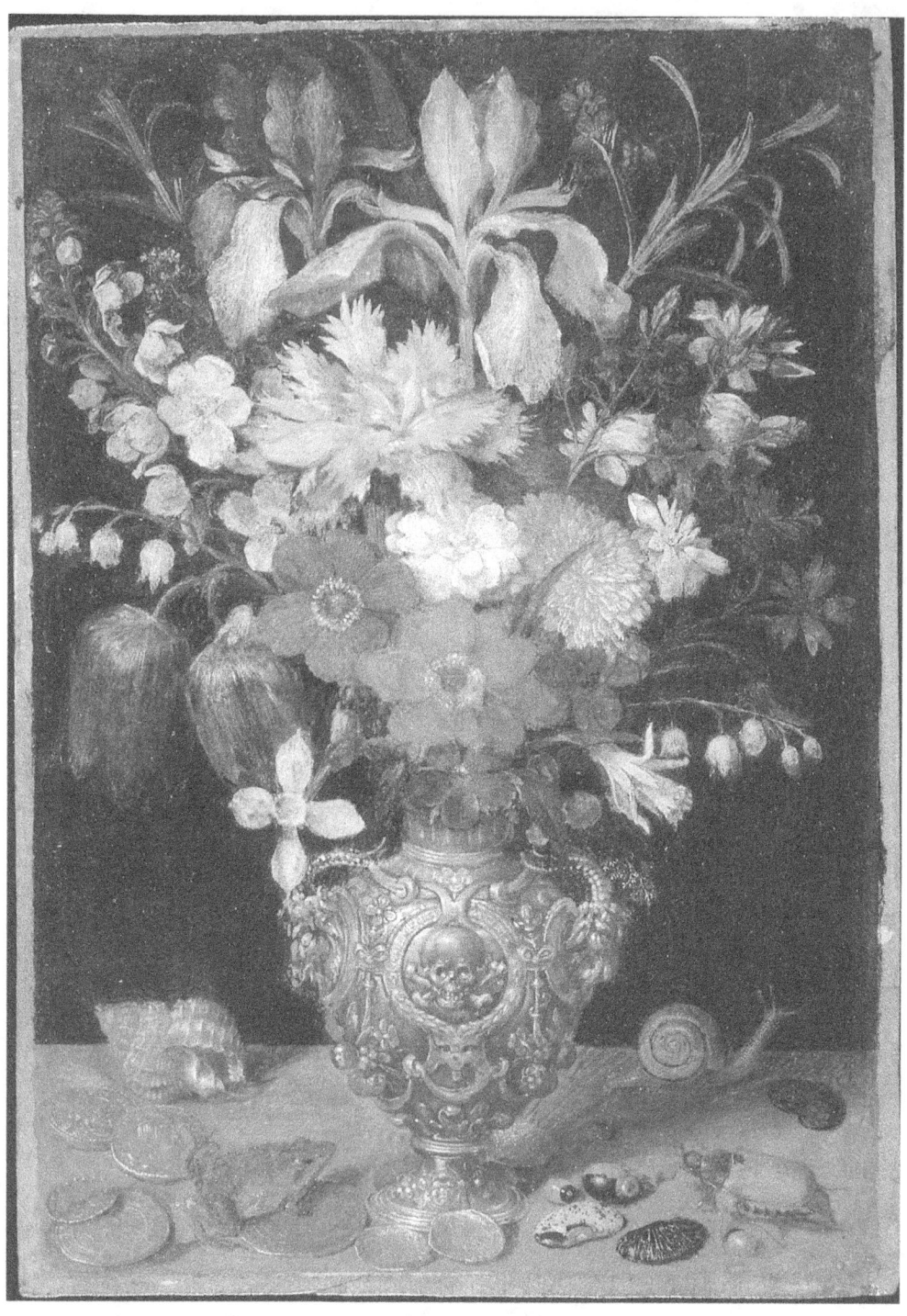

Figure 3.7. Georg Flegel, *Still Life with
Flowers*, c. 1604. Photograph courtesy
of Fitzwilliam Museum, University of
Cambridge.

and Katharine Park have shown, encyclopedic collections did not necessarily aim at universality but were instead concerned with achieving a particular type of broad coverage. According to Daston and Park, collectors of this period sought to amass large numbers of objects and images that adhered to the criteria of wonder or rarity.[33] This predilection for wonders and rare objects was part of the broader shift in attitudes toward curiosity and its role in the investigation of nature in the early modern period. While in earlier times curiosity was considered a vice or sin, in the sixteenth century it was increasingly understood as a virtue.[34] Princely collectors of the sixteenth century favored extraordinary specimens—or "curiosities"—over ordinary ones, and one of the organizing concepts of such a collection was that of a hierarchy, according to Kaufmann, "by which the world of nature no less than the world of states is organized into a coherent system."[35] The prince's collection, or kunstkammer, was aimed at celebrating the magnificence of the ruler, and as such it was intentionally mysterious rather than being designed for the masses. Paralleling this appeal to elite audiences, the realignment of curiosity during the sixteenth and seventeenth centuries directed attention toward a select group of objects to the exclusion of others. Curious objects were those that were rare, displayed fine craft in their construction, and were small and intricate, thereby revealing new details to those who spent the time to inspect them carefully and thoroughly. Daston has argued that cabinets and the objects in them embodied an "aesthetic of rarity, variety, and, especially, 'Delicacy and Excellency of the Workmanship,' [that] marked objects as worthy of inquiry as well as worthy of acquisition . . . The early modern psychology of curiosity reinforced this preference for the novel and the bizarre, and also channeled it toward the small, the intricate, and above all the hidden."[36] Insects—or at least certain types of insects—fit these criteria for curious objects, just as shells and coins did. When still life painters chose to include these items in their paintings, they did so to appeal to the tastes of collectors and patrons immersed in the late sixteenth-century culture of curiosity. Objects of small size were ideal for collectors, as were those that could withstand the rigors of transportation, preparation, and the passage of time, since smaller objects could be amassed in large quantities but stored and displayed in smaller spaces. Although some collectors and artists may have had particular interests in insects, the decision to include them in cabinets and still life paintings was also based on the similarities between insects and other curious objects. Artists working in the emerging field of still life painting at the end of the sixteenth century made works almost exclusively for elite patrons such as these, and in developing a visual language for representing objects in still life these artists drew from existing cultural practices associated with collecting and displaying wondrous and rare objects in order to appeal to the tastes of their patrons.

As Krzysztof Pomian points out, when dealing with the collections of this era we are entering into "the realm of small things."[37] While the terms "kunstkammer" and "cabinet" can refer to both collections in general and the specific item of furniture used to store and display items in a collection, the latter meaning is significant for understanding the relationship between collecting and still life painting. Material connections between still life, the practices of collecting, and the cabinets used to store and display collections are evident in some of the earliest examples of floral still life painting. The German artist Ludger Tom Ring the Younger is credited with painting the earliest independent floral still lifes in Europe—two panels from 1562 (Figure 3.8 and Figure 3.9). Each panel presents an alabaster vase, one containing irises and another containing irises and lilies, against a dark background resting upon what appears to be a tabletop of burled wood in a narrow, rectangular space. The Latin motto "In verbis in herbis et in la [pidibus deus]," or "God is in the words, the plants, and the stones," is inscribed diagonally across the surfaces of the vases.[38] The narrow, rectangular format of the panels would make them suitable for use as doors on a cabinet, while the motto could make reference to the types of items—plants and stones—that might be stored in an apothecary's cabinet. It has been proposed that the paintings were originally used as doors for such a cabinet, and if this is the case then the two alabaster vases could also represent the small vessels used to store medicinal preparations.[39] Alabaster itself would have been considered a precious substance, and objects fashioned out of alabaster, such as the vases pictured in Ludger Tom Ring the Younger's panels, would have been considered valuable by early modern collectors.[40]

Paintings depicting collections, also known as cabinet paintings, feature both physical cabinets and the curious objects they contained. The Flemish artist Frans Francken II is best known for these types of paintings, and his works usually present idealized versions of the spaces in which collectors assembled paintings, natural curiosities, and precious objects. An example of one of Francken's paintings from about 1618 presents the contents of a collection in a room with a window opening out onto a landscape (Figure 3.10). Landscape is the subject of five of the six paintings hanging on the wall of the room, and a wide shelf in the foreground shows additional paintings with books, exotic shells, coins, and medals, and a vase with an elaborate bouquet. Francken has painted the flowers in a manner echoing early still life compositions, especially in his mixing of flowers that bloom in different seasons and the bouquet's flattened, squared off appearance. The artist's decision to portray the bouquet as if it occupies the three-dimensional space of the collector's room rather than presenting it as a reproduction of a painting on display in that space is an indication of the ways that still life painting could traverse the boundary between image and object.

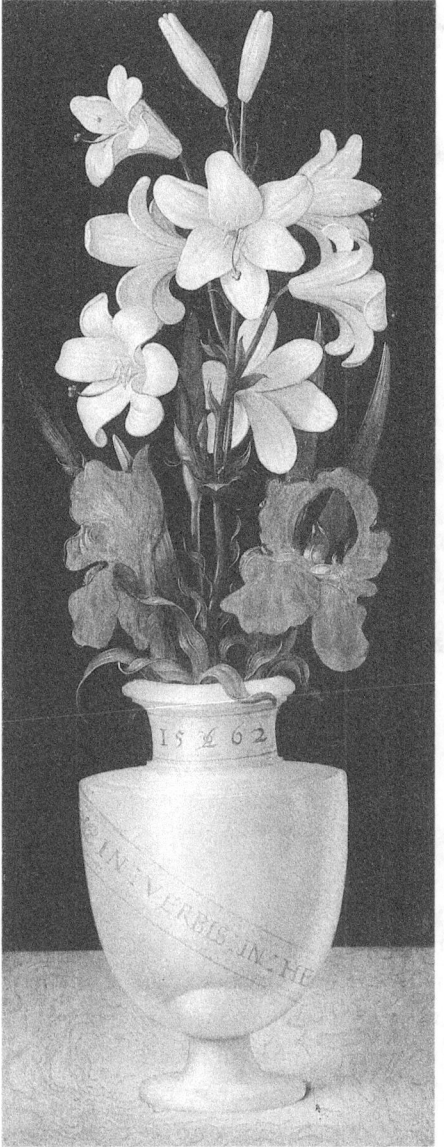

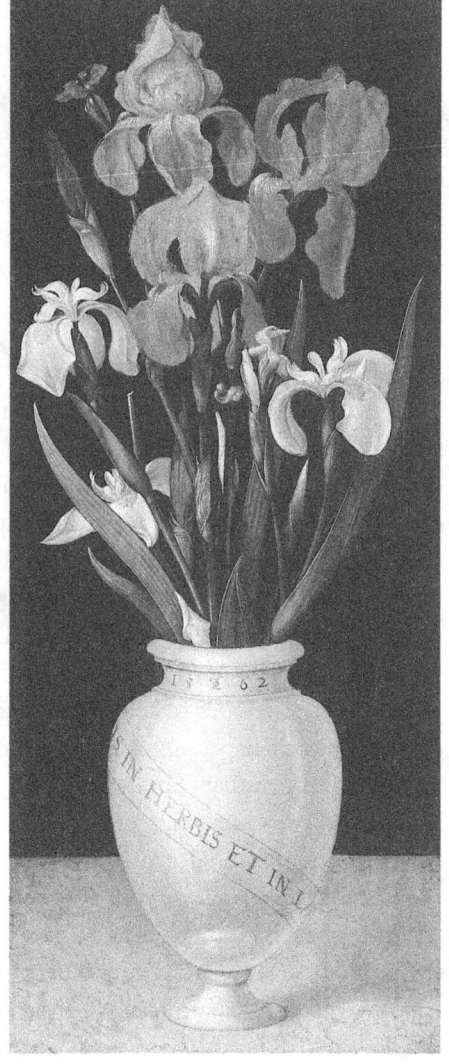

Figure 3.8. Ludger Tom Ring the Younger, *Vase with Lilies and Irises,* 1562. LWL–Landesmuseum für Kunst- und Kulturgeschichte, on permanent loan from Westfälischen Kunstverein. Photograph by Sabine Ahlbrand-Dornseif.

Figure 3.9. Ludger Tom Ring the Younger, *Vase with Irises,* 1562. LWL–Landesmuseum für Kunst- und Kulturgeschichte, on permanent loan from Westfälischen Kunstverein. Photograph by Sabine Ahlbrand-Dornseif.

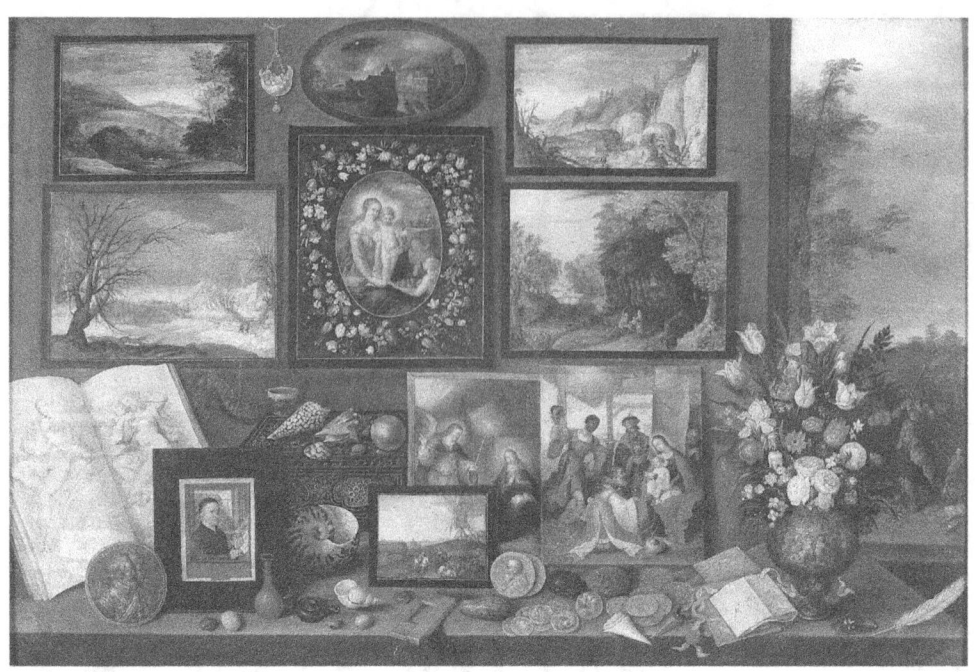

Figure 3.10. Frans Francken II,
A Collection, c. 1618. Copyright Royal
Museum of Fine Arts Antwerp. Copyright
Lukas-Art in Flanders.

In the late sixteenth and seventeenth centuries, cabinets ranged from small cases containing smaller boxes or drawers to elaborately decorated pieces of furniture. Both types of cabinets gained popularity across a range of social groups during the sixteenth century. Apothecary's cabinets were used by medical practitioners to store plants, minerals, and other preparations, and merchants traveling within Europe and abroad used writing cases for storing and transporting papers and other small items. Larger cabinets such as those depicted in other paintings by Francken were made specifically for wealthy collectors, but what all of these types of cabinets had in common was their function of storing, and sometimes displaying, small, portable objects. The designs and decorations on sixteenth-century cabinets thus suggest that the material practices of collecting and the emerging genre of still life painting were informed by a shared set of representational strategies. Like the apothecary's cabinets that preceded them, cabinets made for travelers often consisted of a simple design of a rectangular chest containing small drawers, sometimes with a fall-front panel. The surfaces of the drawer fronts and panel doors lent themselves to decoration, and in some cases these surfaces were carved in relief, but they could also be decorated with inlaid mother-of-pearl and other materials, as well as painted scenes.

In the middle of the sixteenth century Augsburg became the leading European center for the production of cabinets, and Augsburg cabinetmakers utilized printed designs for decorating the surfaces of these cabinets.[41] Furniture makers drew widely from printed sources for the decorative motifs on cabinets, but they could also turn to publications specifically intended for this purpose, such as Jacob Guckeisen and Hans Jacob Ebelmann's *Schweyfbuch* of 1599 (Figure 3.11 and Figure 3.12). One of Guckeisen and Ebelmann's designs presents a bouquet of flowers in a vase flanked by mice and butterflies, a symmetrically arranged image that echoes the format, composition, and subject matter of early floral still life paintings.

When purchasing a cabinet European consumers could choose from designs such as Guckeisen and Ebelmann's, but they could also select from imported wares. Cabinets inlaid with ivory were a specialty of western India during this period and were produced in large numbers for European buyers from the time of the Portuguese presence in India.[42] European travelers to India in the early seventeenth century noted the skilled handiwork and use of luxury materials in goods available in trading centers in Gujarat, where the export market had been firmly established by 1620.[43] Although the form of the rectangular fall-front cabinet was most likely based on European prototypes, the decoration of these cabinets often merged Indian and European motifs. Indian artisans made use of European printed designs and prototypes, producing objects that one traveler described as being done "in the fashion of those in Germany."[44] Except for a few very clear examples, it is difficult for scholars to determine with certainty the direction of influence between Indian and European designs. However, it is interesting to note the similarities between Indian and European cabinet decoration and their links to early still life compositions. As in the example from Guckeisen and Ebelmann's book discussed above, Indian cabinet decoration often featured symmetrically arranged elements around a central vegetal motif. An early seventeenth-century cabinet with decoration typical from this period for cabinets produced in Gujarat and Sindh features a front panel with three vases containing stylized, circular bouquets (Figure 3.13). Inside the cabinet are multiple drawers decorated with a series of stylized flowering shrubs, with the central drawer showing a flowering shrub in a vase with birds in the upper and lower corners. Although Indian cabinets are very different from European still life painting in terms of materials and style, the imagery on the cabinet discussed here employs the same basic compositional structure as an early floral still life by Roelant Savery or Ambrosius Bosschaert the Elder. In both the Indian and the European compositions, small creatures or objects flank a central vase that appears against a dark background, be it the space of a niche or the wood surface of a cabinet.

Another example that reflects the shared set of concerns that link collecting with still life is a nature-themed decorated box by Wenzel Jamnitzer. This

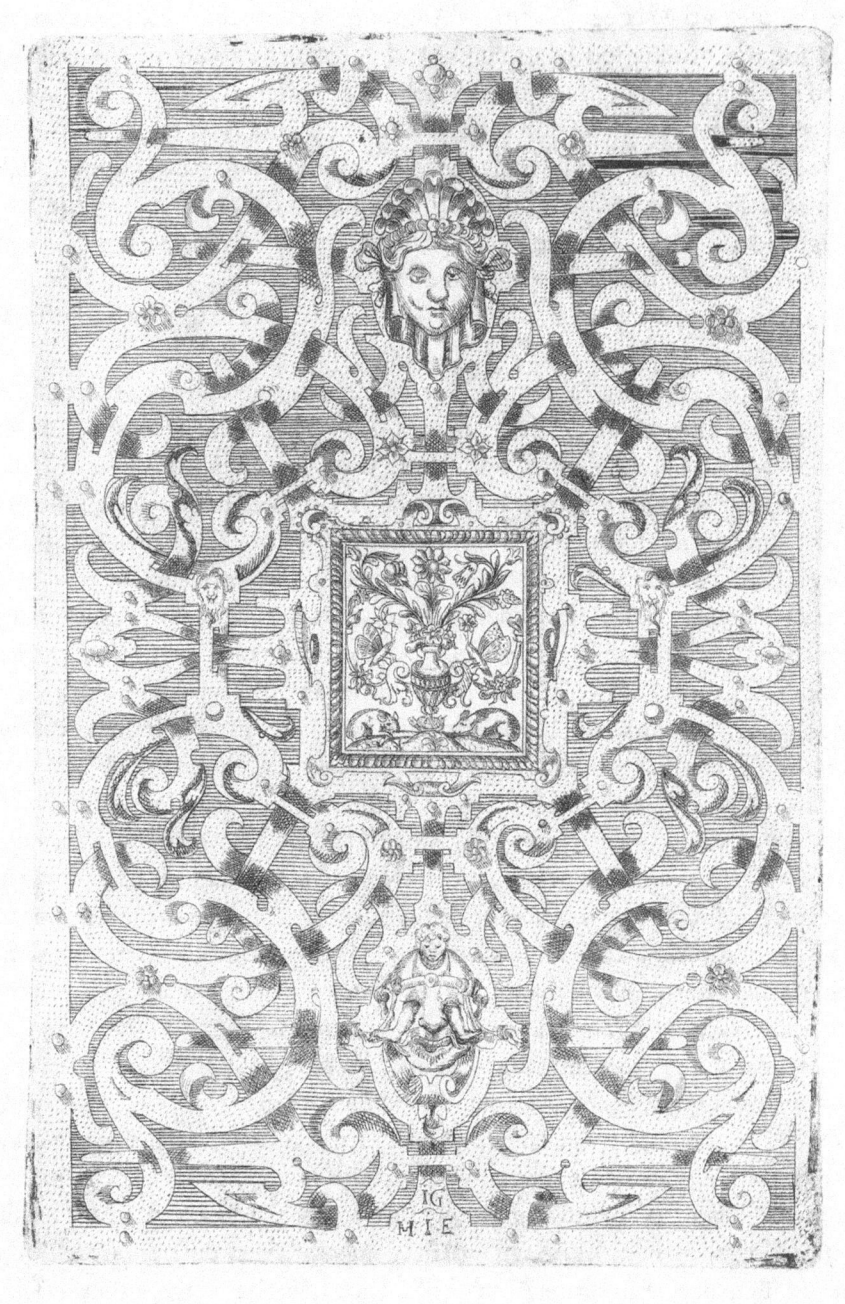

Figure 3.11. Jacob Guckeisen and
Hans Jacob Ebelmann, *Intarsia Design with
Butterflies,* Etching, Plate 1, *Das Schweyfbuch*
(Cologne: Johann Bussemacher, 1599).
Museum of Fine Arts, Boston. William A.
Sargent Fund, 46.787. Photograph copyright
2011 Museum of Fine Arts, Boston.

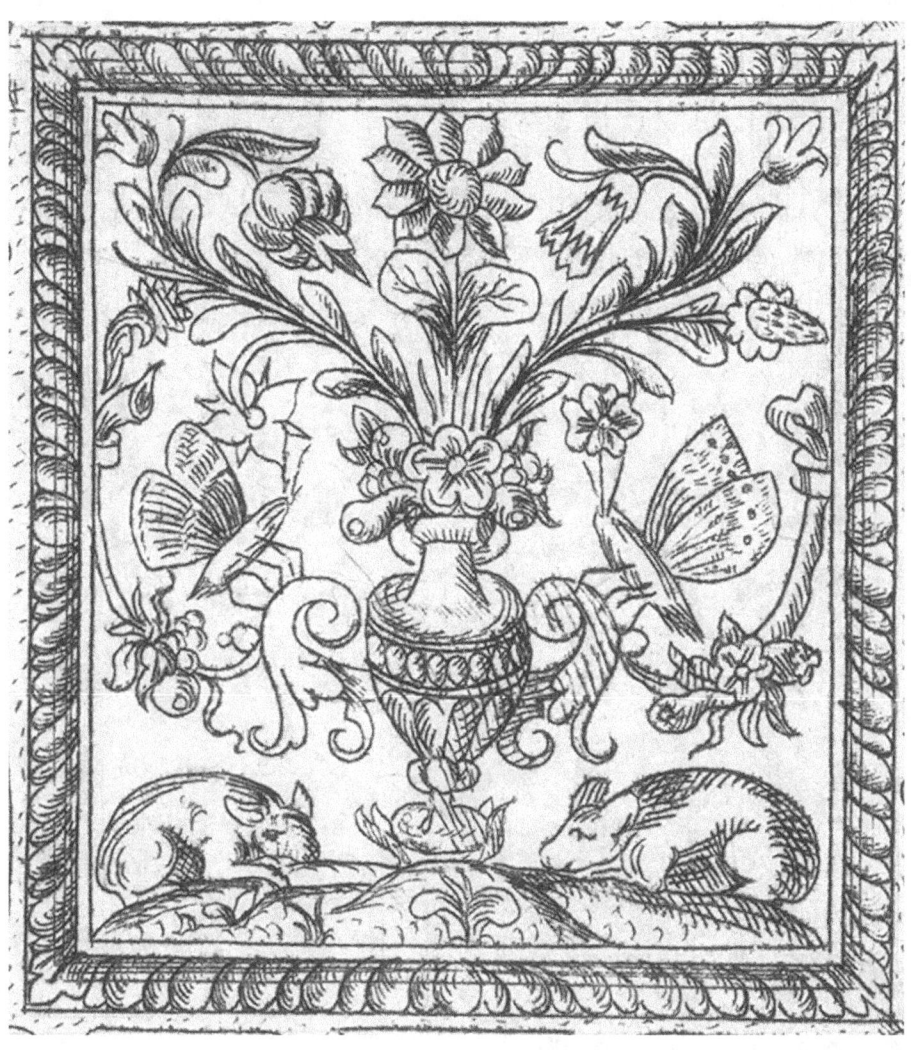

much-admired silver writing box from circa 1560–70 is a supreme achieve-ment of the goldsmith's art (Figure 3.14). The box is adorned with objects and animals cast from life along with a dense vegetal garland encircling its sides. The top lid of the box displays larger specimens arranged in a gridlike forma-tion. Although the box was designed to store writing implements, the life-cast decorations also suggest other types of objects that could be stored in the inte-rior compartments of the box. As the art historian Mark Meadow has shown, Jamnitzer's writing box is essential for understanding the relationships between the seemingly disparate objects of a sixteenth-century collection. Meadow argues that the writing box represents the different forms of knowledge collected within a wunderkammer, and that the wunderkammer as an institution "illustrates how

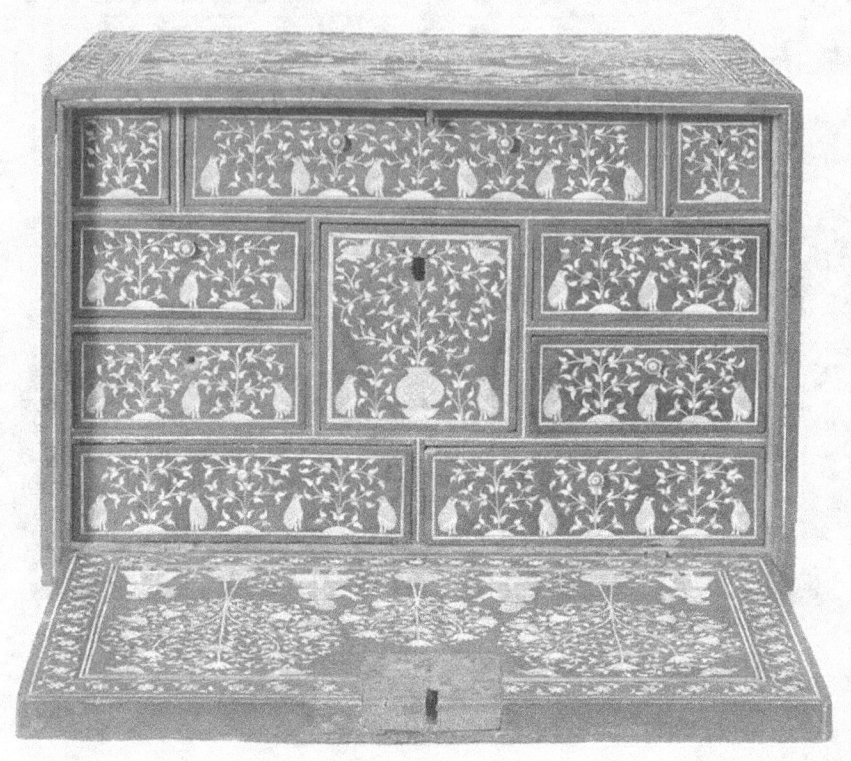

Figure 3.13. Fall-front cabinet of
shisham wood, veneered with rosewood
and inlaid with various tropical woods,
Gujarat or Sindh, early seventeenth century.
Photograph copyright Victoria and Albert
Museum, London.

objects function within such sites of multiple, simultaneous, and dynamic systems of value."[45] Jamnitzer's writing box also reveals parallels between collecting
practices, still life painting, and the culture of curiosity in its relatively limited
range of subject matter. Jamnitzer's choices of objects were constrained by
the multiple factors of medium, material, technique, and context. Only objects
and animals that could withstand the casting process could be used, and these
had to be of a size that could fit the dimensions of a small box. Still life artists
were not subject to exactly the same physical limitations as Jamnitzer, but for
them the culture of collecting and patrons' tastes for rare and exotic objects
that merged natural and human artifice exerted a similar set of limitations. The
insects, shells, and other small creatures that adorn Jamnitzer's box would be
equally at home in the foreground of a still life painting of this era. Early floral
still life paintings, in functioning as both virtual collections and collectable

objects, were embedded within this nexus of visual and material knowledge that characterized the culture of collecting in late sixteenth-century Europe.

Jan van Kessel's Insect Still Lifes

After early still life painters established insects as subject matter, they continued to be included in still life painting throughout the seventeenth century. Insects usually appeared as accompaniments to floral bouquets and banquet scenes, but one seventeenth-century still life painter made them the focus of many of his compositions. Jan van Kessel the Elder's small paintings on panel and copper often featured insects as their primary subject matter, in arrangements that closely followed those in Hoefnagel's *Archetypa*. Van Kessel usually chose butterflies, beetles, and caterpillars for his insect paintings, and in several instances he used caterpillars to spell out his name. Although insects were Van Kessel's primary focus in these compositions, the insects cavort among flowering sprigs and sometimes among shells. In some ways, these small-scale paintings can be seen as concentrating attention on the objects found in the foreground of the early floral still lifes discussed in this chapter. Van Kessel's paintings also connect to this earlier history of still life in the ways that they engage with the context of cabinets and collecting; most scholars agree that Van Kessel's small paintings were originally mounted on cabinet panels.[46] These paintings functioned as

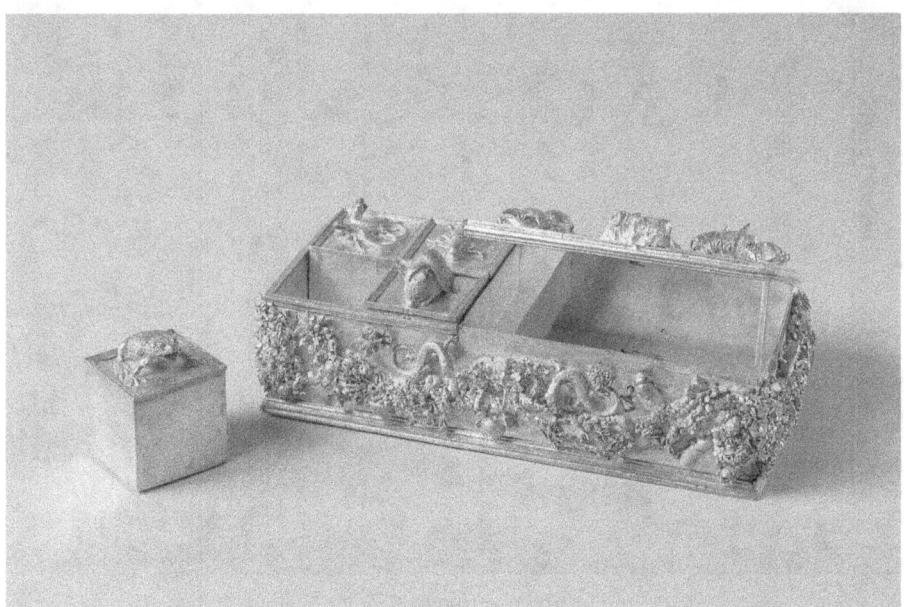

Figure 3.14. Wenzel Jamnitzer, writing box with life cast insects and shells, 1560–70. Silver. Kunsthistorisches Museum, Vienna.

decoration, but they may also have served to indicate the contents of the cabinets they decorated. A series of nine insect still lifes attributed to Van Kessel from the Natural History Museum, London, retains what appears to be the original configuration of the panels, although they are no longer attached to a cabinet (Figure 3.15). It is easy to imagine these panels as part of a specimen cabinet housing insects, parsed into the various categories of beetles and butterflies that would also be used to organize the drawers of specimens within the cabinet.

Van Kessel's paintings incorporate the convention of the specimen drawing, with many of the panels depicting individual insects in isolation against a pale background. The effect is that of individual specimens laid out in a drawer while at the same time presenting them as vibrant living creatures. This combination of the conventions of specimen drawing and still life painting results in compositions where space and perspective shift imperceptibly from insect to insect, from flat surface to vertical surface. Unlike specimen drawings, however, Van Kessel's insect images are unquestionably still life paintings made using oil paint and conveying a sense of life and vitality. In his choice of objects and medium, Van Kessel combines the sensibility of the cabinet with that of still

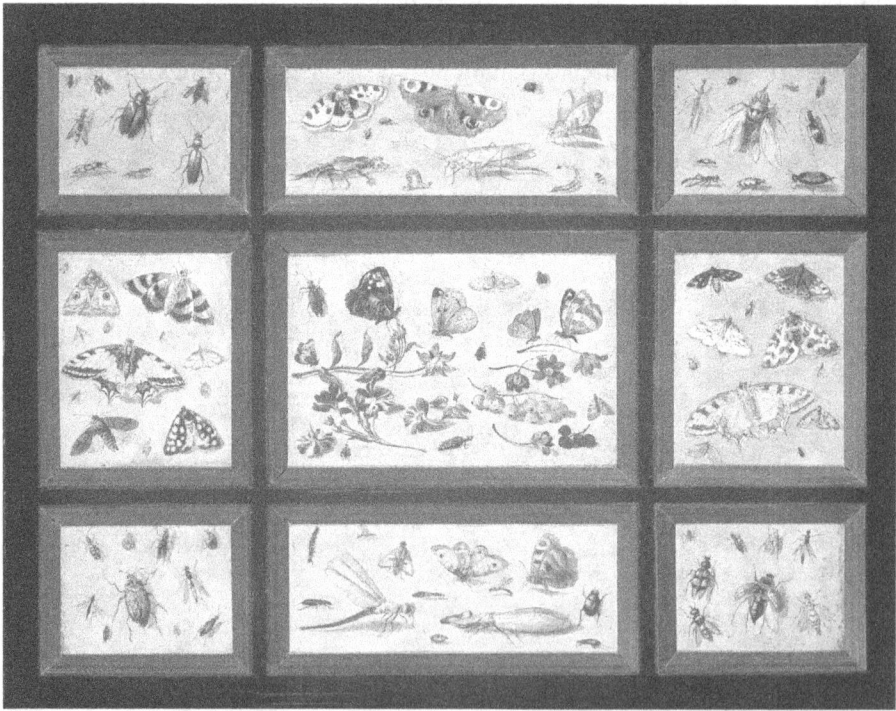

Figure 3.15. Attributed to Jan van Kessel the Elder, *Nine Panels of Insects and Flowers,* n.d. Oil on copper. Photograph copyright The Natural History Museum, London.

life painting. As in earlier still life paintings, part of the value of the image resides in the artist's skill in imbuing the forms with life. But the medium of still life painting allows objects in a collection—especially insects—to be seen as living creatures rather than preserved specimens.

By the later seventeenth century, insects were well established as subject matter for still life painting. Jan van Kessel's insect still lifes show that insects were not only acceptable subjects for still life but also are examples of how insects could move from the margins to the center of such compositions. Like Joris Hoefnagel before him, Van Kessel made a name for himself by specializing in insects and offering his viewers a pleasing spectacle of an insect world filled with beauties and wonders. Lest his viewers forget the connection between the ingenious artist and his fascinating subject matter, he spelled it out for them by signing his name with insects. The process of establishing insects as subject matter for still life took place within the late sixteenth-century culture of curiosity and collecting, in which the boundary between objects and images was increasingly blurred and the value of both was more and more a function of both the exchange value within the economies of collecting and an artist's skill at rendering lifelike appearances. Insects were no longer novel subjects by the late seventeenth century, but they continued to serve as focal points for new techniques of observation and new strategies for representing the natural world in visual form. Insects also continued to provide artists and other image makers with rich material around which professional identities could be constructed and imagined. In the next two chapters I will explore the roles that insects and images of insects played in fashioning the persona of the trustworthy observer of nature and the continuing relationship between insects and the cultures of collecting and exchange in early modern Europe.

II

NEW WORLDS AND NEW SELVES

FOUR

BETWEEN OBSERVATION
AND IMAGE

*Representations of Insects in
Robert Hooke's Micrographia*

The English polymath Robert Hooke (1635–1703) can be credited with a number of mathematical and mechanical inventions, including originating the term "cell" in biology and composing the equation describing elasticity known as Hooke's Law. But it is Hooke's *Micrographia,* published in London in 1665, that is considered a landmark in the history of scientific illustration.[1] His spectacular illustrations of plants, insects, astronomical bodies, and mechanical objects have long been praised for their artistic merit, scrupulous accuracy, and careful attention to detail. With its thirty-eight copperplate engravings, *Micrographia* stands as a testament to Hooke's talents as an observer and illustrator. Little attention has been paid, however, to the visual techniques and traditions Hooke used to translate into two-dimensional images what he saw through the lens of the microscope.

In examining the sequence of events leading up to the publication of *Micrographia* as well as the circumstances of the book's production, in this chapter I will reveal how Hooke labored to present himself as an authoritative and trustworthy observer. This persona, like Hooke's images and specimens, was carefully crafted to support the authority and authenticity of Hooke's observations and, in turn, to create a sense of trust among readers that his images were accurate reflections of those observations.

In addition to considering the published images in *Micrographia,* I also discuss several recently discovered early sketches of microscopical observations of insects by Hooke. These sketches shed important light on his early work with the microscope and his working practices as a scientist and a scholar. In this chapter, I assert that the illustrations in *Micrographia* were central to Hooke's efforts to present the microscope as an instrument that provided access to a hidden world of marvels and wonders. As the gatekeeper to this hidden world, Hooke both controlled access to and shaped perceptions of these wonders through the skilled crafting and manipulation of images. The serenely arranged objects and elegantly posed creatures that are featured in Hooke's images can also be understood within the broader context of the Royal Society's efforts to promote the New Philosophy as a source of rational, useful knowledge that supported, rather than undermined, stability and peace in Restoration society. Hooke's carefully constructed images and authorial persona worked together to produce the impression that *Micrographia* offered readers straightforward, transparent descriptions of natural phenomena. The persona of the distanced, unbiased observer was designed to support Hooke's own status as a trust-worthy member of the Royal Society while supporting the society's desire to avoid associations with contentious political and religious issues.

As the son of the curate of a parish church on the Isle of Wight, Hooke came from a respectable but modest family. His early years were marked by tragedy; his father died when he was thirteen years old. Hooke used the small inheritance he received to finance his move to London around 1648, where he served an apprenticeship in the workshop of the Dutch portrait painter Peter Lely (1618–1680) and soon after attended Westminster School. Because his inheritance would not have been enough to cover all of the expenses he incurred during this time, it is believed that Hooke may have received financial assistance from a family friend to help fund his education. When he later attended Oxford, he also received financial assistance.[2] Thus, although his intelligence and talents were widely recognized from an early age, his humble background would not have gone unnoticed by his peers and teachers. The issue of Hooke's social standing would be an important factor in his later interactions with members of the Royal Society. Furthermore, there is some indication that he suffered a facial disfigurement from a childhood bout with smallpox, which likely would have contributed to his social status as an outsider.

The Origins of *Micrographia*

The origins of *Micrographia* are rooted in a series of microscopical drawings of insects made by Christopher Wren (1632–1723) in the late 1650s or early 1660s, which Wren presented as a gift to King Charles II sometime in early 1661. Contemporary accounts by visitors to the king's cabinet indicate that

Wren's drawings included illustrations of a flea, a louse, and the wing of a fly.[3] The drawings caused a troublesome situation for the Royal Society shortly after Wren made his gift to the king. Charles II was so delighted by the drawings that he requested more of them from Wren. He made his request not directly to Wren, however, but through the Royal Society, of which Wren was a founding member. Henry Powle, another founding member, then wrote to Wren at Oxford informing him of the king's request, indicating that Wren should fulfill the commission without delay: "He doth expect an Account of this from you shortly."[4] On May 17, 1661, society members Paul Neile and Robert Moray wrote to Wren on behalf of the Royal Society with the same request, directing Wren "to delineate by the Help of the Microscope the Figures of all Insects, and small living creatures you can light upon, as you have those you presented to his Majesty."[5]

The king's request had placed the Royal Society in an awkward position. Rather than presenting the original set of drawings through the Royal Society, Wren had made his gift directly to the king. The members had not been given the opportunity to approve or discuss the gift before it was made, nor had they enjoyed the privilege of viewing the drawings before they became the king's property, where access to them was restricted to a select few. Now, however, the Royal Society was responsible for carrying out the king's request that Wren provide him with more drawings. Although the society's members would have been eager to carry out the king's wishes and thereby gain his favor, their ability to induce Wren to complete the project was limited. Although he was a well-respected founding member of the society, Wren's activity and attendance at meetings had become somewhat sporadic after he took up his appointment as the Savilian Professor of Astronomy at Oxford in May 1661. Apparently Wren did not act on the royal assignment, for on July 3 the Royal Society asked Henry Powle to speak with him about his progress on "the pictures of small insects by the microscope."[6]

There is no record of Wren's response to Powle's inquiry, but by August 1661 Wren had indicated to the Royal Society that he would not be able to make the drawings for the king. Wren's withdrawal from the project is recorded in a letter about the matter sent to him by Robert Moray. Rather than leave the king's request unfulfilled, Moray and the other members of the society had decided to reassign the task to Robert Hooke. On August 13, Moray wrote to Wren as follows: "In Compliance with your Desire to be eased of the further Task of drawing the Figures of small Insects by the Help of the Microscope, we have moved his Majesty to lay his demands on another, *Vander Diver*; and we have also persuaded Mr. Hook, to undertake the same thing."[7] Thus Moray's text indicates that two people had been asked to work on the project; unfortunately the identity of the person referred to as "Vander Diver" remains unknown.

The politics of the situation were grave. The Royal Society naturally stood to gain a great deal of favor through Wren, whose work greatly pleased the king, but they also risked disfavor if Wren did not comply with the king's wishes. Furthermore, they could not be assured that they would reap the benefits of Wren's talents if Wren did not choose to associate his gift with the Royal Society; however, they were almost guaranteed to lose favor with the king if they failed to carry out his request. Robert Moray, who acted as the Royal Society's unofficial liaison to the court, seems to have successfully negotiated this potentially damaging situation by convincing the king to pass the task on to Hooke. There is no further mention of the project in the Royal Society's minutes, however, and neither is there any record that the society presented the king with drawings by Hooke.[8] No further reference to microscopical drawings is made in the Royal Society's minutes until March 1663, almost two years after the episode involving Wren's drawings.

Hooke's Role in the Royal Society

At the time he was asked to produce microscopical drawings of insects for the king, Robert Hooke had a markedly different standing in the Royal Society from the more central role he would later occupy as its curator of experiments. Hooke was not yet a member of the society in 1661, and he began attending meetings only after he was appointed its curator in November 1662. He was certainly known to the membership before 1661, however, as he had been employed by Robert Boyle in Oxford since 1657 and had been acquainted with many of the society's founding members since the start of his studies at Oxford in 1655.[9] Hooke's name first appears in the minutes of the society on April 10, 1661, when members decided to discuss his tract on capillary action at their next meeting.[10] Although Hooke had worked closely with Boyle in Oxford on the design and construction of the air pump, it is not likely that he was present in London for Boyle's demonstrations of the device at the Royal Society meetings during the early 1660s. Indeed, Steven Shapin and Simon Schaffer have argued that the problem of the repeated failures of the air-pump demonstrations at these meetings was solved only when Hooke was appointed curator; the frustrations of the failed demonstrations seem to have played a part in the society's decision to appoint a curator of experiments.[11] As Shapin and J. A. Bennett have shown in separate studies, Hooke's status in the Royal Society remained deeply ambiguous throughout his tenure as curator and later as secretary.[12] In later years, Hooke would move on a daily basis between the different social worlds of gentlemen virtuosi and those skilled in trades, and although his ability to conduct experiments and operate instruments was rarely questioned, he was never fully accepted as a social or a professional equal by the other members of the Royal Society.

Hooke's status in the early 1660s was more marginal than it was during his later years with the society, but it was also less ambiguous. As noted earlier, Hooke's family was not wealthy, and he relied on assistance for his attendance at Westminster School and at Oxford. Thus, the fact that Hooke was not a person of independent financial means would have been well known to the members of the Royal Society in the early 1660s. As a paid assistant to Robert Boyle and later a paid employee of the society, he would also have occupied a social position distinct from that of a gentleman. A gentleman, in the context of the Royal Society, could be trusted to offer his unbiased opinion precisely because he was a person of independent financial means. Members of the Royal Society were, for the most part, men from the upper levels of society—gentlemen and nobles who were, as Shapin has shown, "free" men and thus considered trustworthy because of their social standing. "Unfree" men, such as technicians and servants who expected remuneration for their efforts and ideas, were subject to the control of others and therefore considered less trustworthy.[13] When Hooke became an official member of the Royal Society while retaining his paid position, he held an uncertain status among the society's members. His work on *Micrographia* in the early 1660s would coincide with the transition he was making from the relatively stable social position of paid assistant to the more ambiguous and problematic role of paid employee and fellow of the Royal Society. This dual role and its associated problems would come to characterize many of Hooke's later interactions with the society and its members. The procedures that Hooke and the Royal Society devised during this period to bring the *Micrographia* into print also reflected Hooke's increasingly ambiguous status within the organization.

Hooke had been curator for six months when the members of the society decided, in March 1663, that he should "prosecute his microscopical observations, in order to publish them."[14] From this point on, Hooke presented microscopical observations at meetings on an almost weekly basis, showing his first drawing—of "common moss"—on April 8.[15] Hooke continued to present microscopical drawings at meetings throughout 1663, showing his last drawing on December 12.[16] The drawings he presented between March and December 1663 would become the basis for most of the illustrations in *Micrographia*. The society's minutes do not record who actually proposed that a publication on the microscope should be pursued, but it is clear that the members intended to monitor closely the book's progress and, to some extent, its content.

For the most part Hooke's drawings pleased the members, but on occasion they did withhold their approval, as in the case of the drawing of a spider with six eyes. On April 29, Hooke presented two microscopical drawings, "one of a mine of diamonds usual in flints; the other a spider appearing to have six eyes: but the latter was not yet perfectly drawn." The spider with six eyes was not satisfactory because the image Hooke presented was not judged to be "perfectly drawn," and this deficiency does not seem to have been rectified because the insect never did appear in print.[17] Hooke was generally allowed to choose his own subject matter for observations, but on occasion he would be ordered to investigate items of interest to other members. For this reason, much of the time he spent making microscopical investigations during the month of June was devoted to the search for spiders living in sage leaves. On June 3, for example, he was ordered to "observe by a microscope, whether there be any cavities in sage leaves for little spiders to lodge in," and at the next meeting, on June 10, he reported that he had seen none. Despite this negative report, the interest in spiders in sage leaves persisted. On June 17, Hooke was ordered to continue looking for such spiders and, on July 1, the order was issued to him again. No illustration of spiders in sage leaves appears in *Micrographia*, and thus it is probable that Hooke never found the creatures, despite his repeated searches.[18]

It is interesting to note that the meeting at which Hooke was ordered to begin the unfruitful series of sage-leaf investigations was also the meeting at which he was elected a fellow of the society. Thus, at the moment at which Hooke had achieved the recognition and respect of the members of the society in being elected to their ranks, he also found himself under obligation to carry out duties that highlighted his subservient role within the group. It is not difficult to imagine Hooke bridling at the suggestion that he may have failed to see the spiders in the sage leaves as well as resenting the time spent making the additional observations when he had many other pressing duties to fulfill as curator. On the other hand, he may have felt that there was good reason to continue the investigations. Although this example points to the collaborative nature of the activities of the early Royal Society, it is important to note that these collaborations were not always carried out among equals. Unlike other members of the Royal Society, Hooke could be ordered to take a course of action that he might not otherwise have chosen to pursue; conversely, a request that might not have been appropriate to make of a regular member of the society—one that entailed an implicit questioning of that person's judgment and abilities—was appropriate to make of Hooke because he was an employee.

Hooke and Wren

Many scholars have assumed that Robert Hooke and Christopher Wren worked together in making microscopical drawings for *Micrographia* and that the book represents a collaboration between the two men. Hooke and Wren were associated with the natural philosophers, mathematicians, and other scholars in the circle of John Wilkins, Seth Ward, and Robert Boyle in Oxford during the 1650s. Wren had been making microscopical drawings since the early 1650s, and Hooke is believed to have begun his own work with microscopes some time during the late 1650s.[19] Hooke's and Wren's collaborations in the years after 1666 are well documented, as the two worked together closely on a number of architectural projects after the Great Fire of London.[20] Knowledge of their working relationship during the period of the preparation and publication of *Micrographia,* however, is not as firmly established.

Wren's association with *Micrographia* stems primarily from two contemporary sources. The first of these is an account of a conversation that occurred in September 1655 between Wren and the Polish-born scientist and gentleman Samuel Hartlib (c. 1600–1662), in which Wren told Hartlib of a book he was preparing "with pictures of 'Observ[ationes] Microscop[icae].'"[21] The drawings that Wren gave the king in 1661 may have been related to the publication Wren had already been planning in 1655, but no further references to Wren and a book on microscopy occur until 1663. In August 1663, Henry Power informed readers of the preface to his *Experimental Philosophy* that they could "expect shortly from Doctor Wren, and Master Hooke . . . Cuts and Pictures drawn at large, and to the very life of these and other Microscopical Representations."[22] Power (1623–1668) was a corresponding member of the Royal Society who resided in Halifax, and his *Experimental Philosophy* included a section on microscopical observations that contained several small illustrations. In June 1663, Power attended a meeting of the Royal Society in London and presented his own microscopical observations. At the same meeting, Hooke, Wren, and John Wilkins were "appointed to join together for more observations of the like nature."[23] At the time of Power's visit to the Royal Society, Hooke had already made seven presentations of microscopical drawings to the society and was thus well along with his preparations for *Micrographia.* It is possible that Power mistakenly assumed that Wren was working on the book with Hooke, since Wren had been in attendance when Power presented his microscopical observations and was a member of the group assigned to make similar investigations on that day. However, there is no further mention in the minutes of such a collaboration among Hooke, Wren, and Wilkins, or of any presentations of drawings by Wren. Hooke continued to present his microscopical drawings at meetings throughout the year, and he is the only person noted as engaging in such activities during the period preceding *Micrographia*'s publication.

Wren has also been linked to *Micrographia* on the basis of the drawings he presented to Charles II in 1661. Because those drawings are known to have included illustrations of a flea and a louse—two of the more memorable creatures appearing in *Micrographia*—and because there is no record that Hooke presented drawings of these particular insects to the Royal Society, it has been argued that Wren provided the illustrations for these and several other insects in *Micrographia*.[24] But although Wren's early microscopical drawings and observations clearly influenced the decision by the Royal Society's members to direct Hooke to work on a book of microscopical observations, it is unlikely that the drawings Wren made in 1661 were used as the basis for any of the engravings in the book. For one reason, Hooke would not have had access to Wren's drawings in the king's cabinet, and neither is there any record that a representative of the Royal Society approached either Wren or Charles II with such a request. It is more likely that Hooke produced his own versions of these insects for *Micrographia,* perhaps acting upon the advice of Wren or other members of the society, but did not present these drawings at any meetings. Hooke did not present any microscopical drawings after December 1663, but he continued to work on *Micrographia* in the following months. He seems to have been nearly finished with the book in June 1664, at which time it was decided that Lord Brouncker (William Brouncker, 1620–1684; first president of the Royal Society) would review Hooke's manuscript and pass it on to other members of the society for their review before approving it for publication.[25]

Hooke himself referred to Wren's early drawings in the preface to *Micrographia,* writing that "Dr. Wren . . . was the first that attempted any thing of this nature: whose original draughts do now make one of the Ornaments of that great Collection of Rarities in the Kings Closet."[26] It is significant that Hooke does not specifically state in this passage that Wren contributed material to *Micrographia,* although he did take pains to recognize Wren's earlier work in the field. Thus there is no reason to believe that Hooke would not have acknowledged Wren's contributions to the book if any had, in fact, been made. Indeed, it is likely that both Hooke and the Royal Society would have liked very much for Wren to work on the publication, since Wren had already distinguished himself in the field and had been closely associated with the microscope since the 1650s. However, in spite of these circumstances, it is unlikely that Wren was involved with the book in anything more than an advisory capacity. Although Wren had been planning to publish an illustrated book of microscopical observations in 1655, he seems to have ceased working actively in this area after declining the king's request for additional microscopical drawings in 1661. So although *Micrographia* was shaped by the various collaborative relationships that existed between Hooke and the members of the Royal Society, it remains largely the product of Hooke's labor, and as such it reflects many of Hooke's particular experiences, ideas, and beliefs.

Crafting Natural Appearances: Creating Calm and Order in the Microworld

In his preface to *Micrographia,* Robert Hooke wrote that all that was required for studying nature was "a sincere hand, and a faithful eye, to examine and to record the things themselves as they appear."[27] In this well-known passage, Hooke implies that with his words and pictures he acted as the microscope's amanuensis by simply recording whatever appeared through the lens of the instrument. Rather than taking Hooke at his word, however, let us instead consider the techniques he used to craft his images, specimens, and observations to create "natural" appearances. In so doing, we will see that between the moment of looking through the lens and the act of creating the image more was involved than a seamless transfer of information from the eye to the hand of the observer, as Hooke would have us believe. In composing the words and images of *Micrographia,* Hooke not only rendered visible a world that could be seen only by means of the microscope but also filled that world with a tantalizing collection of strange and unusual creatures and objects. From a chaotic and confusing array of disparate observations, Hooke constructed for his readers a quiet, ordered, and comprehensible world of hidden wonders.

In order to present microscopical observations in a familiar format for seventeenth-century audiences, Hooke's illustrations for *Micrographia* drew upon a number of sophisticated image-making traditions such as still life painting, manuscript illumination, and natural history illustration. Although Hooke never worked professionally as an artist, he had been apprenticed for a short time to the royal portrait painter Peter Lely, and he maintained an interest in art throughout his life.[28] In Schema 18 of *Micrographia,* Hooke presents a magnified view of thyme seeds (Figure 4.1). In the corresponding text he describes the seeds as "pretty fruits" that resemble oranges and lemons, writing that "the Grain affords a very pretty object for the *Microscope,* namely, a Dish of Lemmons plac'd in a very little room."[29] Hooke's statement recalls the banquet and "laid-table" compositions typical of seventeenth-century still life painting. The vivid colors and unusual textures and shapes as well as the rarity of citrus fruits made them a favorite subject for seventeenth-century still life painters. Like a seventeenth-century still life painting, Hooke's engraving of thyme seeds arrayed on a tablelike surface has the crisp delineation and well-illuminated composition that facilitates the quiet contemplation of the objects thus portrayed. Still life painting offered Hooke a pictorial strategy for translating fleeting impressions of surfaces and fragments into images of physical objects in a three-dimensional space. Like oranges and lemons in a still life, Hooke's thyme seeds engaged his viewers' eyes in a delightful exploration of the unusual textures and shapes of rare objects.[30]

Throughout *Micrographia,* Hooke described the objects he viewed through the microscope as "pretty" and "beautiful." These aesthetic qualities were important criteria for Hooke in choosing which objects to view with the microscope, and more important, which objects to illustrate in *Micrographia.* Hooke's choices were informed by principles similar to those governing illustrated natural histories and cabinets of curiosities—in particular, the focus on rare, unusual, and beautiful forms shared by many early modern European collectors and artists. In his description of purslane seeds (Figure 4.2), for example, Hooke noted that the seeds had "very notable shapes, appearing through the *Microscope* shap'd somewhat like a *nautilus* or *Porcelane* shell. . . . The order, variety,

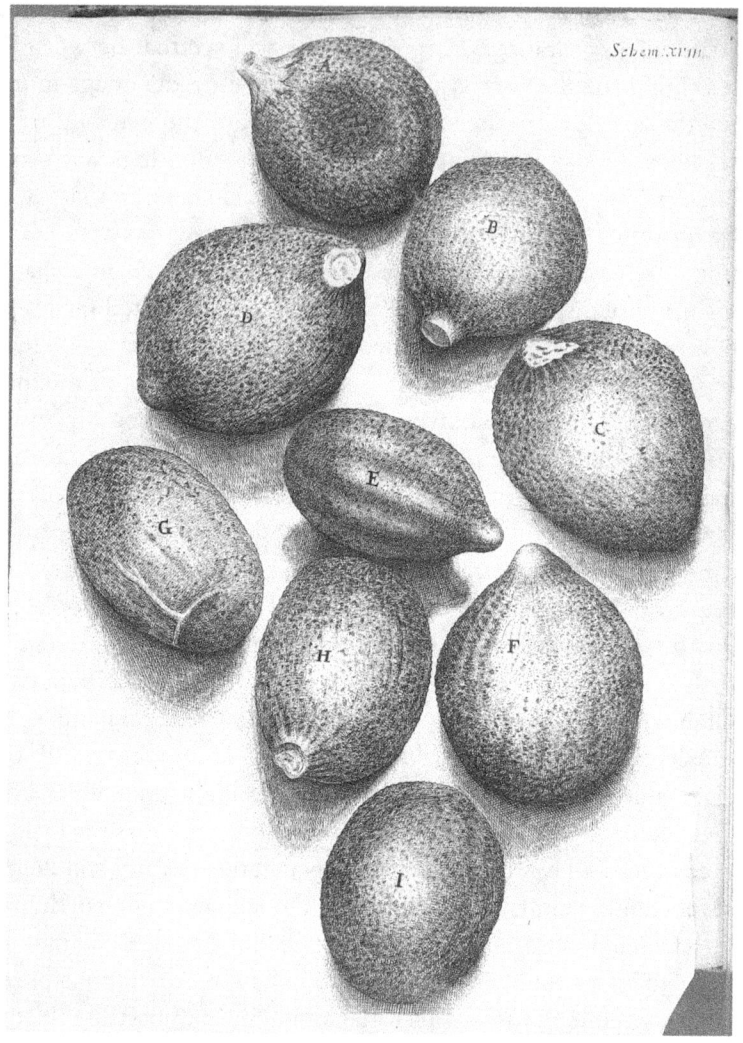

and curiosity in the shape of this little seed, makes it a very pleasant object for the *Microscope.*"[31] Hooke also noted that, when viewed through the microscope, the seeds of the Venus looking glass, or corn violet (Figure 4.3), seemed "to be the Cabinet of Nature, wherein are laid up its Jewels . . . nor indeed is there in any part of the Vegetable so curious carvings, and beautiful adornments, as about the seed."[32] Couched in the language of collecting and natural history, Hooke's text suggests that the world revealed by the microscope was made up of chambers filled with beautiful and unusual objects, much like a cabinet of curiosities. The illustrations in *Micrographia* also recall the visual conventions and "specimen logic" of natural history and botanical illustration. Many of the images present objects against white backgrounds containing few contextual elements, aside from an occasional shadow, as in the illustration of an ant in Schema 32 (Figure 4.4).

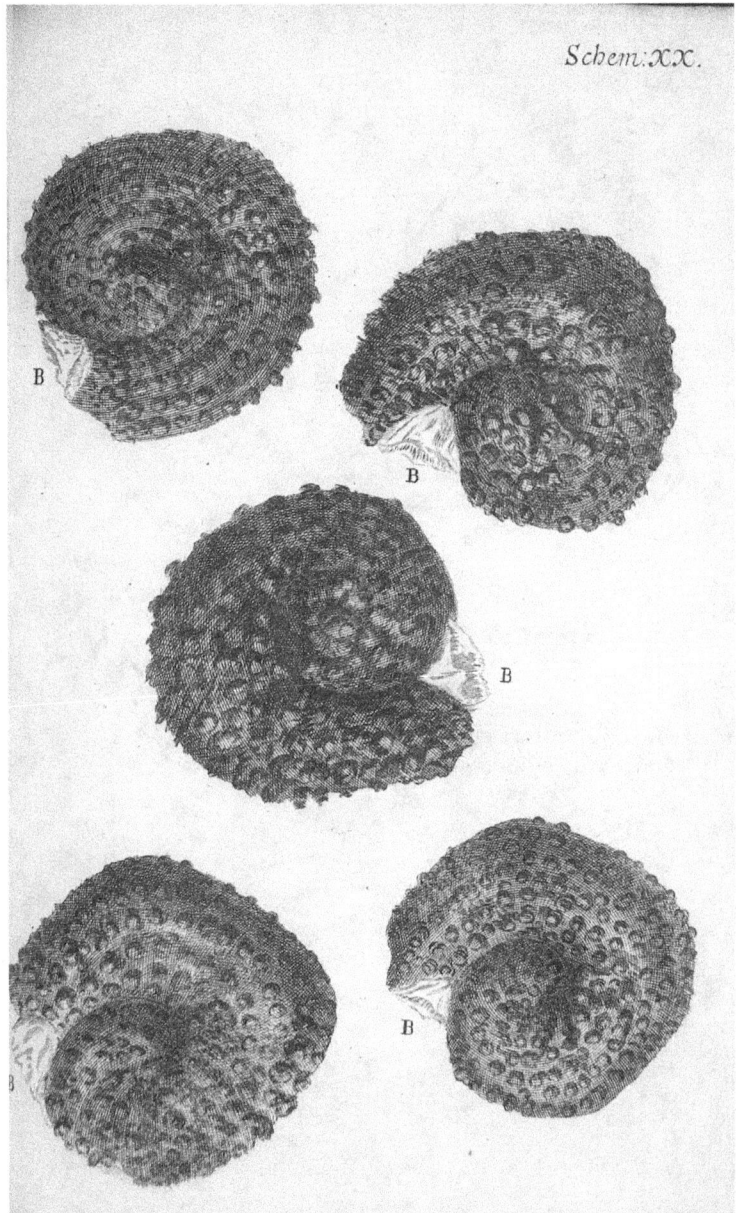

Schem:XX.

Figure 4.2. Robert Hooke, *Micrographia* (London, 1665), Schema 20 (purslane seeds). Courtesy of History of Science Collections, University of Oklahoma Libraries.

Although *Micrographia* was not intended as a treatise on insects, such creatures occupied much of Hooke's attention, appearing in fifteen of the thirty-eight plates, or "schemas," of the book. Hooke's use of the term "schema" to identify his plates implies the study or visual dissection of the objects portrayed, and as such indicates that he approached his images in a diagrammatic manner. Insects were closely associated with the microscope from the time of its invention in the early seventeenth century. Indeed, they appear in the earliest published microscopical image from 1630—an engraving of a bee and bee parts by

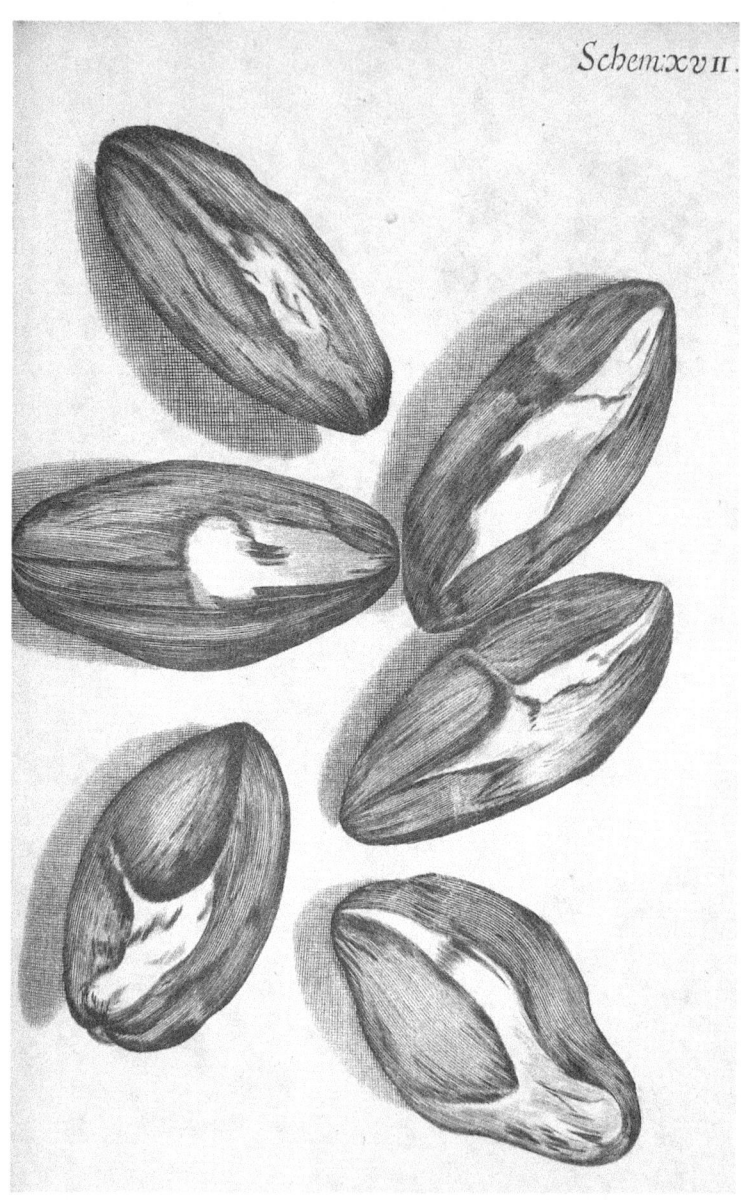

Figure 4.3. Robert Hooke, *Micrographia* (London, 1665), Schema 17 (corn violet seeds). Courtesy of History of Science Collections, University of Oklahoma Libraries.

Schem:xv II.

Francesco Stelluti (1577–1652), a member of the Accademia dei Lincei (Figure I.3). By the 1660s insects had long been accepted as appropriate subject matter for early modern European artists and naturalists, and they had become frequent subjects of study for microscopists.[33] Prior to the Royal Society's decision to have Hooke publish a book of microscopical investigations, several members had met on a regular basis to discuss the topic of the generation of insects.[34] The intricate structures of insects, as well as their size, made them ideal subjects for viewing through the microscope, but these small, delicate creatures also posed a great challenge to Hooke. Insects were easily damaged and killed in the

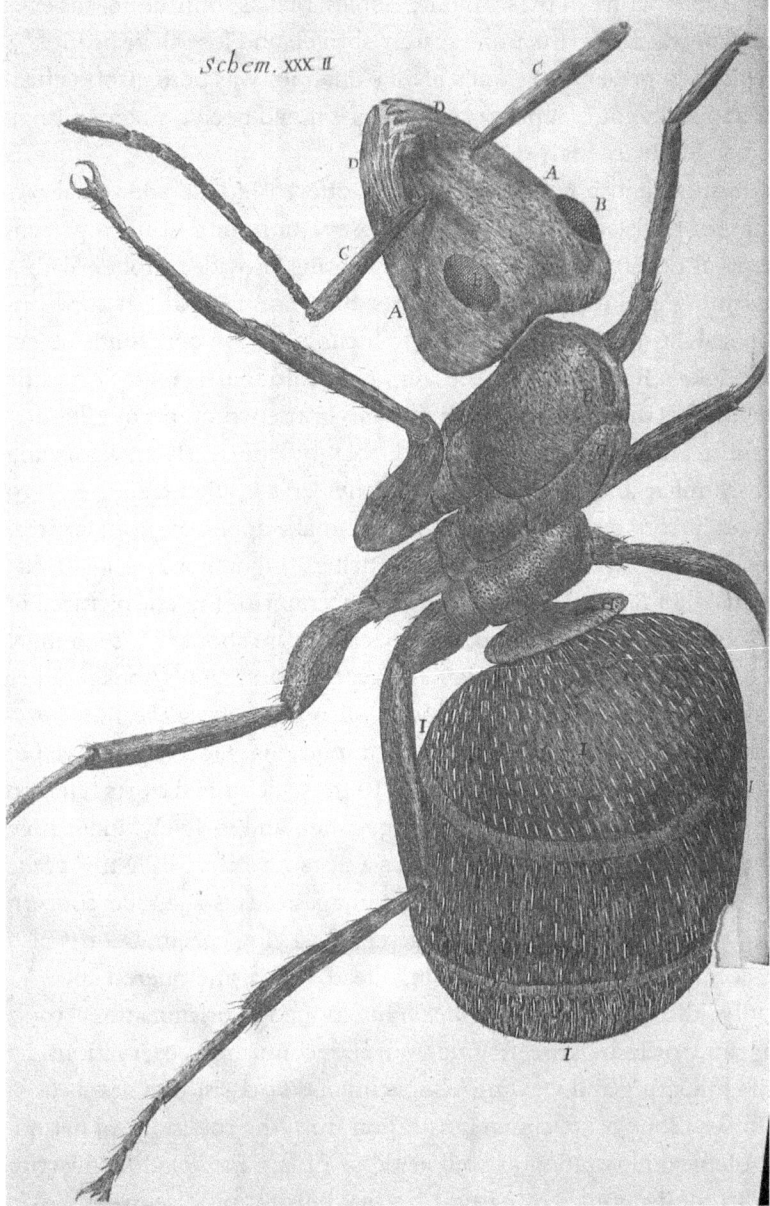

Schem. XXX II

process of preparing them for viewing, and for this reason they were difficult to represent in "natural" poses, as was the case with the delicate body of the ant pictured by Hooke in Schema 32 (Figure 4.4). Hooke wrote that the ant was "a creature, more troublesome to be drawn, then any of the rest, for I could not, for a good while, think of a way to make it suffer its body to ly quiet in a natural posture."[35] Hooke eventually discovered that he could incapacitate the ant by submerging it in spirits, a technique he favored because "if I killed it, its body was so little, that I did often spoile the shape of it, before I could thoroughly view it: for this is the nature of these minute Bodies, that as soon, almost as ever their life is destroyed, their parts immediately shrivel, and lose their beauty."[36] Using this method to preserve the ant's lifelike qualities was perhaps too effective, for after about an hour, "upon a sudden, as if it had been awaken out of a drunken sleep, it suddenly reviv'd and ran away."[37]

Natural history and botanical illustration offered Hooke additional pictorial strategies for translating microscopical observations into visual representations of physical objects. Whereas still life painting provided Hooke with a means of presenting multiple objects in a three-dimensional space, natural history and botanical illustration allowed him to focus attention on a single object or creature. Hooke's illustration of the ant, at a scale that is many times its actual size, highlights the insect's spiky limbs and brittle structures by silhouetting them against a stark white background, thereby effectively transforming the seemingly familiar ant into an exotic and mysterious alien creature. Two works of natural history contemporary to Hooke make appearances in his text: Willem Piso's *Historia naturalis Brasiliae* and Richard Ligon's *A True and Exact History of the Island of Barbados*.[38] In his written account of the ant pictured in Schema 32, Hooke refers readers who wish to learn more about the "seemingly rational actions" of ants to Ligon's book.[39] His reference to Ligon's book—rather than to a book dealing specifically with insects, such as Moffet's *Theatrum insectorum* or Aldrovandi's *De animalibus insectis*[40]—implies that Hooke conceived of his insects as exotic and strange creatures akin to those described by travelers to the New World. With its focus on a single specimen and its lively, inquisitive stance, Hooke's ant can also be understood as a distant relative of Dürer's *Stag Beetle* of 1505 (Figure I.2). But while Hooke's image reflects a similar concern with conveying the movements and character of a living specimen, it is also based on the combined observation of living, dead, and dismembered ants.

For Hooke, the images in *Micrographia* functioned as organizational tools for arranging and ordering the fragmented and confusing observations he amassed while preparing and viewing his specimens. To assist him in this task, he adopted conventions of specimen illustration from the traditions of natural history and botanical illustration as well as those of late medieval manuscript illumination. In the examples of natural history illustration discussed above, individual specimens are presented against plain backgrounds, a technique that encourages both artist and viewer to understand the plant or animal depicted as an isolated physical object. Similarly, in the illustrated borders of illuminated manuscripts, such as the Book of Hours of Catherine of Cleves (c. 1440), the surface of the page serves as a repository for mussel shells and a small crab (Figure 4.5), which are presented in much the same manner as precious, jewel-like objects in a curiosity cabinet. Each of these pictorial traditions offered Hooke varying means of using images to organize information and to transform obscured forms and uncertain knowledge into gleaming objects and solid bodies.

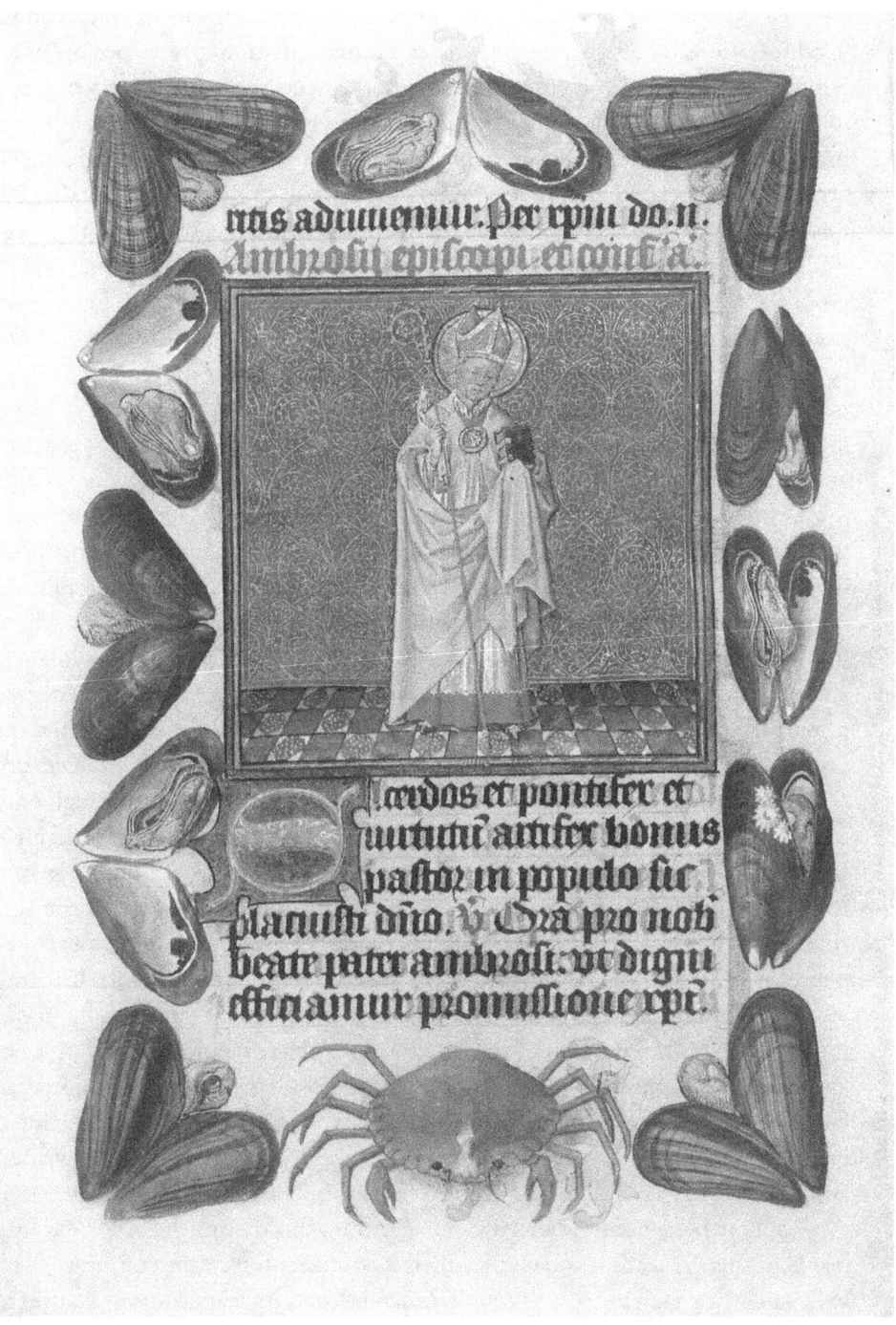

Figure 4.5. Hours of Catherine of Cleves, c. 1440, page 244. The Pierpont Morgan Library, New York. Purchased on the Belle da Costa Greene Fund with the assistance of the Fellows, 1963. MS M. 917.

As the example of Hooke's ant reveals, making observations and producing illustrations for *Micrographia* involved far more than simply recording, as he
put it, "the things themselves as they appear." Insects served as "pretty objects"
for the microscope only if they could be made visible to the instrument; "natural appearances" required careful crafting of both specimen and image. I have
argued thus far that Hooke used a variety of pictorial strategies to show that the
microscope provided access to a "hidden world" of strange, unusual, and beautiful objects and creatures. Presenting the microworld *as* a world was one of
Hooke's central aims in constructing the images in *Micrographia.* The illustrations were crucial to his overall effort to portray the microworld as a place of
order, quiet, and calm rather than a confusing array of incomprehensible fragments. Consequently, the images minimize the disjointed experience of viewing objects through the microscope and betray none of the messiness, tedium,
and uncertainty experienced by early users of microscopes.

Hooke built up his images from numerous observations made from multiple vantage points, under varying lighting conditions, and with lenses of differing powers. Similarly, his specimens required a great deal of manipulation and
preparation in order to make them visible through the microscope. Nowhere
in *Micrographia,* however, do we find visual traces of mutilation or violence,
although insects were of course dismembered, pinned, and killed in the course
of preparing and dissecting specimens. Instead Hooke almost always presents
his insects as whole, living creatures. In Schema 24, for example, viewers are
confronted with the intense stare of the eye of a fly (Figure 4.6). Although only
a small part of the fly's body is visible, one does not sense that this is merely a
fragment of a body. The fly seems to be peering out from the darkness in a
lively and compelling manner, suggesting that its body lies hidden in the surrounding shadows. As with the other illustrations in the *Micrographia,* an atmos-
phere of calm and quiet prevails as the insect serenely presents itself to the
viewer for study and contemplation, and as such the image successfully masks
the violence of the removal of the fly's head for microscopical observation. In
another illustration of a fly (Figure 4.7), Hooke presents the insect alongside an
enlarged view of one of its wings. The fly is shown as an intact, living specimen,
gently cradling its own severed wing in a peaceful embrace. Hooke's written
account describes his dissection of the fly and his observations of its internal
organs, but these observations are not illustrated in the image: "Nor was the
inside of this creature less beautiful then its outside, for cutting off part of the
belly, and then viewing it . . . I found, much beyond my expectation, that there
were abundance of branchings of Milk-white vessels, no less curious then the
branchings of veins and arteries in bigger terrestrial Animals."[41] Again, Hooke
has crafted the image to reveal his specimen as a beautiful and noble creature,
in part by carefully concealing the destruction of its body that was necessary to
observe it.

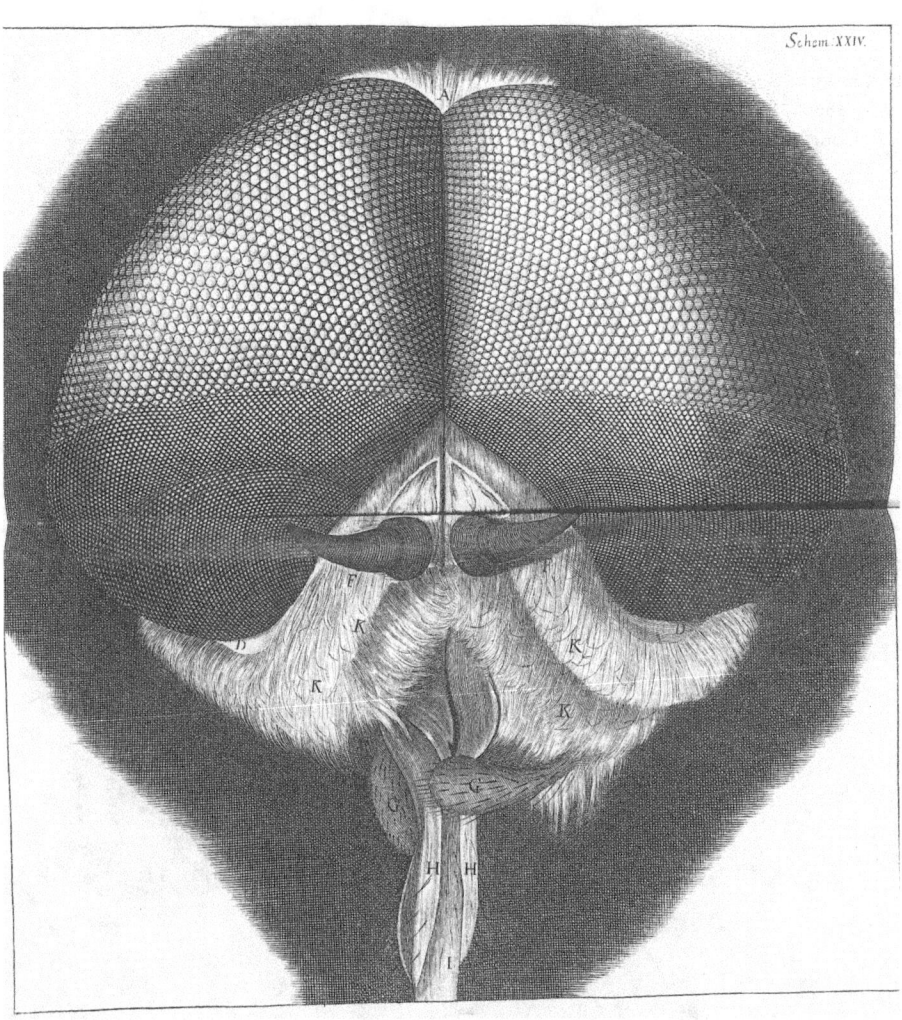

Figure 4.6. Robert Hooke, *Micrographia* (London, 1665), Schema 24 (eye of fly). Reproduced by permission of The Huntington Library, San Marino, California.

Hooke's desire to show the microworld as a place of calm and order paralleled the aim of the members of the Royal Society to be perceived as purveyors of useful knowledge gained through reasoned inquiry. The Royal Society was eager to show that it contributed to the orderly workings of society and that its activities would not undermine the smooth functioning of the newly restored monarchy. John Harwood has argued that *Micrographia* was "designed to persuade Restoration society that the New Philosophy addressed important intellectual and social problems."[42] The project emerged from the social and political concerns of the Royal Society at a time when the group was becoming

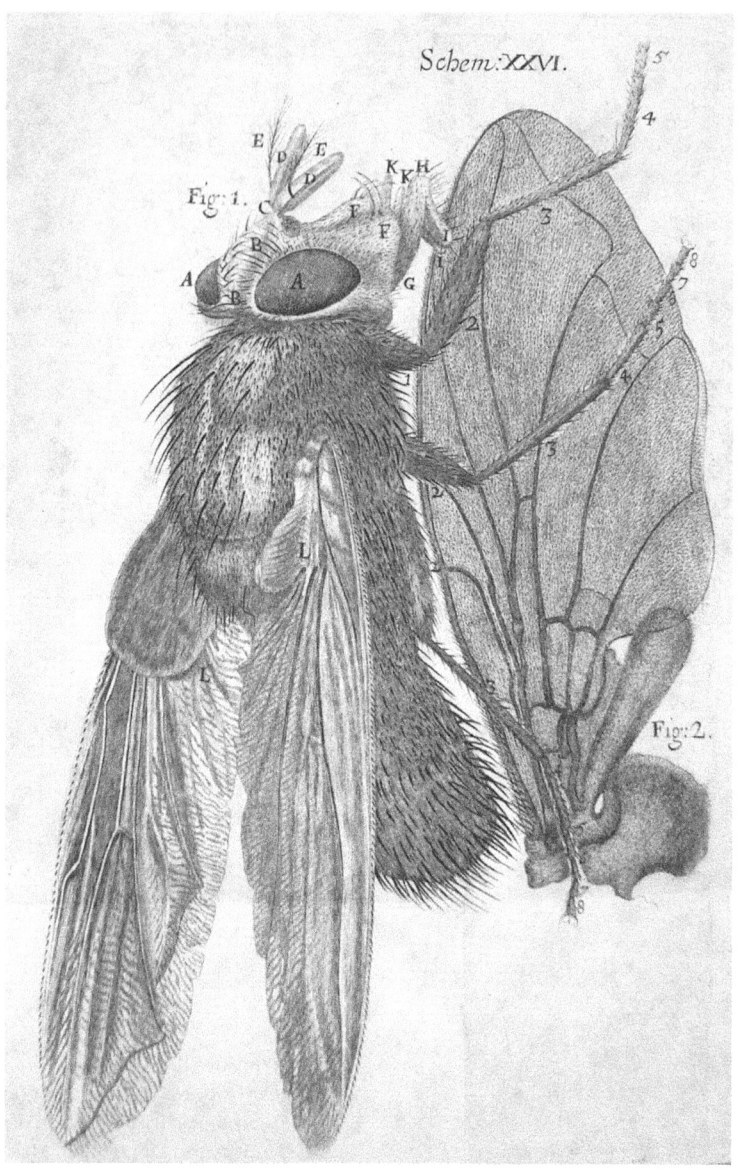

increasingly aware of the printing medium as a way to establish and protect its public image.[43] In *Micrographia,* Harwood argues, Hooke presented an idealized picture of himself and the New Philosophy and emphasized "the existence of an interpretive community committed to serious questions and likely to achieve significant, highly visible results."[44] Michael Aaron Dennis has also argued that political concerns shaped the form and content of *Micrographia,* particularly the problems of public indifference and ecclesiastical criticism that the early Royal Society faced: "Responding to the charge that the experimental philosophy was part of the atheism, materialism, and enthusiasm unleashed during the Civil

War and Interregnum, [Micrographia] demonstrated that the new optical instruments generated 'safe' knowledge clearly distinguished from the claims of atheists and enthusiasts, the elusive outcasts of Restoration society."[45] Dennis argues that interpretation was a central concern of both the Reformation and the New Philosophy. While Protestant reformers were involved in a conflict over the meaning and significance of the Bible, the New Philosophy was involved in a contest over the meaning and theological status of the Book of Nature. Religious enthusiasts and atheists alike were attacked by the restored episcopacy and monarchy because they were believed to bear most of the responsibility for the chaos of the Civil War and Interregnum. As Dennis writes: "Cast with such suspect company, the experimental philosophers vigorously attempted to differentiate themselves from these 'subversive' readings of the Book of Nature while demonstrating the ability of their own interpretive practice—representation—to undermine the claims of both atheists and enthusiasts. A 'reformation in Philosophy' promised a philosophy appropriate to the needs of church and state, one that would aid the 'Mighty King' in maintaining 'an Empire over the best Invisible Things of this World, the Minds of Men.'"[46]

Dennis sees the interpretive practice of representation that Hooke advocated in Micrographia as the visual expression of disciplined seeing: "Observers practicing disciplined seeing were transparent conduits through which an audience *re-viewed* the observer's private experience. With disciplined seeing, Hooke denied that there was a translation of private experience into public knowledge. . . . In Micrographia the reader sees through Hooke to 'the things themselves,' forever *re-presented* on the printed page."[47] In denying that communicating microscopical observations required an act of translation on the part of the author, Hooke sought to avoid the troublesome political and religious issues raised by critics of the New Philosophy. As we have seen, however, he used numerous pictorial strategies to translate his observations into comprehensible visual images for his audience. He carefully crafted the deceptively straightforward illustrations in Micrographia by using sophisticated image-making techniques designed to convince viewers that the objects and creatures found in the microworld were similar to those found in the macroworld.

Making Images and Making Observations:
Hooke's Early Drawings of Insects

Let us now turn to Hooke's early microscopical observations in order to investigate some of the processes by which he created his written and visual accounts and how his published accounts worked to conceal the very circumstances of their production. Hooke's early activity with the microscope and his preparatory work for Micrographia are known primarily through the minutes of the Royal Society meetings at which he presented drawings and observations. Although a listing for "Dr Rob. Hookes *Micrographia* with Fig. And the Original Drawing" was included in the sale catalogue of Hooke's library published after his death, the drawings are not known to have survived, and it has been suggested that most of them were destroyed in the Great Fire of London in 1666.[48] Only one preparatory drawing for *Micrographia* is known to exist, a drawing of figures found in frozen urine and water, which is published as figures 1 and 6 of Schema 8.[49] Outside of the detailed accounts given in the text of *Micrographia*, little has been known about Hooke's working methods during the period he was making observations and creating the images for the book. Recently, however, I have found evidence for attributing to Robert Hooke a series of sketches that sheds important light on his early working practices as well as on the development of his visual and verbal observations in *Micrographia*.

The sketches appear in a notebook, now in the collection of the British Library, that once belonged to Dr. John Covel (1638–1722), an English naturalist who served as the chaplain to the English ambassador to Constantinople from 1670 to 1677.[50] The notebook contains numerous drawings by Covel of the plants and insects he observed while living in Turkey.[51] Many years after his return to England, Covel passed on his notebook to the noted collector and botanist James Petiver (1663/1664–1718). Covel seems to have given the notebook to Petiver in order to obtain his help in identifying the Turkish plants and insects. Petiver's written comments, which appear throughout the notebook, usually signed "Petiver" or "Pet.," can be dated to sometime after 1707 but most likely were written after his trip to the Netherlands in 1711.[52] The notebook is a fascinating example of the role of visual images in facilitating the exchange of information about the natural world in early modern Europe. In his annotations Petiver often addresses Covel directly in a conversational manner, and he offers praise and encouragement for Covel's work. In commenting on a plant Covel drew on April 10, 1675, Petiver wrote: "You have been very happy in the accurate delineation of the different Leaves of this peculiar and to me unknown plant. PET."[53] Petiver's comments show that he gained his knowledge of Middle Eastern plants and insects from reading published works by travelers and botanists who had visited the area, as well as through the study of specimens and drawings he obtained by way of his broad network of correspondents. He writes of receiving drawings of plants that grow in Tuscany from the botanist Bruno Tozzi (1656–1743) and of receiving specimens of insects from the Levant.[54] The written entries and annotations show that both Covel and Petiver shared Hooke's affinity for unusual and beautiful specimens. Covel includes the following written description with a drawing of a beetle: "The English in Turkey call this a Buffalo bugge: it is of a shining Azure or blewish green like the feathers of a drakes neck, bravely checker'd or fretted on the back. . . . It is set in gold, and silver and worn as a great amulet by the Turkes. The first that ever I found (june 4.72 at belgrade) as I was peeping upon it pist or squirted on my left eye. . . . This I found at Pera. May 20th. It lived till this day the 30th. I perct its 2d & 3d trunk with a pin to hold him fast, and yet he lived." Petiver concurs with Covel's assessment of the beetle's appearance by writing, "I have observed of this wingless genus 3 or 4 sorts in England one of a copperish green [pocked] like this but not so beautiful. Pet."[55]

The earliest entries in the notebook, however, date from 1660 and 1661, thereby predating Covel's trip to Turkey by ten years. These entries consist of sketches of insects and inscriptions in Hooke's hand and include several unmistakable links to Micrographia. It is not presently known how or why the notebook passed from Hooke to Covel; the bindings appear to be original and it does not appear that any pages have been added or removed. Covel and Hooke were students at Cambridge and Oxford, respectively, during the 1650s, but there is no record of their knowing or working with one another, and Covel was never a member of the Royal Society. Although the early provenance of the notebook is as yet unclear, the written and visual records it contains are nevertheless important documents of Hooke's working practices. Hooke's entries in the notebook reveal how closely the making of images and the practice of observation were linked in his early work with the microscope. They also show how he worked and reworked his written and visual observations over the course of several years.

Most of Hooke's contributions to the notebook appear on folio 113v (Figure 4.8), a page that contains sketches of magnified views of insects accompanied by written entries.[56] Several of the sketches are dated, and from these dates it appears that the entries on folio 113v were made over a one-year period, between July 28, 1660, and July 17, 1661. The insect pictured in the lower-right corner of the page, labeled "A Kind of Teek," is accompanied by an inscription that is signed and dated "April 11. 1661. R.H." The entries on this page are written in a different hand from Covel's and Petiver's entries and are consistent with known examples of Hooke's handwriting.[57] In the lower-left corner of the page, the legs of the insect referred to as "another sort of mite creeping on rotten wood" are labeled with a small letter *s* in two places. These letters are used to direct readers to a particular aspect of the drawing referred to in the written inscription, which reads, "I supplyed the forme of the leggs all being double joynted, and forked at the end like S.S. they not being in theire naturall position." This technique of labeling similar body parts with the same key letter is used quite frequently by Hooke in Micrographia, as seen, for example, in figure 1 of Schema 27, an illustration "Of the Water-Insect or Gnat" (Figure 4.9). Figure 2 of Schema 27 illustrates another pictorial technique used on folio 113v, that of the dotted or "prickt" line. Hooke had used a dotted line to represent the leg of the dissected insect pictured at the upper right of folio 113v (see Figure 4.8), in order to show the position of the legs after it had died.[58] Conversely, in figure 2 of Schema 27 of Micrographia, this same dotted line is used to indicate

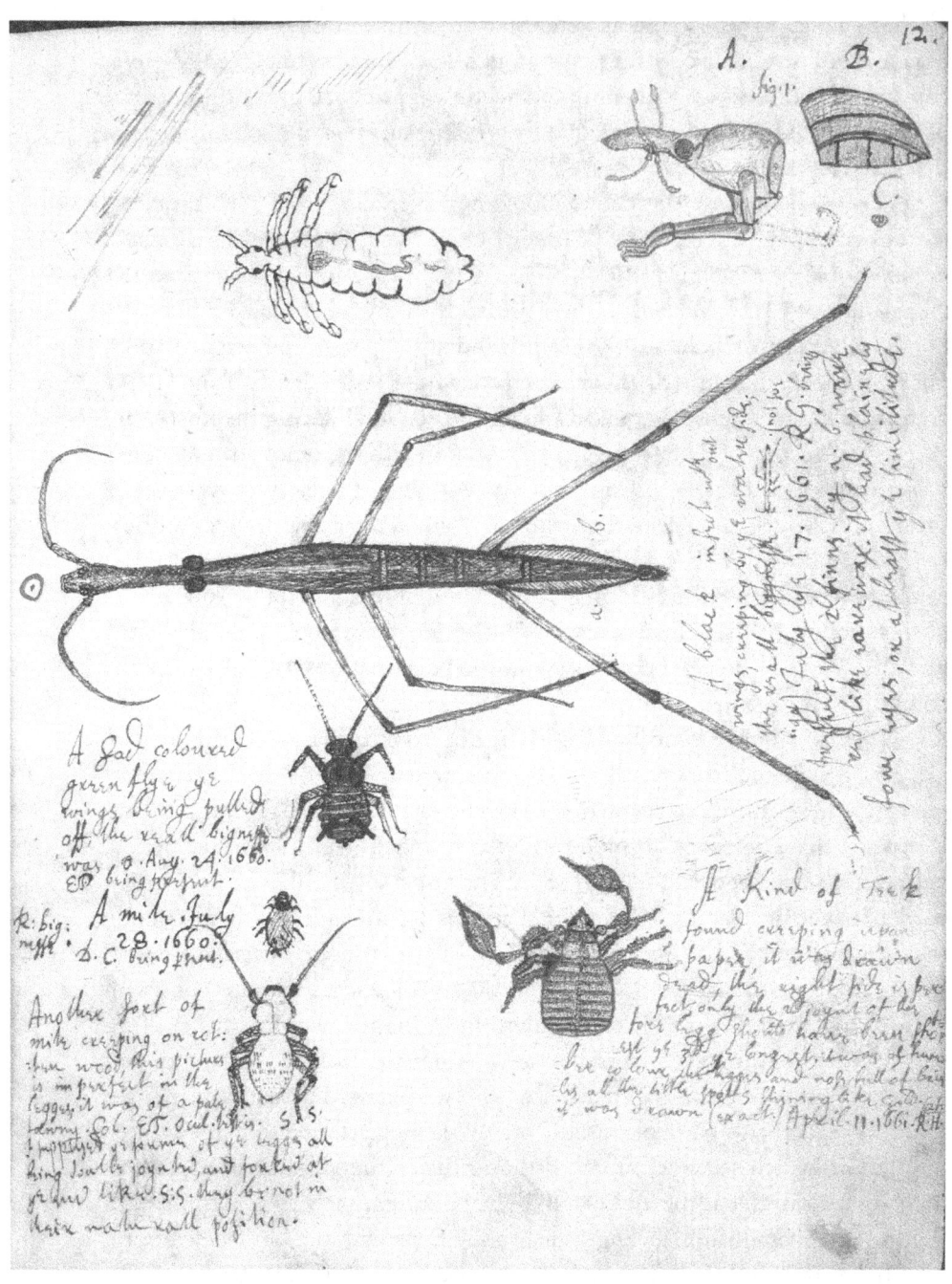

the movement of the living insect's swimming appendage as it propels itself through water, rather than the position of its body when dead.[59]

The sketches on folio 113v afford a glimpse into the development of Hooke's visual strategies for representing the microworld that would eventually be fully elaborated in *Micrographia*. One of the fundamental visual problems Hooke needed to work out was that of scale. Because there were no stationary markers or recognizable landmarks in the microworld, early users of microscopes were limited in their ability to communicate the relative sizes of objects and the distances between them when viewed through the microscope. In the sketches in the Covel notebook, this problem was solved by including a small drawing or mark to indicate the insect's actual size, or "real bignesse." For example, to the left of the drawing labeled "A mite. July 28. 1660" is a small dot

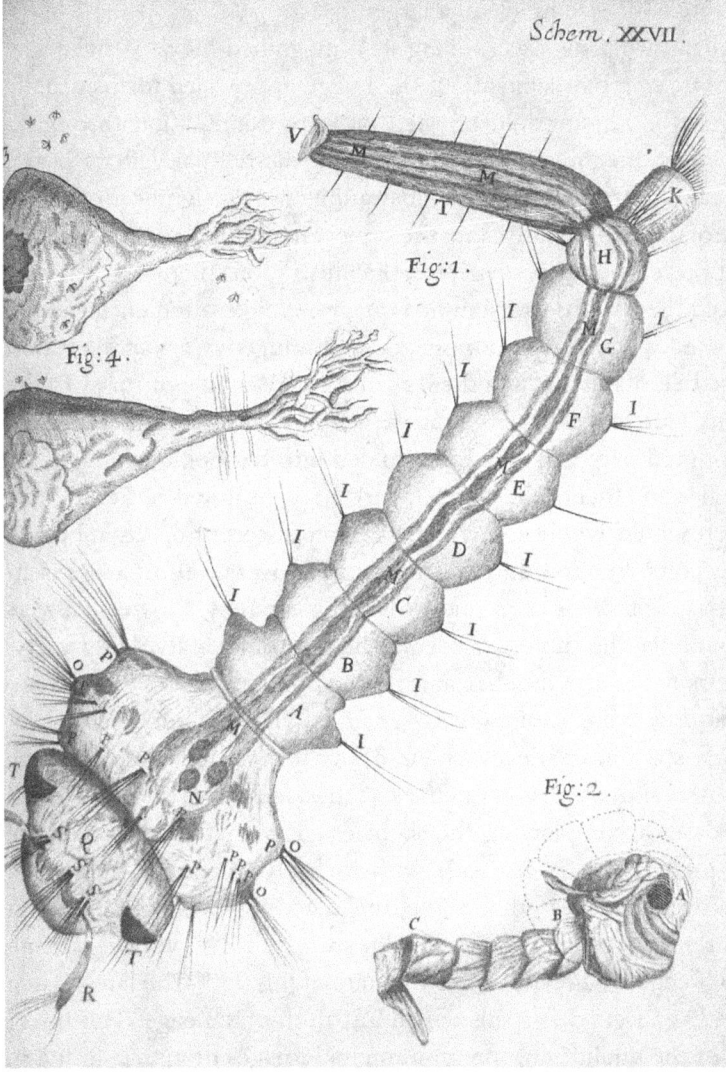

Figure 4.9. Robert Hooke, *Micrographia* (London, 1665), Schema 27 (water-insect). Courtesy of History of Science Collections, University of Oklahoma Libraries.

marked "R. bignesse." Similarly, within the descriptive text for the insect pictured in the center of the page, following the phrase "the reall bignesse was," is a small drawing of the insect at actual scale. This schematic strategy seems to have been lost on James Petiver, since his annotation for the drawing indicates that he may not have been aware that he was viewing a magnified image. Petiver's comment, which appears opposite the sketch on folio 114, is directed to Covel, who he assumed was the artist. In this passage, Petiver refers Covel to his own work on English beetles that appears in *The Monthly Miscellany: or, Memoirs for the Curious* . . . (London, 1707–1709). "A Tipula or water spider, this has been observed at London but very rare. PET. vid. Scarab. Anglic. in the Memoirs for the Curious. id." Petiver obviously did not read the inscription accompanying the drawing and thus missed the basic information that the image represents a magnified view of a very small insect, rather than a life-size or slightly larger than life-size image of a larger insect.

In not realizing that he was looking at a magnified view, Petiver also missed what was most remarkable about the image: it revealed forms usually hidden from the eye, which could be seen only by examination through a microscope or with a magnifying lens. In *Micrographia*, Hooke left no doubt that what his readers were seeing in the illustrations were magnifications. Not only did the title of the book clearly state the subject matter and the text include detailed descriptions of the microscope, but the illustrations themselves almost always announced their status as magnified images. Hooke often enclosed the objects he presented within a round frame, thus offering viewers an evocation of the experience of looking through the lens of a microscope. For many of the insect illustrations, Hooke created large-format images that sometimes exceeded the size of the printed page and had to be folded into the book, as he did for his image of the flea and that of the louse (Figure 4.10). As noted earlier, Hooke almost always presented whole insects rather than dissected or disembodied parts of insects. Thus, in addition to furthering his overall effort to present the microworld as a place of calm and order, this strategy allowed viewers to comprehend at once the subject matter of the illustrations. By deftly weaving together elements of the familiar and the unfamiliar, Hooke signaled to his audience that they were seeing greatly enlarged views of very small creatures. A very early stage in the development of this technique for representing microscopical observations is evident in the sketches on folio 113v of the Covel notebook. The dates accompanying the sketches show that the insects were depicted on an increasingly larger scale with each observation. The earliest dated image is also the smallest, that of the mite dated July 28, 1660. The next dated image on the page is that of "a sad coloured green flye, ye wings being pulled off," which appears directly above the mite of July 28.[60] The latest dated image is the sticklike insect that spans the entire width of the page. This insect possesses some of the qualities of the monumental images of insects found in

Micrographia, but judging from Petiver's annotation it was not quite large enough to signal its status as a magnified image.

The most direct link between the sketches in the Covel notebook and the plates in *Micrographia* is the insect pictured in the lower-right corner of folio 113v (Figure 4.8), which is accompanied by an inscription dated April 11, 1661, and signed with the initials "R. H." The inscription reads as follows: "A Kind of Teek found creeping upon paper, it was drawn dead, the right side is perfect, only the 2d joynt of the four legg should have been shorter the 3d the longest. it was of humber coloure the legges and nose full of bristles, all the little spotts shining like Gold. [Cat] it was drawn (exact) April 11. 1661. R. H." The insect in this sketch corresponds to the "crab-like insect" that appears as figure 2 of Schema 33 of *Micrographia* (Figure 4.11). In Hooke's longer and more detailed account of this insect in *Micrographia,* he notes that he had observed the insect

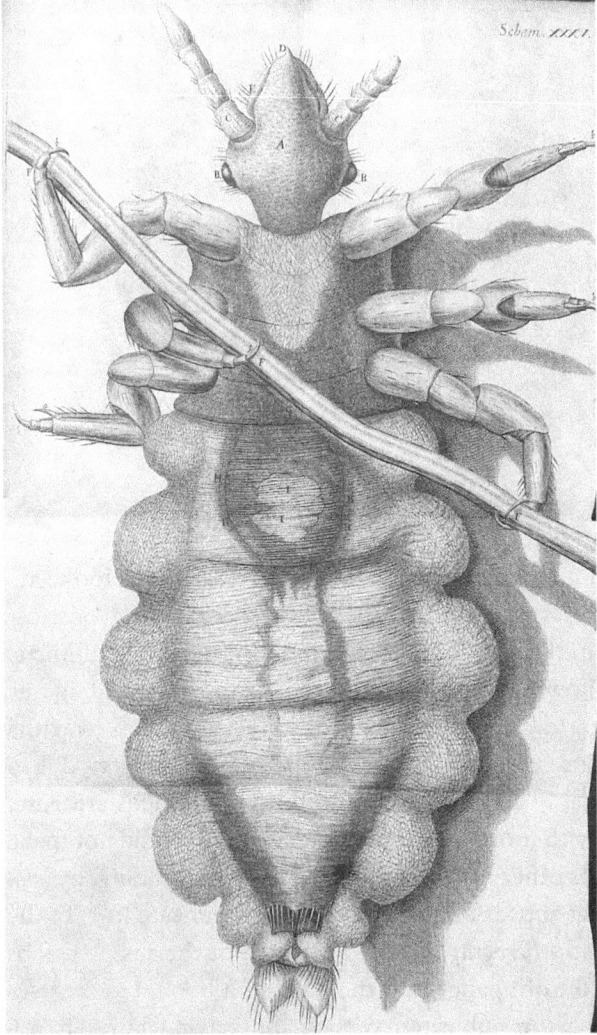

Figure 4.10. Robert Hooke, *Micrographia* (London, 1665), Schema 35 (louse). Courtesy of History of Science Collections, University of Oklahoma Libraries.

Figure 4.11. Robert Hooke, *Micrographia* (London, 1665), Schema 33 (crab-like insect). Reproduced by permission of The Huntington Library, San Marino, California.

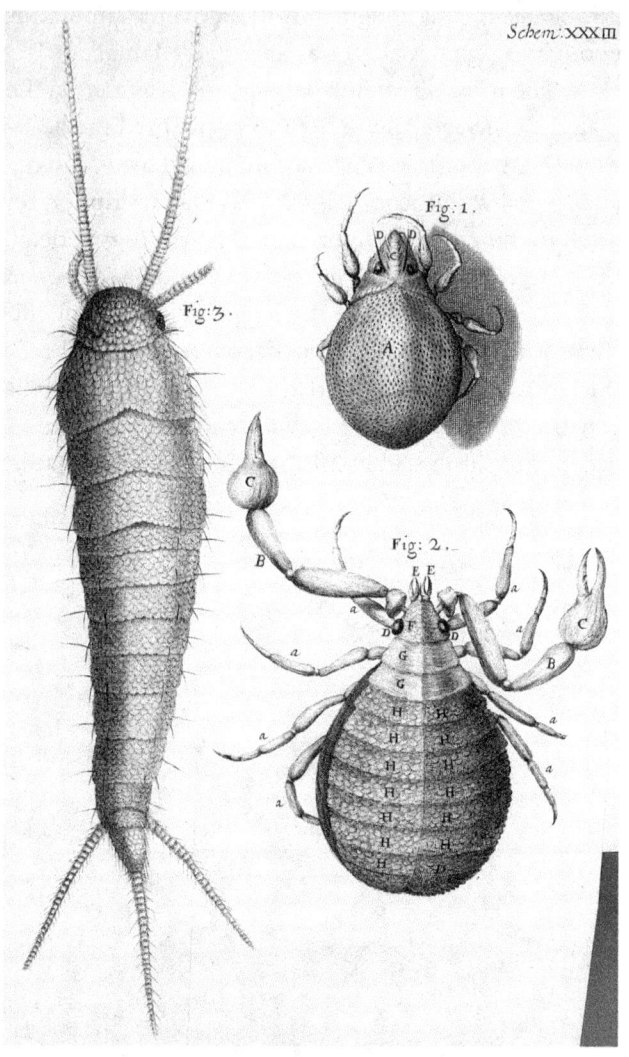

on only one occasion. From this statement, then, it is clear that Hooke reworked his visual and written accounts of the insect from memory and possibly by consulting the sketches in the Covel notebook. In Observation 51 of *Micrographia,* Hooke gives the following account of the crablike insect: "Reading one day in September I chanced to observe a very small creature creep over the Book I was reading, very slowly; having a Microscope by me, I observ'd it to be a creature of very unusual form." He continues by stating: "This creature, though I could never meet with more then one of them, and so could not make so many examinations of it as otherwise I would, I did notwithstanding, by reason of the great curiosity that appear'd to me in its shape, delineate it."[61] In the written account published in *Micrographia,* the surface that the insect creeps across has been transformed from paper into the page of a book. Hooke uses the word "reading" twice in the published version, portraying himself as a scholar sitting quietly in his study, immersed in the world of learning but at the same time alert to the workings of nature. With his microscope at his side, Hooke presents himself as ready to conduct observations carefully and thoughtfully whenever an interesting phenomenon might present itself.

The fully realized portrait of the crablike insect that appears in the engraving has also been substantially reworked from the version that appears in the notebook sketch, which was presumably made at some point closer in time to Hooke's actual observation of the insect. In the sketch, the insect is shown to possess only six legs, but in the published version it has eight legs. As noted above, Hooke did not make any further observations of the insect and thus may have included the additional legs on the basis of his knowledge of the anatomy of other types of insects, such as spiders. As is typical of other such illustrations in *Micrographia,* the crablike insect is shown on a large scale, and every surface of its body is described in minute detail. Hooke's early sketch does not contain the same level of surface detail, so it is likely that he again filled in the "missing" information from his knowledge of other insects. Another source he may have consulted in composing the published version of the image was Willem Piso's illustrated book on the natural history of Brazil, *Historia naturalis Brasiliae,* which Hooke referred to and quoted from in Observation 43, "Of the Water-Insect or Gnat." Hooke may have used the illustrations of crabs that appear in Piso's book as a model for constructing his image of the crablike insect, for the insect that is represented in Schema 33 is shown using the same convention of depicting the "claws" in similar positions and as slightly open (Figure 4.12).[62]

Constructing the Persona of a Scientific Observer: Quietly Peeping in at the Window

The differences between the early sketch and notes and the published version in *Micrographia* show that just as Hooke labored to create specimens with "natural appearances" so too did he labor over the crafting of his visual and verbal accounts. In presenting himself as a solitary observer, however, Hooke not only substantially reworked his early observations but also eliminated certain pieces of information from them. The written inscriptions accompanying the sketches in the Covel notebook reveal that Hooke did not conduct his observations alone. Several colleagues, or co-observers, are noted as having witnessed the observations with Hooke. A person referred to only as "DC" was present for the earliest observation, that of the mite dated July 28, 1660, while two other people—referred to as "ET" and "RG"—are listed as being present for observations made on August 24, 1660, and July 17, 1661. The undated sketch of a "mite creeping on rotten wood" also lists "ET" as a witness. Hooke makes no mention of conducting microscopical investigations with co-observers in the text of *Micrographia.*

ditum, uti & forceps: finiftrum parvum. Color teftæ fuperius olivaceus, flavo in extremitate intermixto. Corpus inferius pallide flavefcit. Omnia crura & brachia caftanei funt coloris, forcipes etiam fuperius, cætera pallide flavefcentia una cum tenaculis. Crura omnia inferius veftita funt copiofis pilis longiufculis obfcure caftanei coloris. Carne funt bona.

Uca Gvacv Brafilienfibus, figura & conformatione fua plane convenit cum antecedente, excepta magnitudine & colore, hic enim longe eft major.

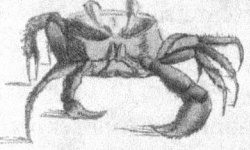

Cvnvrv Brafilienfibus, fœmella eft Uca una, figura & colore conveniens, folum magnitudine corporis & brachiorum differt, minor enim eft quam mas & brachia habet parva, brachio maris finiftro vix æqualia, finiftrumque paulo grandius dextro; raros quoque habet pilos, cum Uca una tota fit hirfuta.

Gvanhvmi Brafilienfibus: Cancer terreftris: corpus illi rotundum, fed paulum compreffum: magnitudine Auriaci mali. Pedes habet octo, quinque digitos longos, quatuor internodiis: crura longis pilis veftita: os magnum & ad latera oris

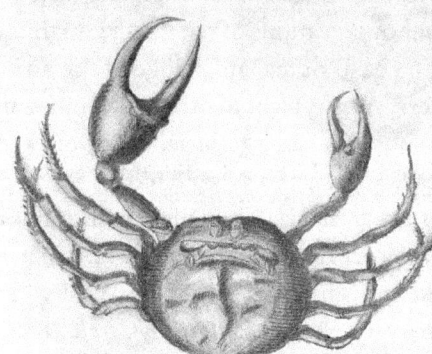

utrimque latitudine unius digiti hirfutum, ut & corpus: Brachia duo, dextrum magnum, finiftrum parvum; magnum longum octo digitos, latum plus duobus; parvum vix quartam partem illius æquat. Oculi bini, quos inftar parvæ pilæ erigit & iterum in longa cavitate abfcõdit: circa os quafi duo brachiola quæ recondere poteft & iterum depromere, & ufque ad oculos deducere. Celerrime currit, & quidem tranfverfim, ita ut oculos ad latus convertat & branchium majus elevatum geftet. Maximo numero in filvis paludofis oberrant. Carnem habent bonam.

Aratv & Aratv pinima Brafilienfibus: Cancer terreftris, quadratæ figuræ, corpore haud magno, tefta multiplici colore picta, brunno, cœruleo, albo, rubro punctatim varie intermixtis. Oculi prominentes longe diftantes, nigri, in angulis roftri pofiti. Crura octo, quatuor internodiis, compreffa, ruffa, ac per totum maculis purpureis, nigris ac albis diftincta: Chelæ duæ non ita magnæ, æquales, ruffæ, in extremitate ex albo flavefcentes. In ventre etiam flavefcit. Crura raris pilis nigris veftiuntur.

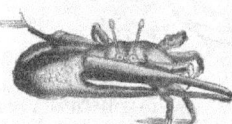

Ciecie Ete Brafilienfibus: *Cranguerfinho des Manges* Lufitanis: corpus ad quadratam figuram accedit, & totum vix fuperat majoris avellanæ magnitudinem: oculi prominentes longiufculi, quos recondere poteft. Crura octo: brachium finiftrum ipfi maximum, non tamen craffum, forcipis tenacula tenuia & longa, æqualia, fine dentibus: dextrum brachium tenuiffimum. Tefta coloris hepatici: reliquum totũ corpus cum cruribus obfcure flavum ex pallido. Edulis eft; & Brafiliani illo curant morbum quem vocant Mia.

Ciecie Panema Brafilienfibus, cancer præcedenti plane fimilis, excepto quod forcipis tenaculum inferius brevius eft fuperiori.

Potioviqviya Brafilienfibus. Locufta marina, Belgis Ƶeekreeft. Longitudo corporis à fronte ad exortum caudæ feptem digitorum; caudæ fex: latitudo teftæ dorfi feptem, ventris duorum & femis: totius corporis ambitus novem digitorum & femis; caudæ ambitus quinque, quæ conftat fex tabellis, feptem juncturis fibi invicem appofitis: & habet cauda ad quodlibet latus inferius quatuor pinnas, unum digitum longas: caudæ autem finis definit in quinque pinnas, fefquidigitum longas, unum latas: extremitates laterales cujufque tabellæ definunt in cornu acutum: in quolibet latere quinque habet crura, quinque internodiis conftantia; primum par crurum fex digitos longum, alterum novem, tertium pedem, quartum feptem digitos, quintum quinque: Quodlibet autem unguem habet incurvum, acutum, pilis multis flavefcentibus hirtum, inftar penicilli pictorum: crus anterius craffitie digitum æquat, reliqua graciliora. Tefta corporis varie tuberculata, frequentibus eminentiis inftar corni-

A a

Figure 4.12. Gulielmus Piso, *Historia naturalis Brasiliae . . .* (Amsterdam, 1648), page 185. Courtesy of Missouri Botanical Garden. http://www.botanicus.org.

Steven Shapin has noted the appeal of the persona of the solitary observer for Hooke and other early modern intellectuals: "The presentation of the philosopher's persona as hermit was [not only] a way of understanding . . . who the philosopher was and what might be expected of him, but also a way of warranting claims to knowledge. A man so abstracted from the world was a man free of the hold of its idols and in immediate contact with reality, divine or mundane."[63] As Shapin also points out, the image of the scholarly hermit did not necessarily reflect Hooke's actual experience or social status. However, to gain the respect and trust of his fellow members of the Royal Society, Hooke may have found it prudent to present himself as a solitary and authoritative observer who did not need to rely on others to confirm or support his observations. As noted earlier, Hooke occupied an unusual social position within the Royal Society, and while in some ways he embodied the ideal of the mechanical philosopher in his ability to conduct experiments and construct and operate instruments, he was never fully accepted as a social or professional equal by the society's other members. A possible candidate for the "RG" listed as a witness to the observation of the sticklike insect in the Covel notebook is Ralph Greatorex (1625–1712), an instrument maker from London who worked with Hooke on building the air pump for Robert Boyle during the late 1650s.[64] Greatorex and other instrument makers had many dealings with members of the Royal Society but were never elected to their ranks.[65] Hooke may not have wanted to be identified with the world of instrument makers as he endeavored to prove his trustworthiness and intellectual capability as a newly elected member of the society.

Another reason why Hooke may have minimized the presence of his co-observers in his written accounts was his desire to avoid difficult questions about whose observations were being presented and how the viewpoints of multiple observers were or could be reconciled. Potentially disruptive issues such as these could have undermined Hooke's credibility and that of the microscope itself, and had to be avoided if Hooke were to be perceived as the microscope's amanuensis and if his images were to be accepted as exact transcriptions of his observations. Shapin has shown that "virtual witnessing" was both central to Robert Boyle's experimental program during the late 1650s and early 1660s and an essential component in the process by which matters of fact were generated within the community of natural philosophers. Virtual witnessing was made possible through the dissemination of experimental reports that were rich in circumstantial detail, thereby allowing readers to witness an experiment without directly experiencing it or replicating it themselves. Although Shapin describes virtual witnessing as a "literary technology"—generated primarily through written reports—visual images also played a role in the process. Shapin argues that visual representations functioned as mimetic devices in Boyle's texts by facilitating virtual witnessing in their density of circumstantial detail and because they "imitated reality and gave the viewer a vivid impression of the experimental scene."[66] For Hooke, visual images also created a vivid impression for viewers through, among other things, their density of circumstantial detail. But unlike Boyle's illustrations, Hooke's images of his microscopical observations were not meant to function as mimetic devices for his audiences by creating a mental image of a demonstration of a device such as the air pump. Instead, the drawings Hooke presented at meetings of the Royal Society and later published in *Micrographia* functioned as demonstrations in themselves.

Whereas the air pump could be brought into the public rooms of the Royal Society and experiments with it performed for all to see, demonstrations of the microscope could not be "performed" in this manner. Because of the nature of the instrument, microscopical observations were conducted by individuals in private settings. The public nature of demonstrations at the Royal Society meant that Hooke and others generally presented "completed" phenomena—in other words, experiments or instruments they were sure would work. The potential for embarrassment was great if an instrument or demonstration were to fail. Experiments in which the outcome was uncertain thus usually took place in private.[67] The physical setting of the Royal Society's meetings would also have precluded individuals making their own observations with a microscope. The president sat at a central table with members sitting in rows surrounding him. Instruments were set up in advance on the central table or at a table to the side of the room. It is unlikely that members would have wished to make their own observations during meetings even if it were physically feasible to do so, because receiving instruction on the use of the microscope might have been viewed as an activity inappropriate to their elevated social status. Gentlemen did not engage in manual labor or crafts, precisely the activities that Hooke was hired to handle. In the context of the microscope, therefore, images were more effective than verbal or written descriptions in facilitating the process of virtual witnessing that was so central to generating facts in Restoration science.

For Hooke's images to be trusted as exact descriptions of his observations, there could be no doubt that he had made the observations himself and that the images represented precisely what he himself had seen. If Hooke's co-observers in making the early sketches were indeed fellow technicians and instrument makers, their very presence in Hooke's "official" published accounts might have cast doubt upon Hooke's authority and his capacity to act as a trustworthy conduit of microscopical observations. For, as Shapin has also pointed out, although technicians and operators enjoyed the type of direct and unmediated access to experimental phenomena that was highly valued by natural philosophers, "directness of experience did not in itself confer epistemic value on technicians' understandings." Whether such experience was recognized by employers depended on the "moral texture of social relations."[68] Given Hooke's ambiguous social and professional standing within the Royal Society, it was important for him to eliminate elements from his visual and written accounts that might suggest unreliability. The entries in the Covel notebook show that, in his earlier accounts, Hooke was more likely to acknowledge the uncertainty of knowledge gained through microscopical observation. In describing the insect pictured in the center of folio 113v, he writes that "it had plainely some eyes, or at least the similitude." Such uncertainty is rarely present in the written accounts published in *Micrographia*. Instead, Hooke shifts uncertain knowledge from himself to less experienced or less skilled practitioners. In the preface to *Micrographia,* he notes that the microscope can produce confusing images and that in viewing objects through the microscope "there is much more difficulty to discover the true shape, then of those visible to the naked eye."[69] Hooke goes on to describe how careful he is in making observations, unlike other users of microscopes such as Henry Power: "The Eyes of a Fly in one kind of light appear almost like a Lattice, drilld through with an abundance of small holes; which probably may be the Reason, why the Ingenious Dr. Power seems to suppose them such." Hooke found the eye of a fly to instead be "shap'd into a multitude of small Hemispheres."[70] In *Micrographia*, Hooke presents himself as an authority on the microscope, fully in command of the instrument and in possession of all the skills necessary for obtaining accurate observations.

The sketches in the Covel notebook record some of the earliest microscopical observations Hooke made, working with several co-observers. Though in some cases his sketches are but a hasty set of pen scratches, they are nevertheless crucial to our understanding of the images in *Micrographia*. Hooke's illustrations for *Micrographia* are intricately constructed compositions that were the products of several years of work and many reworkings of the material; indeed, they are not simply direct transcriptions of observations but rather complex translations of those observations. The sketches, on the other hand, record an important early stage in the process of making microscopical observations and demonstrate that this process was inseparable from making the final images.

These twin processes can also be seen at work in Hooke's illustration of the louse, one of the most striking and best-known images from *Micrographia* (see Figure 4.10). Folio 113v of the Covel notebook contains a sketch that is perhaps one of Hooke's first efforts at depicting the louse and suggests that he may have used an earlier image of a louse as a model. In its size and format, Hooke's sketch resembles an illustration that appears in Thomas Moffet's *Theatrum insectorum* of 1634 (Figure 4.13). Where the Moffet louse hints at the presence of internal organs with a jagged line along its abdomen, Hooke's sketch renders these forms in greater detail. In addition, folio 112v of the Covel notebook includes a separate sketch of the internal organs of an insect that could be those of the louse (Figure 4.14). If this is the case, the sketch of the

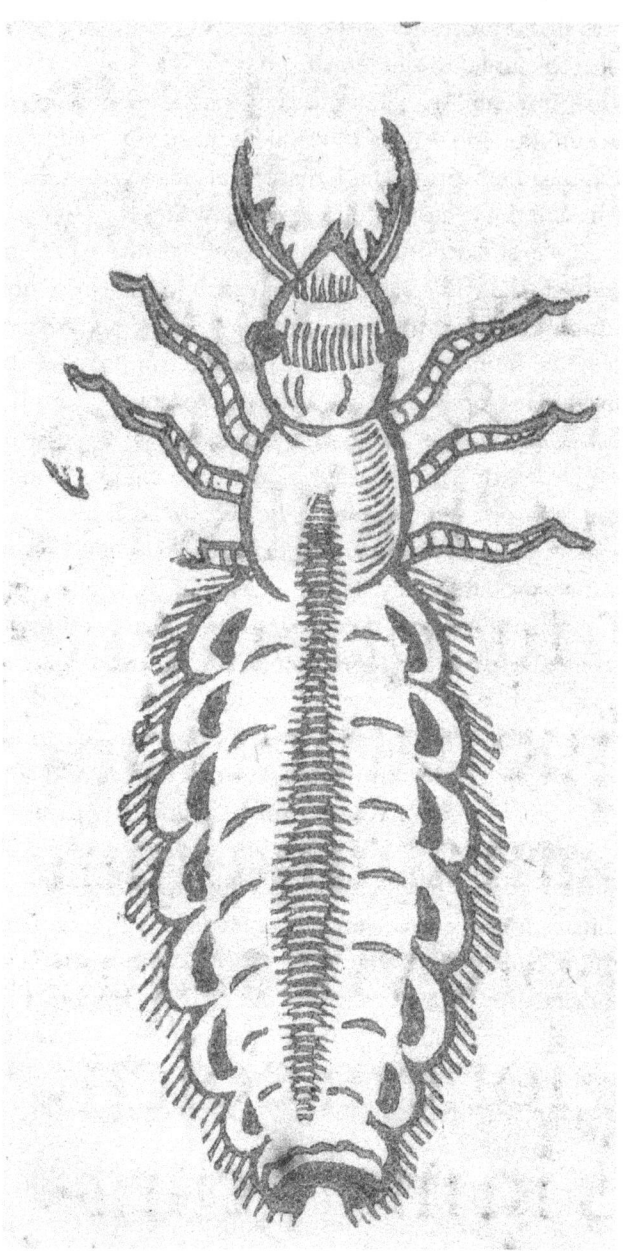

louse on folio 113v shows that Hooke combined his observations of the dissected internal organs with the Moffet model to produce an image that begins to approximate the finished version in Schema 35 in *Micrographia*.

In the engraving, Hooke uses the transparent membrane of the insect's body as a window to its internal structures, thereby uniting in a single image information gained through dissection with observations of the creature's external appearance. The louse's anthropomorphized forelegs curl around the

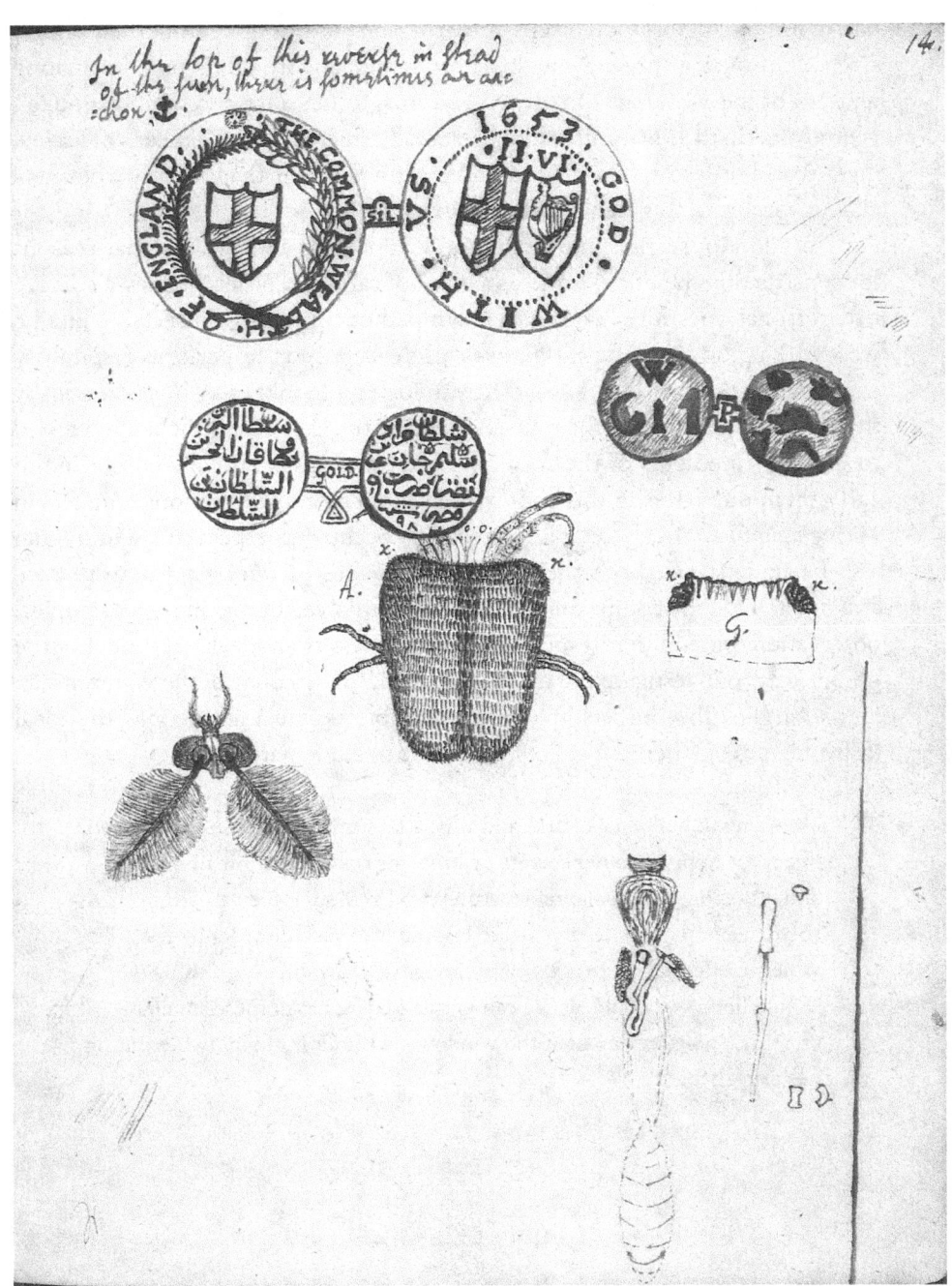

Figure 4.14. Robert Hooke and others, folio with drawings of various coins and insect parts, from John Covel, *Natural History and Commonplace Notebook* (London, c. 1660–1713). Copyright 2011 the British Library Board; all rights reserved. MS Add. 57495, folio 112v.

hair of its human host with biceps, forearm, and pincers, and rather than aggressively confronting the viewer, this stoic sentry quietly offers us a fascinating glimpse of the workings of nature. The image thus supports one of Hooke's central aims in the *Micrographia*: to present the microscope as a means of accessing a world of fascinating objects and creatures rather than merely offering a chaotic jumble of fragments and dead insects.

While Hooke had important social, political, and professional reasons for constructing the microworld as a place of calm and order, he may also have had personal ones for creating such sympathetic images of insects. While he was working on *Micrographia,* he was also called upon to perform respiratory experiments on live dogs for the Royal Society. Hooke was so distressed by the animals' suffering that he refused to repeat the experiments when requested to do so by members of the society, though he was eventually compelled to carry them out.[71] It is in this context that Hooke's representation of insects in *Micrographia* as whole—as living creatures occupying a peaceful world—can also be understood. Both the words and images of *Micrographia* were constructed to hide the confusion and messiness involved in making microscopical observations, and in that respect the book served as an idealized model of the proper way to investigate the natural world. In his account of the water insect, Hooke argues that insects are the ideal subjects and microscopes the ideal instruments; together, they allow observers to study nature

> according to her usual course and way, undisturbed, whereas, when we endeavor to pry into her secrets by breaking open the doors upon her, and dissecting and mangling creatures whil'st there is life yet within them, we find her indeed at work, but put into such disorder by the violence offer'd, as it may be easily imagin'd, how differing a thing we should find if we could, as we can with a *Microscope,* in these smaller creatures, quietly peep in at the windows, without frightening her out of her usual byas.[72]

FIVE

STITCHES, SPECIMENS, AND PICTURES

Maria Sibylla Merian and the Processing of the Natural World

By all accounts, including those written during her own lifetime, Maria Sibylla Merian was a remarkable woman who led an extraordinary life. Born in Frankfurt, Germany, in 1647 into the eminent artistic and publishing family of Matthäus Merian the Elder, Maria Sibylla demonstrated an early passion and talent for the subjects that would come to dominate her professional life. As a child she was fascinated by insects, and she raised silkworms in order to observe the stages of their development. She received artistic training in the workshop of her stepfather Jacob Marrel, where she became skilled in the depiction of flowers and other natural subjects, and like many young women of her social group she also learned embroidery and needlework from an early age. She was accomplished enough to offer lessons in these subjects to the daughters of several wealthy families in Nuremberg, where she and her husband Johan Andreas Graf settled in 1670. Merian referred to these students as her "company of maidens," and she worked with them on several projects that made use of the techniques she devised for colorfast painting on satin, linen, and silk—most notably a tent for an army general who "desired to have his field quarters designed to give him the illusion of living in a garden house full of birds and flowers."[1] It was for her company of maidens and others like them that Merian is believed to have published her *Neues Blumenbuch* in 1675, the first of a three-volume series of floral designs for use as embroidery and needlework patterns, and the first of the artist's many illustrated publications. Merian was also actively engaged in the study of insects during this early period in Nuremberg, where she developed her unique approach to the visual representation of insect life cycles and collected and traded insect specimens, an activity that would come to serve as an important source of income for her later in life. Merian's *Raupenbuch* of 1680 presented innovative illustrations of butterflies, moths, and caterpillars that included picturing the stages of the insects' life cycles along with their food plants. The *Raupenbuch* represented the culmination of Merian's early research on insects and was the first of what would eventually form a three-volume series on European moths and butterflies. The study of insect life cycles, the art of painting and drawing, and the decorative concerns of embroidery and fabric design were the foundations upon which Merian would base her professional and commercial activities and would play an important role in her approach to the natural world over the course of her long and unusual career.

In 1685 Merian made a decision that profoundly affected both her personal and professional life. It was in this year that she left her husband and, along with her two daughters and her mother, joined the Labadist religious community at Waltha Castle near Wieuwerd in the Dutch province of Friesland. The Labadists maintained strict rules aimed at providing their members with complete separation from the outside world, and included among their precepts was the belief that marriages within their sect were valid only if both parties were Labadists. Merian never reconciled with her husband Graf, and he later filed for divorce and remarried. Merian and her daughters left the Labadists in 1691 and moved to Amsterdam, where Merian supported her family through the sale of drawings, insect specimens, and art supplies, and by offering lessons in painting and drawing. In 1699 Merian and her daughter Johanna Helena embarked on a journey to the Dutch colony of Surinam in order to observe, collect, and record the life cycles of South American insects. When they returned to Amsterdam in 1701 they brought with them a large number of specimens and drawings that served as the basis of Merian's most famous work, the *Metamorphosis insectorum Surinamensium,* published in Amsterdam in 1705. This illustrated volume of sixty copperplate engravings, accompanied by Merian's written accounts, was the high point of Merian's artistic career and established her reputation as a gifted artist and naturalist well into the eighteenth century.[2]

Although Maria Sibylla Merian has been the subject of extensive scholarly research, her far from ordinary life experiences have until recently posed a barrier to understanding the artist within her specific historical context. As Natalie Zemon Davis has shown, the topos of the remarkable woman has functioned as the lens through which Merian has been viewed from the earliest accounts of her life, and Merian herself was at times complicit in supporting such an interpretation. Davis's biography provides a richly detailed picture of the social and cultural world Merian inhabited while also offering an important analysis of the relationship between Merian's gender identity and her focus on the organic interactions between plants and insects—or as Davis describes it, Merian's "ecological" approach.[3] More recently, Tomomi Kinukawa has explored Merian's place within the culture of early modern natural history, and has shown that Merian's many-faceted approach was shared by the local networks of burghers whose interest in the natural world stemmed from ideas related to religious reform, women's domestic roles, and a desire for knowledge of the individual things of the world through empirical observation. Merian's immense artistic talent has to some extent also been a barrier to understanding her visual imagery within its historical, cultural, and artistic contexts. While Merian is universally recognized as an extraordinary artistic talent, the meaning of her work beyond this extraordinary talent has been somewhat more difficult to pin down. Although Merian's illustrations of insect life cycles were innovative, she was not the first artist to depict the stages of insect development, nor was she the first artist to raise the portrayal of insects to "great art," nor was her research into insect development an important factor in resolving the heated debates among natural philosophers about the existence of spontaneous generation and the myriad associated philosophical and religious questions raised by this debate.[4] Analyses of Merian's imagery have often focused on distinguishing between the aspects of Merian's work that can be categorized as "art" and those that can be categorized as "science," with each of these understood as opposing forces engaged in a perpetual battle for dominance. Kurt Wettengl, for example, describes Merian's approach to illustration in *Metamorphosis* as "aesthetic" and "painterly" and sees other illustrations as exemplifying a "systematic, classifying methodological approach."[5] In this chapter I contend that there is little to suggest that Merian's approach in the *Metamorphosis* was not systematic or methodological, or that her approach in other visual work was not aesthetic.

In order to gain a better understanding of Merian's visual strategies and the significance of her illustrations it is necessary to look beyond questions of their artistic or scientific "value." Art and science were not separate concerns in Merian's work, nor is this the only axis by which Merian's imagery should be approached. I argue that the specific visual requirements of decorative arts practices such as embroidery and needlework played an essential role in the formation of Merian's approach to creating images of the natural world, and that Merian's involvement in the trade and exchange of natural history specimens within the community of collectors in Amsterdam during the 1690s was another important influence on her artistic development. It was in this milieu of preparing, circulating, and selling *naturalia* that Merian learned to see the natural world as composed of beautiful objects that could be bought and sold for profit. This chapter examines Merian's illustrations and drawings for publications she produced prior to *Metamorphosis* in order to show how these various strands of her interests developed, and to show how they came together in the Surinam book. Merian's combination of skills in embroidery and needlework design, specimen preparation, and the observation of insect life cycles allowed her to create a book that offered European audiences an elegant and exotic vision of nature in the New World. In the illustrations for this work Merian used her wide-ranging talents to transform plants and insects into objects of exchange that could travel between continents and within cabinets; the techniques she developed for processing the natural world were the outcome of her immersion in the practices and culture of seventeenth-century natural history, but were also embedded within the practices of the global trade in commodities that came to serve as the foundation of the economic prosperity of early modern Europe.

Insects were a personal interest of Merian's, but they also suited her professional ambitions as an artist, author, and broker of natural history specimens. As discussed in part 1, by the later seventeenth century insects were well-established as subject matter for studying, collecting, and picturing in a variety of contexts. Although Merian's professional activities put her at risk of losing the respect of the international networks of collectors who preferred to obtain their specimens, books, and drawings through the exchange of gifts, the inherent interest and value of her subject matter was not at all in question. Rather, in order to garner interest in her insects, Merian needed to set herself apart. To do this, she not only cultivated the persona of the remarkable woman, but also ventured farther afield than Europe to find new insects that would satisfy the evolving tastes for the exotic among collectors and naturalists. In her culminating work, the *Metamorphosis,* Merian utilized all of the tools at her disposal to render nature newly exotic and collectable for this audience.

Blumenbuch Series: The Foundations of Merian's Visual Style

Merian's *Blumenbuch* series has received comparatively little attention from scholars in favor of the better-known *Metamorphosis* volume and the *Raupenbuch* series, but the *Blumenbuch* illustrations are important for understanding the concerns that would shape Merian's approach to creating images of the natural world throughout her career. The *Blumenbuch* series was published in Nuremberg in three parts between 1675 and 1680; in 1680 Merian issued a second edition of the series that contained all three volumes together under the new title *Neues Blumenbuch.*[6] Each of the three volumes consists of twelve copperplate engravings featuring a single flower type, with the title pages composed of floral garlands. At this early stage in her career Merian relied heavily on existing pictorial works as models for her own illustrations, but she adapted these models to the unique pictorial demands of embroidery and needlework design. The pictorial requirements of the decorative arts would come to play an important role in Merian's illustrations, in particular her approach to representing relationships between insects and plants.

As the name implies, the *Blumenbuch* series focused primarily on illustrations of flowers. Merian created clear, uncluttered compositions of flowers and insects with bold, crisply outlined forms. Color played an important role in these compositions, as it would in her later work. Merian used broad areas of deeply saturated hues to block out and define the compositions. She stated in the preface to the second edition that the illustrations were "to be of use and pleasure to people who know and love art as [models] for drawing and painting and to women for sewing."[7] An essential component of a respectable young woman's education and upbringing was the acquisition of skills in embroidery and other types of crafts, and pattern books such as Merian's provided an essential source of motifs for women engaged in these activities. The earliest European pattern books were published in Italy beginning in the fifteenth century and focused on patterns for lace. With the spread of printing, however, came an increase in the type and number of pattern books, and by the later seventeenth century needleworkers had many books to choose from when composing their designs. Pattern books were primarily intended for women whose involvement with embroidery and needlework were leisure-time pursuits rather than professional activities, but the work of such "amateurs" was often very sophisticated and exhibited high levels of skill.[8] In composing her own book of models Merian drew extensively on the work of others, and as Sam Segal has noted, "although the title pages state that the motifs were painted from nature, the claim cannot be taken literally."[9] Copying models was a regular component of an artist's training and, like other artists engaged in the production of images "from life," Merian learned to create lifelike illustrations through the twinned processes of observing nature and following established examples. Segal has identified many of the sources Merian used for the *Blumenbuch,* and primary among these was Nicolas Robert's *Variae ac multiformes florum species.* . . .[10]

An example of the ways in which Merian adapted her models for the purposes of needlework and other types of handwork is reflected in a succession of images of an iris. A bearded iris and swallowtail butterfly appear as plate 8 of the first volume of the *Blumenbuch* (Figure 5.1). Several scholars have noted that Merian based this composition on a plate from Nicolas Robert's *Variae ac multiformes florum species* . . . (Figure 5.2).[11] Merian's *Blumenbuch* copy is reversed, and she made two other changes to Robert's composition: the tulip was replaced with an iris bud, and a swallowtail butterfly was added to the lower-left corner.[12] A vestige of the original tulip remains in the form of the short, truncated leaf that appears in the lower-left corner, where Merian incorporated Robert's tulip leaf into the leaves of the iris. Merian retained the compositional balance provided by the tulip while at the same time simplifying the overall arrangement by eliminating the tulip's stem, which in Robert's composition ran diagonally

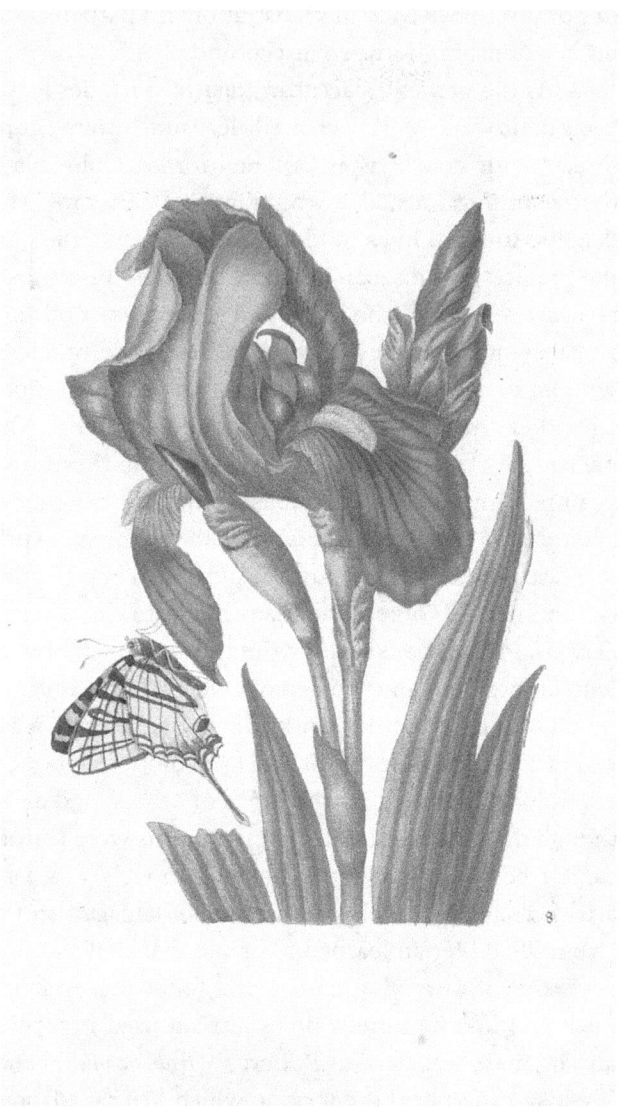

Figure 5.1. Maria Sibylla Merian, *Neues Blumenbuch* (Nuremberg, 1680), plate 8. Photograph copyright The Natural History Museum, London.

through the center of the image. In the *Blumenbuch* engraving, Merian further simplified the image by creating even more distance between the individual elements of the composition. The leaves have been separated from one another and the butterfly has been lowered in order to distinguish it from the petal upon which it rests, and the central structures of the flower are silhouetted against a small area of white space that has been carved out of the center of the bloom.

Commentators on Merian's early work have generally understood the deviations she made from her models as evidence of her burgeoning artistic talent. Segal has described the changes Merian made to Robert's composition as "improvements" that gave the composition "a more natural appearance," and

Thomas Bürger's opinion of the effect of these changes is that they "breathe artistic life into floral pictures."[13] While this emphasis on Merian's originality and artistic talent is certainly justified, Merian's modifications can also be understood as a translation of Robert's original design into the medium of embroidery, and thus devised in response to the specific needs of the genre in which she worked. One of the foremost pictorial requirements of such pattern book

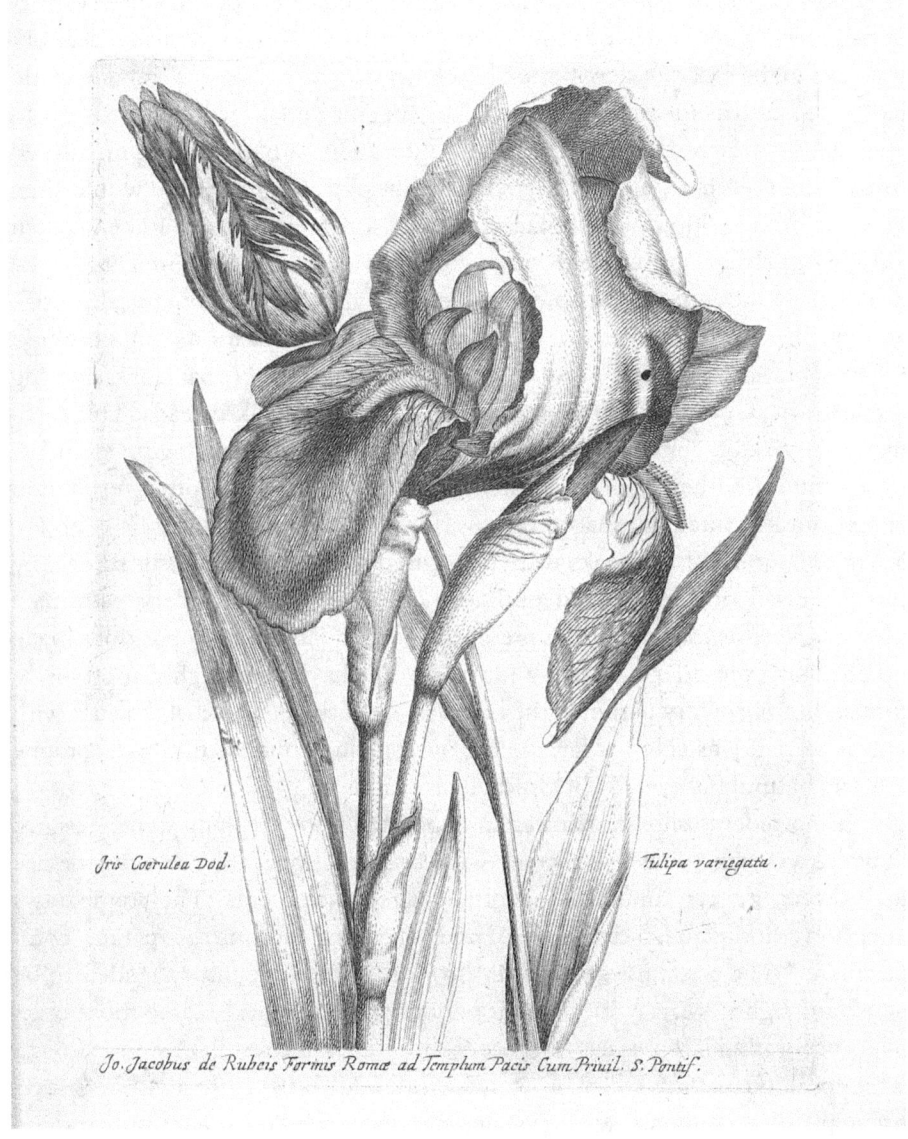

Iris Coerulea Dod.

Tulipa variegata.

Jo. Jacobus de Rubeis Formis Romæ ad Templum Pacis Cum Priuil. S. Pontif.

Figure 5.2. Nicolas Robert, *Variæ ac multiformes florum species . . .* (Rome, 1665). Photograph courtesy of Brown University Library.

images was that they be easily copied; the more distinctly defined the individual elements of a given pattern, the better the chances that a needleworker would be successful at translating the printed image into the medium of fabric and thread. The elegant interplay between negative and positive space demonstrated in Merian's iris and swallowtail composition was an important feature of her embroidery and needlework patterns, and it would later play an important role in her illustrations of the insects of Surinam.

Merian's *Blumenbuch* illustrations are also indicative of the close connections among botanical illustration, natural history imagery, and needlework practices in the early modern period. For example, the *Blumenbuch* images could have been used as models for making slips, an embroidery technique similar to appliqué, in which designs are stitched onto linen canvas backing in colored thread and then cut out (Figure 5.3). The cut out design, or slip, would then be sewn onto another piece of fabric, usually a more expensive material such as satin or velvet (Figures 5.4 through 5.7). This technique was often employed in the early modern period to create designs with floral or botanical motifs, and the term "slip" derives from an early gardening term for a plant cutting.[14] Often lighter-colored slips were sewn onto black or dark blue backings, creating a sharp contrast and accentuating the bright colors of silk threads. This technique provided a convenient method for arranging and rearranging elements of a composition before permanently attaching the slips, and often resulted in compositions containing sharp juxtapositions of scale. Illustrations from botanical and natural history books were often used as sources for motifs since they too possessed the uncluttered outlines necessary for embroidery patterns.[15] Merian's *Blumenbuch* illustrations present elegant and unified compositions made up of distinct pictorial elements and would thus have been highly appropriate for use as embroidery patterns; they are also an early indicator of the ease with which Merian was able to traverse the overlapping contexts of the decorative arts and natural history and botanical illustration.

Embroidered slips were often incorporated into "scrolling stem" designs, in which a curving vine or trellis served as both a support and a framing device for flowers, insects, animals, and other pictorial elements. The stem draws attention to individual pictorial elements by isolating each into a separate "compartment." The scrolling stem patterns of the sixteenth century tended to be dense and tightly woven, but in later periods these designs became more open by incorporating larger areas of space.[16] Although later scrolling stem patterns were loosely constructed, plant stems continued to be used to separate and frame pictorial elements. Merian's illustrations in the *Blumenbuch* utilize some of the same visual strategies as those found in the tradition of scrolling stem embroidery design. In one illustration, the ribbonlike leaf of a hyacinth provides a perch for a damselfly (Figure 5.8); in another, a damselfly is encased within the intricate patterns formed by the overlapping stems and petals of a bouquet of

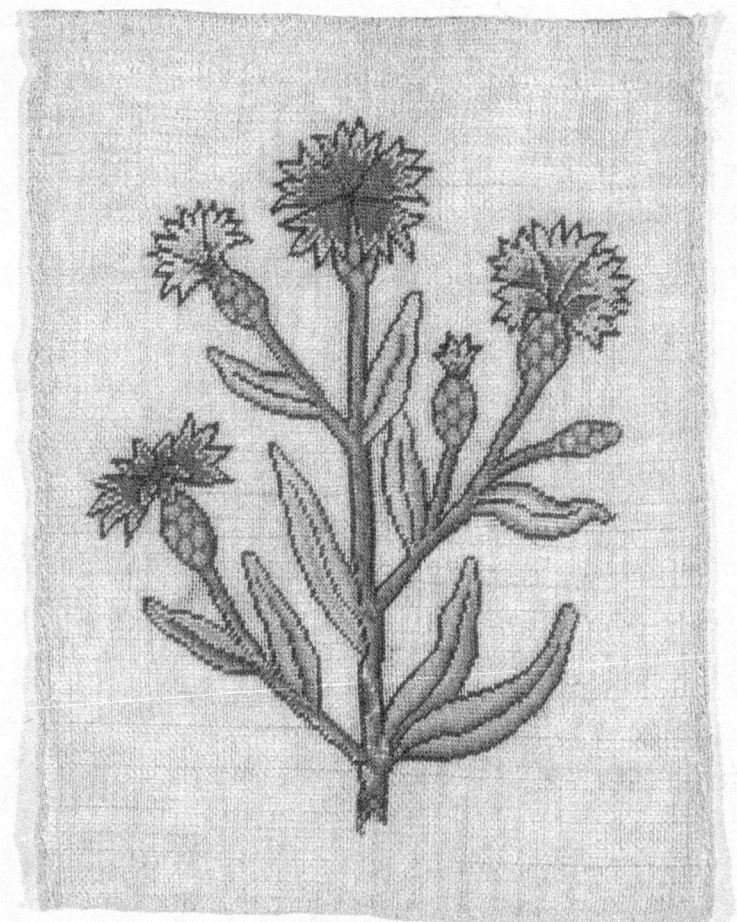

Figure 5.3. Embroidered slip depicting cornflowers (uncut), English, c. 1600. Photograph copyright Victoria and Albert Museum, London.

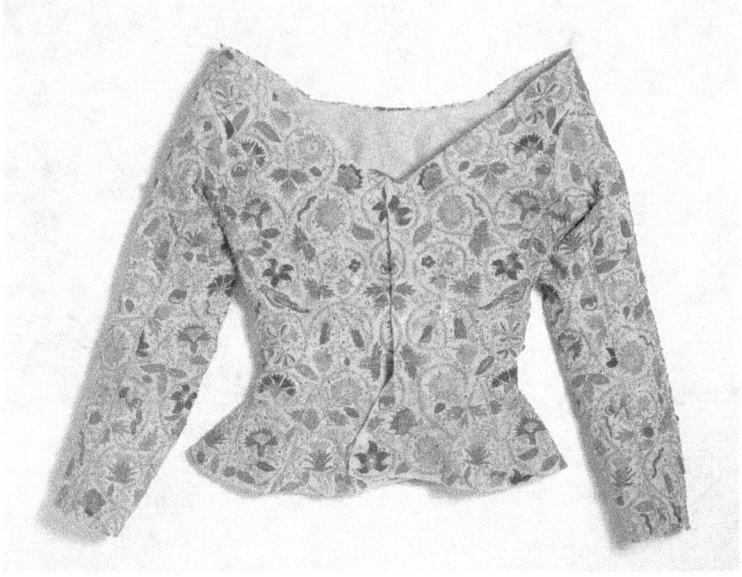

Figure 5.4. Jacket, linen, embroidered with silk and metal thread, England, c. 1600–1625. Photograph copyright Victoria and Albert Museum, London.

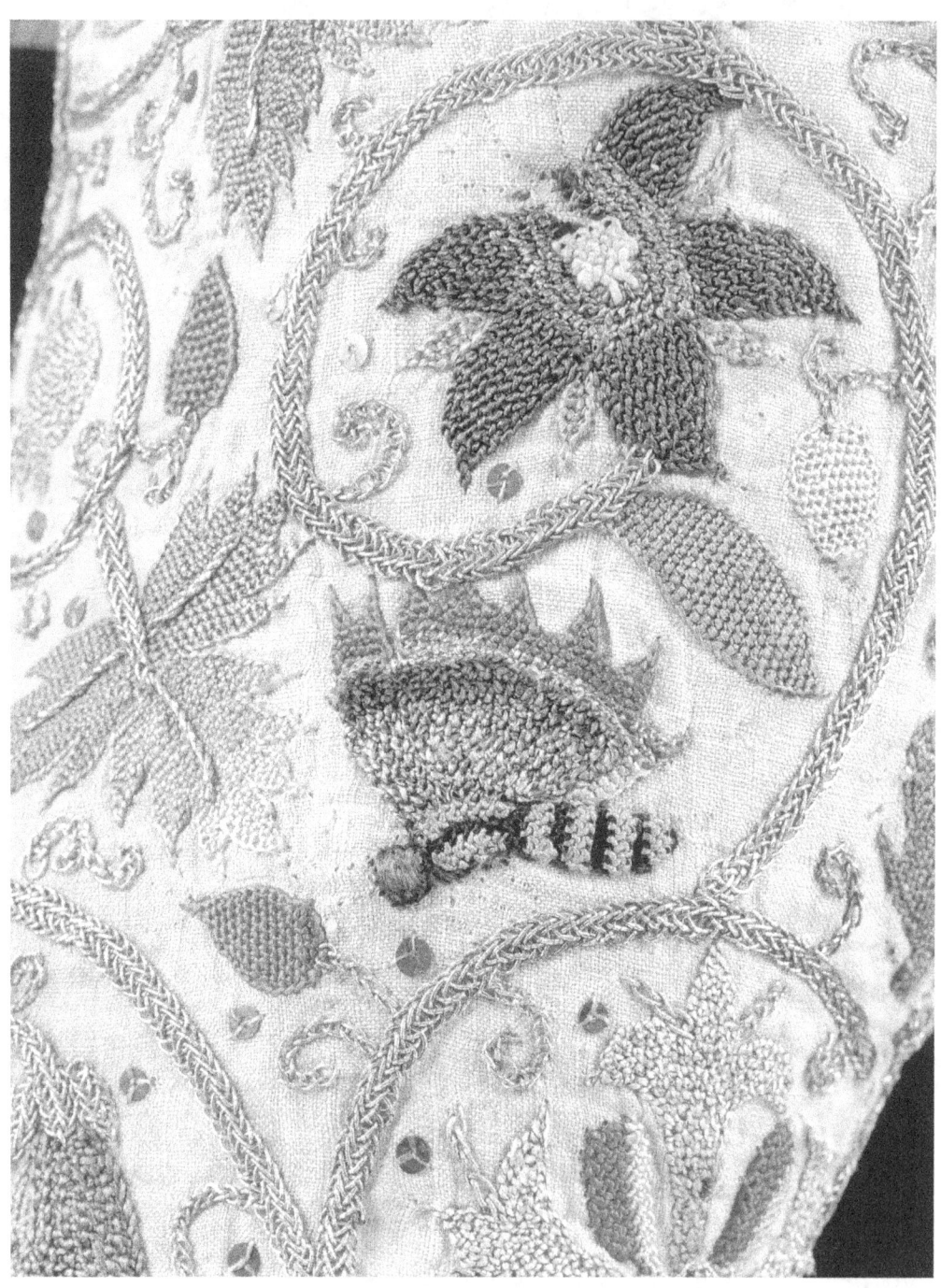

Figure 5.5. Detail of embroidered jacket.
Photograph copyright Victoria and Albert
Museum, London.

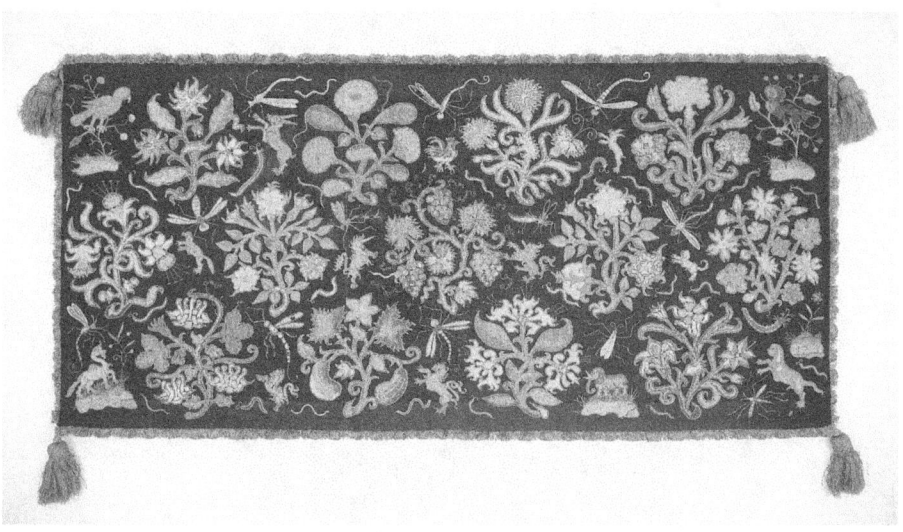

Figure 5.6. Cushion cover, silk velvet, with applied linen canvas embroidered with silk and metal thread in tent stitch and laid and couched work, England, c. 1600. Photograph copyright Victoria and Albert Museum, London.

Figure 5.7. Detail of cushion cover. Photograph copyright Victoria and Albert Museum, London.

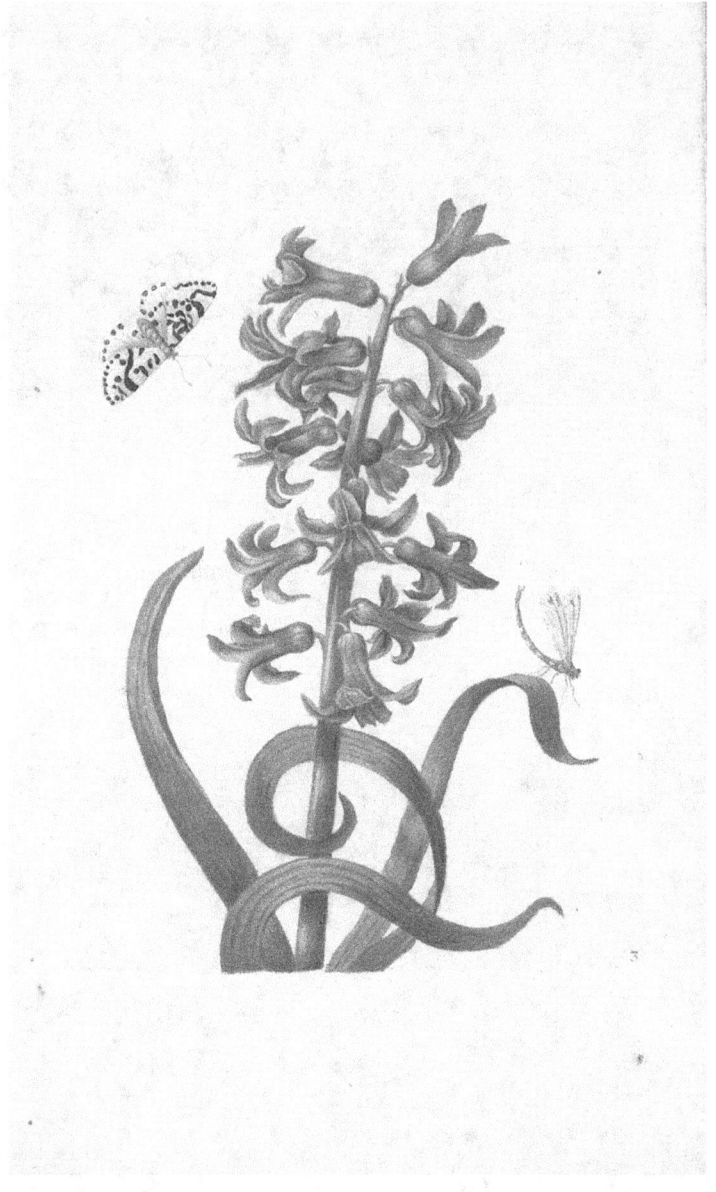

pansies (Figure 5.9). In the iris and swallowtail illustration (Figure 5.1), Merian's
placement of the insect within the greenery of the plant also recalls the compartments formed by scrolling stem embroidery designs. Merian based this
swallowtail butterfly, as well as the damselfly pictured in Figure 5.9, on insects
found in Hoefnagel's *Archetypa* engravings.[17] In Hoefnagel's composition, the
swallowtail butterfly serves as the basis of an exploration and elaboration upon
the theme of slender, elongated natural forms. For Merian, the interplay between
the butterfly and the plant also serves as a motif around which to organize
her composition. Merian frames the butterfly by placing it between the petal

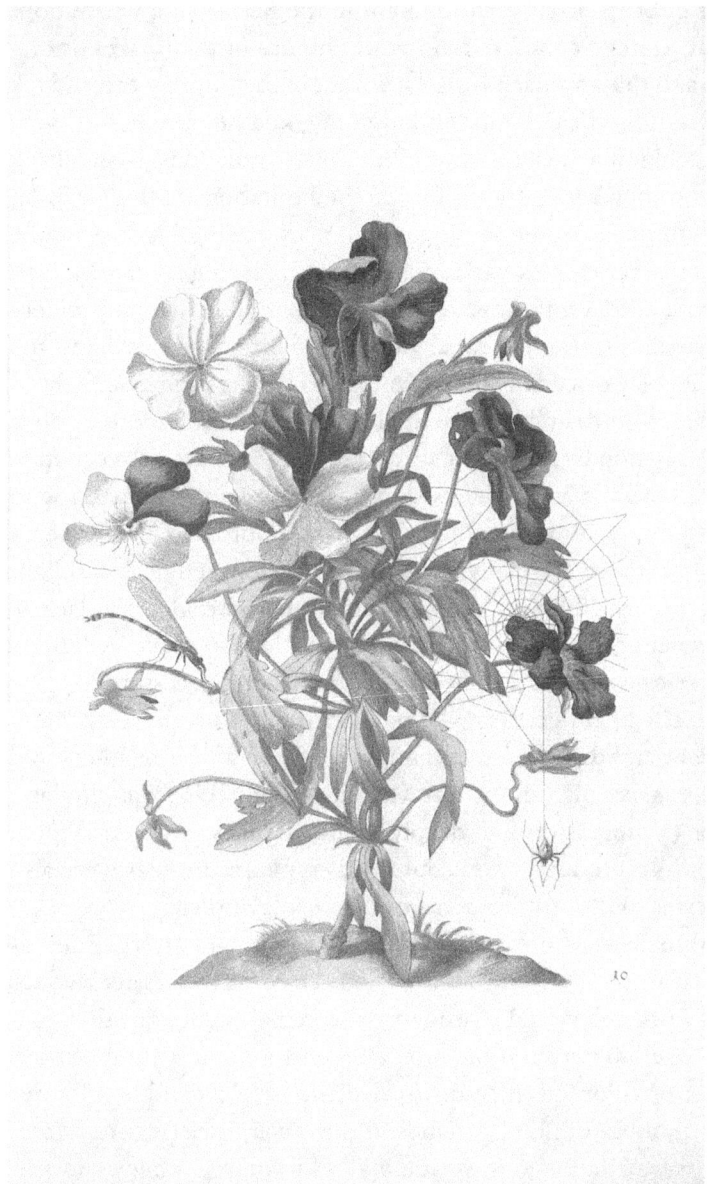

Figure 5.9. Maria Sibylla Merian, *Neues Blumenbuch* (Nuremberg, 1680), plate 10. Photograph copyright The Natural History Museum, London.

and leaves of the iris, and the spaces created by the two leaves surrounding the insect visually extend the elongated form of the wing to the bottom of the page. In departing from Robert's original, Merian has fashioned the tip of the petal that the butterfly rests upon into a curved hook that echoes its own "tail." The theme of elongated hooked forms extends over the lower half of the composition, with the curving tips of each of the leaves forming a pattern that unifies the overall composition. Merian has further adjusted the leaf fragment in the lower-left corner to form a serrated edge that mimics the serrated edge and crescent patterning of the butterfly's wing.

Merian's use of broad tonal areas of color in the *Blumenbuch* illustrations was another visual strategy aimed at facilitating the use of the images as embroidery patterns, and this approach is also apparent in preliminary pen and ink drawings done after Robert's designs. Although talented needleworkers were by no means incapable of achieving very subtle tonal gradations, the embroiderer's palette was somewhat restricted due to the limitations of the medium. Unlike a painter, who can darken or lighten paint by adding black or white pigment, the embroiderer creates variations in tone by laying stitches side by side or on top of one another. For this reason, it is preferable that embroidery designs employ a smaller amount of tonal variation in order not to exhaust the number of different color threads available. A regular progression from light to dark, or vice versa, is also desirable since it allows the embroiderer to achieve the effect of tonal variation by increasing or decreasing the space between individual stitches. Merian's iris and swallowtail image reflects this concern with mapping out tonal areas according to the needs of an embroiderer. In Robert's engraving (Figure 5.2) there are many sharp contrasts between dark and light that would require many different colors of thread to embroider; the lack of regular transitions between light and dark areas would also make it difficult for the embroiderer to achieve the subtle variation in tone necessary for producing naturalistic effects. In contrast, Merian's compositions present clearly defined areas of light and dark and an even range of tones. The hand-colored editions of the *Blumenbuch* also show that Merian's use of color was similarly restrained by relying primarily on four deeply saturated hues—lavender, violet, tan, and dark green—to produce simple but vivid compositions that were also very practical in terms of the requirements of the needleworker.

Fabric and thread were not the only media in which patterns such as Merian's were used in the early modern period. Decorative arts techniques such as cut paper work, inlay, and marquetry also relied on designs that contained little or no overlap between pictorial elements and built compositions out of broad areas of color. In addition to depictions of individual blooms, Merian included a number of floral garlands in the *Blumenbuch* (Figure 5.10).[18] Merian's garlands were conceived in such a way as to offer complex and varied designs that could be easily translated into media such as stone, wood, and paper. An inlaid tabletop of engraved mother-of-pearl, stone, and ebony made by Dirck van Rijswijck provides an example of a type of object for which Merian's floral garlands could have served as patterns (Figure 5.11). The art of paper cutting was another medium in which designs such as Merian's would have been used. Johanna Koerten, the wife of a wealthy Amsterdam merchant who may have been an acquaintance of Merian's, was well known for her work in paper cutting. Koerten's compositions were remarkable for the artist's rendering of multiple tones and intricate forms solely through the juxtaposition of white paper against a dark background. Artisans such as Van Rijswijck and

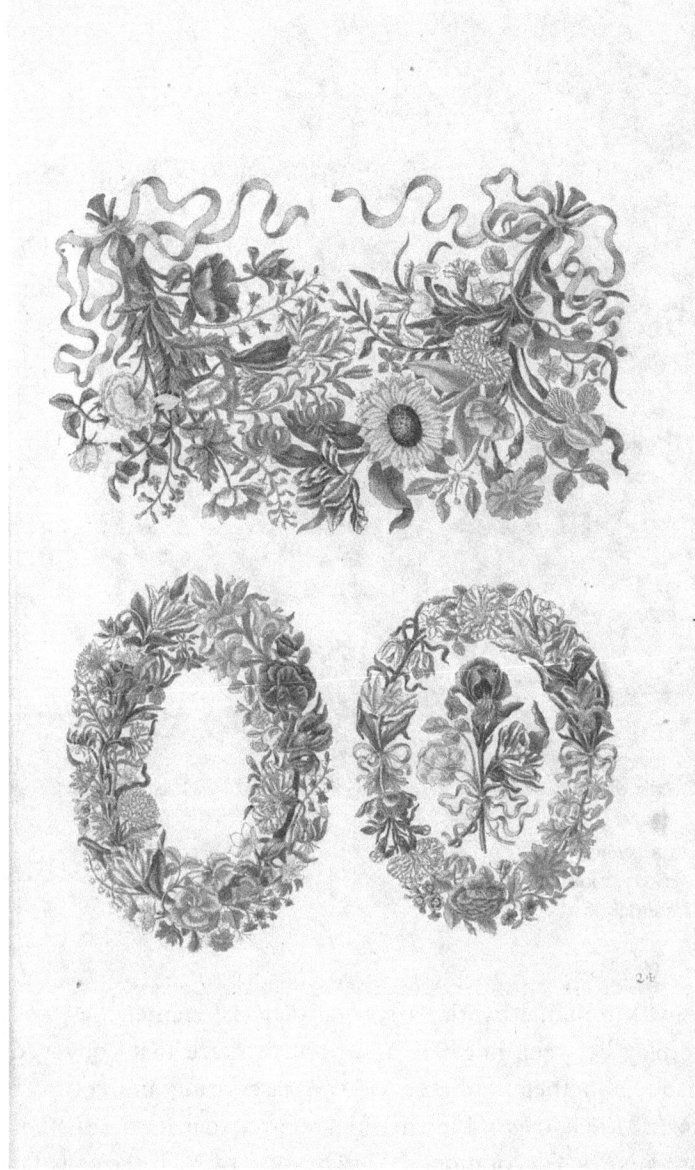

Koerten conveyed form and volume in their inlaid panels and paper cuts through incising the surface of the mother-of-pearl lamellae, or by varying the size of the cuts in the paper, but both relied on designs with clearly defined pictorial elements that could be worked in the rigid substances of paper, shell, and stone.[19]

Although the *Blumenbuch* was Merian's only publication that was explicitly intended for use in a craft or artisanal context, the concerns of the decorative arts would play an important role in the images she created for her other publications. In the *Raupenbuch* and *Metamorphosis* volumes, Merian would

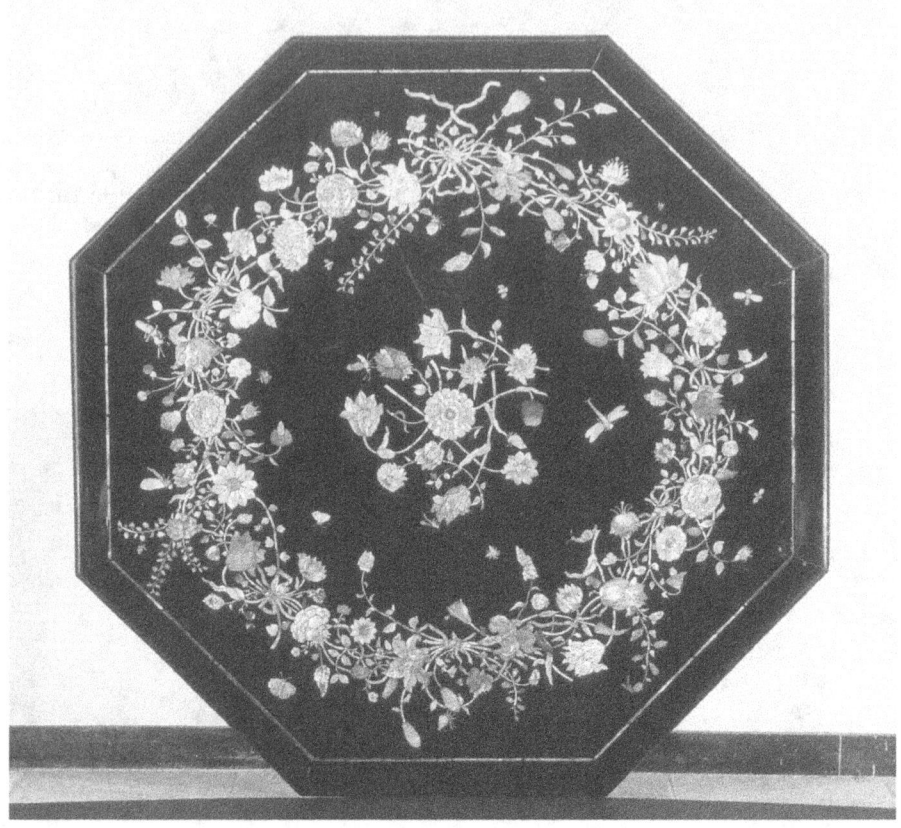

Figure 5.11. Dirck van Rijswijck,
tabletop, 1655. Marble, ebony, mother-of-
pearl. Rijksmuseum Amsterdam.

create illustrations with minimal overlap between pictorial elements as well as an intricate interplay between negative and positive space that conveyed little spatial depth but nonetheless offered viewers fascinating and complex spectacles of nature. In *Metamorphosis* Merian often referred to insects and other creatures as "ornaments" or "decorations." Merian's use of such terms is far from an indication that these elements played a secondary or inferior role in the images but rather is instead evidence both of a continuing concern with the visual logic of the decorative arts and of the centrality of ornament and decoration to Merian's approach to constructing relationships between otherwise disparate elements of the natural world.

Raupenbuch Series: Representing the Life Cycles of Insects

In 1679 Merian published the first volume of her second major project, the *Raupenbuch* series, in which she presented her illustrations and research on the life cycles of European moths and butterflies. The *Raupenbuch,* more formally known as *Der Raupen wunderbare . . .* , is considered by most scholars to be Merian's first serious work due to the more "scientific" approach and subject matter it contains, in comparison with the emphasis on "art" found in the *Blumenbuch.* However, Merian and her audience did not see these works as radically different from one another, particularly with respect to the illustrations. The popularity of the *Raupenbuch* seems to have generated renewed interest in the *Blumenbuch,* and this prompted Merian to issue the second edition of the *Blumenbuch* in 1680.[20] Merian's preface to the 1680 *Blumenbuch* indicates that the illustrations in the *Raupenbuch* were being used in the manner of a pattern book. She writes that the new edition was intended "to be of use and pleasure to people who know and love art as [models] for drawing and painting and to women for sewing," and then goes on to thank these same people for the "favour with which they have clearly been pleased to receive the recently published little *Raupenbuch.*"[21] In the *Raupenbuch* Merian developed and established the visual forms for which she would gain renown in natural history circles, but the pictorial requirements and design imperatives of the decorative arts, as well as a continuing interest in the material, decorative, and aesthetic qualities of the natural world, remained a strong influence in her imagery.

In the *Raupenbuch* illustrations Merian paired insects and the stages of their development with their food plants, which she describes in the title as "The Caterpillar's Wondrous Metamorphosis and Particular Nourishment from Flowers."[22] Each of the three volumes contain fifty copperplate engravings accompanied by Merian's written descriptions of her observations of the breeding habits, behavior, and transformation of the insects pictured in the plates.[23] As noted above, Merian often copied motifs directly from other artists, but in the *Raupenbuch* she incorporated their visual strategies for representing insects rather than copying their images. Like Joris Hoefnagel, Merian depicted butterflies and moths in either the flat or profile view, a technique that highlighted the patterns and colors of the insects' wings. And Merian also manipulated size and scale in order to increase the visibility of surface features and the markings of the minute creatures pictured. In plate 35 from the first *Raupenbuch* volume, the caterpillar is nearly twice the size of the adult moth of the same species pictured in the upper-left corner (Figure 5.12). The positioning of the caterpillar is reminiscent of the paradoxical configurations of two- and three-dimensional space often found in Hoefnagel's compositions; Merian's caterpillar rests upon the surface of one leaf, which is contiguous with the surface of the page, while simultaneously receding into the depth of the picture plane as it traverses the edge of another leaf. Another technique Merian often employed in her illustrations of insects was the placement of a flat or open-winged moth or butterfly in one of the upper corners of the composition, as seen in Figure 5.12. According to Segal, Merian took this idea from Robert, and it remained one of her

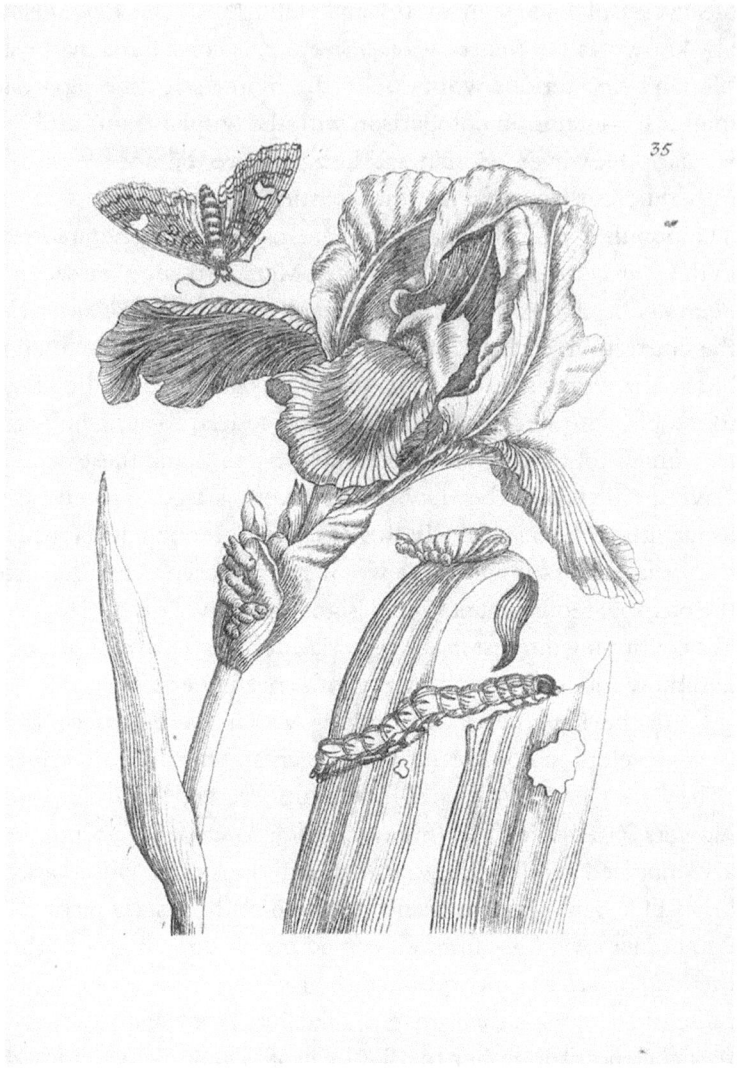

Figure 5.12. Maria Sibylla Merian, *Der Raupen wunderbare . . . [Raupenbuch]* (Nuremberg, 1679), plate 35. Heidelberg University Library.

35

most frequently used pictorial devices in both the *Raupenbuch* and *Metamorphosis* illustrations.[24] Heidrun Ludwig has argued that in the *Raupenbuch* Merian merged the tradition of flower painting with that of the insect piece by reversing "the conventional roles of central motif and secondary elements, so that the plants that now occupied central positions in the composition were employed primarily in support of what had formerly been subordinate to them."[25] Merian's practice of placing flattened butterflies and moths in the corners of her compositions can be understood as evidence of the role reversal Ludwig describes, but it is also indicative of her interest in the symbiotic relationships between insects and plants. By giving each element in the composition equal weight and attention, Merian both highlighted the importance of the insect as subject matter and used it as a framing device for the overall composition.

Merian's depiction of insect metamorphosis was strongly influenced by another artist, Johannes Goedaert, and his illustrated book on insects, *Metamorphosis Naturalis,* which was published in three volumes between 1662 and 1669.[26] Goedaert studied and collected insects for over thirty years and recorded his observations in illustrations that presented insect metamorphosis as a series of distinct stages arranged in vertical format, in which the caterpillar appears in the upper area of the page, followed by the pupa in the center and the adult below. Merian followed Goedaert's format closely in several of her preparatory drawings and watercolors for the *Raupenbuch,* and in the published illustrations she also presented several compositions using this vertical arrangement.[27] However, in most of her illustrations in the *Raupenbuch* (and later in *Metamorphosis*) Merian utilized Goedaert's method but did away with the vertical arrangement, opting instead to present the stages of development among the leaves and petals of the insects' food plants. This technique placed more emphasis on the material qualities and characteristics of the insects' forms, and it was best suited to depicting the life cycles of insects that experience dramatic changes in their physical form during their development. Both Goedaert and Merian focused on the phases of the insect life cycle that are most visibly distinct from one another, and as such they preferred to illustrate butterflies and moths because they exhibit dramatically different visual appearances as they proceed through their development cycles. Butterflies and moths undergo "complete" metamorphosis in which the successive stages of development are very different from one another, but for insects that undergo "incomplete" metamorphosis the changes in form appear more gradually. Typically, the immature, or nymph, forms of insects that experience incomplete metamorphosis resemble the adult form but lack wings and reproductive organs. The different stages of development in these insects are more difficult to distinguish from one another than in insects that experience complete metamorphosis, and thus they are not well suited to the techniques for representing insect life cycles developed by Goedaert. Insects that experience incomplete metamorphosis are not necessarily lacking in visual interest, however, and Merian included a number of dragonflies and damselflies in the *Blumenbuch*. But in the *Raupenbuch* she abandoned dragonflies and damselflies in favor of moths and butterflies. This shift in subject matter was the result of Merian turning her attention to insect metamorphosis and of her decision to concentrate on species that were best suited to the method of visually representing insect development as a series of distinct stages. Merian usually depicted only one larval stage in her illustrations, and she generally chose the stage where the insect displayed the most vibrant colors or intricate patterns, even though butterflies and moths experience between four and nine changes during the larval phase.[28] During her stay in Surinam, Merian would become interested in the life cycles of beetles, another insect that experiences complete metamorphosis.

Merian's major innovation in the *Raupenbuch* was the pairing of insects with their food plants, and this has been the primary reason why she has been described as taking an "ecological" approach to observing, interpreting, and representing the natural world. Merian was not the only early modern European artist whose work can be described as having an interest in ecological relationships—loosely defined as a concern with depicting the habits of living organisms, their mode of existence, and their relations to their surroundings. The Dutch still life painter Otto Marseus van Schrieck, for example, specialized in subject matter similar to Merian's, although he did not depict insect metamorphosis. Van Schrieck's approach to representing the natural world can also be described as "ecological" in that his primary focus was on interactions between living organisms, but his compositions differ dramatically from Merian's in their tone and character. Merian's clear, uncluttered presentations of plants and insects bear little visual resemblance to Van Schrieck's dark and mysterious paintings, a difference that stems in part to Merian's visual roots in the pattern book and decorative arts traditions.

Although it is often noted that Merian produced the first volume of the *Raupenbuch* during the same period she was working on the *Blumenbuch* series, in emphasizing the innovative aspects of Merian's work in the *Raupenbuch* scholars have tended to overlook some of the continuities between the two works. Merian employed almost identical visual styles in both works, with the difference between the two lying mainly in their subject matter. As noted above, the images in the *Raupenbuch* were used by some readers as patterns for painting, drawing, and sewing. And like the *Blumenbuch,* the *Raupenbuch* illustrations present insects as isolated elements framed by the elegantly curving stems and branches of plants, thereby forming complex patterns of negative and positive space. There is no doubt that the *Raupenbuch* signaled a major shift in Merian's career that coincided with her increased involvement in raising and observing insects. But in devising methods for representing the life cycles of insects, Merian did not abandon the design principles of the decorative arts in favor of a "scientific" approach based on empirical observation. Instead, these concerns emerged as inseparable components of Merian's understanding and approach to the natural world. Her interest in exploring the similarities and connections between visual forms and patterns found in nature would play an important role in her illustrations of insect life cycles in the Surinam book, as well as in another area of Merian's interest—the collection and trade of insect specimens.

Collecting, Preparing, and Trading Specimens

It was while preparing the first volume of the *Raupenbuch* in Nuremberg during the late 1670s that Merian began collecting and observing insect specimens systematically. Merian searched for caterpillars within the city, in gardens, and outside the city walls in the surrounding meadows and woodland areas. Merian's techniques for collecting, raising, and observing caterpillars have been summarized by Wettengl as follows: "Merian ordinarily collected the insects herself, determined through close observation which plants provided them nourishment, procured their food, bred the small creatures in little boxes, observed their successive developmental phases, and described the place in which they had been found—their natural living environment—as well as their appearance and behaviour."[29] Merian recorded her observations in a journal now known as the *Studienbuch,* in which she collected written and visual records of her observations of insect metamorphosis over the course of thirty years.[30] The entries in the *Studienbuch* reveal the extent to which concerns with the decorative and material aspects of the natural world informed Merian's approach. In one entry, Merian describes an encounter with "a great web upon which all of 70 caterpillars, which were yet very small, lay in a round circle very close together; but they looked just like a round, black, velvet patch."[31] Merian's description of the caterpillars recalls the black velvet fabric upon which embroidery slips were often sewn; glistening upon a spider's web in the morning dew, the patch of caterpillars must have produced a similarly dazzling effect.

Merian used the same basic methods for collecting and preparing specimens throughout her career. Her particular interest in insect life cycles necessitated that she obtain for her research living specimens rather than dead ones. Merian sometimes received caterpillar specimens through the mail as gifts from friends, but these proved difficult for her to raise to maturity if the correspondent did not provide sufficient information about the insect's diet and behavior. In addition, specimens sent through the mail were also subject to uncontrolled conditions that sometimes resulted in the insect dying in transit.[32] When James Petiver sent her a gift of insect specimens in 1705, Merian refused them because they were not appropriate for her research. Merian explained to Petiver that she could not use specimens that were not alive, especially if he did not provide her with information about the insects' behavior, habitat, and diet:

> I have . . . received animals [insects] from the gentleman [Petiver] on two
> occasions. The first time they were brought by Doctor Reuss, but
> because I was not in need of such creatures I gave them back and thanked
> him, requesting that he write to the gentleman, telling him that I have no
> use for such animals and did not know what to do with them. For the
> kind of animals I am looking for are quite different. I was in search of
> no other animals, but only [wished to study] the generation and
> reproduction and transformation of the animals, and how one emerges
> from the other, and the properties of their food, as the gentleman can see
> in my book [the *Raupenbuch*]. Therefore, I would ask the gentleman not
> to send me any more animals, for I have no use for them.[33]

While Merian's repeated refusals of Petiver's insect specimens would have done little to endear her to him, her insistence upon studying the life cycles of insects through observation of living specimens demonstrates her commitment to the methods and techniques established early in her career.

Merian's introduction to the culture of collecting and exchanging exotic specimens may well have occurred during her stay at the Labadist community at Waltha Castle. The property was owned by Anna, Maria, and Lucia van Aerssen van Sommelsdijk, three sisters who were followers of the sect's leader, Jean de Labadie, and Pierre Yvon, Labadie's successor. The women's brother, Cornelis van Aerssen van Sommelsdijk, served as governor of Surinam from 1683 until his death in 1688. He was interested in natural history, gardening, and collecting exotic specimens, and he maintained ties with botanical gardens in Holland by sending specimens of plants and seeds from Surinam.[34] He also sent exotic specimens as gifts to his sisters, who seem to have displayed them at Waltha Castle for the members of the Labadist community and their visitors. The historian Trevor Saxby notes that the display of specimens included "a 23-foot aboma (tree-snake), stuffed by indians and shipped as a memento by the governor to his sister." In addition to the tree snake, the display contained "a collection of large and brilliantly colored butterflies sent . . . by governor van Sommelsdijk."[35] It is believed by most scholars that Merian's encounter with these butterflies eventually inspired her to travel to Surinam herself to collect and study insects. Another important factor in Merian's decision to travel to Surinam was her experience viewing and studying specimens in collections in Amsterdam.

The exotic flora and fauna of Surinam may have been unknown to the majority of the population of Europe during the early modern period, but in the late seventeenth century many Dutch naturalists and collectors were not only familiar with such items but owned and displayed them in their cabinets and museums. By the time Merian had left the Labadist community and settled in Amsterdam with her daughters in 1691, Surinam had been under Dutch control for almost twenty-five years.[36] Many collectors maintained contacts with and received regular shipments from travelers to Dutch colonies in the East and West Indies. Nicolaas Witsen, Frederick Ruysch, and Levinus Vincent were some of the figures whose collections were among those Merian visited in Amsterdam. In the preface to *Metamorphosis* Merian states that she visited the collections of Witsen, Ruysch, and Vincent before her journey to Surinam, and she describes how she "saw with wonderment the beautiful creatures brought back from the East and West Indies."[37]

These collections not only inspired Merian to travel to Surinam to study insects but also provided her with practical information about the preparation and preservation of specimens. Elizabeth Rücker notes that Merian "was well versed in the technique of preserving specimens in alcohol, either in sugar jars or bottles. Merian had presumably become acquainted with this process in the course of visits to the many and diverse exhibits of curiosities that Amsterdam had to offer, acquiring the necessary skills in order to prepare herself for research in the South American tropics."[38] Merian's specimens would presumably have been prepared using a technique similar to the one Merian described in a letter to Clara Regina Imhoff, a former student and friend from Nuremberg: "If one wishes to . . . kill butterflies quickly, then one must hold the point of a darning needle in a flame, thus making it hot or glowing red, and stick it into the butterfly. They die immediately with no damage to their wings, and the little boxes in which they are then placed can be coated with lavender oil first, so that no worms can get in and feed on them."[39] The technique Merian describes for killing a butterfly was designed to preserve the insect's delicate structures from the violence of its own death. Once safely subdued and killed, the insect was placed in a small box to protect it from damage while being transported, stored, or displayed. A coat of lavender oil in the interior of the box protected the specimen from infestation. Depending on the type of insect being prepared, Merian may then have coated the insect with turpentine oil to further protect it and to give it a glossy finish, much like varnish on an oil painting.[40] It is ironic that Merian's success at crafting well-preserved and attractive insect specimens depended on her ability to prevent the life cycles of destructive insects from taking place, because these skills were at odds with those she cultivated in her research on insect metamorphosis in which the central goal was the collection and preservation of living specimens. As in her artistic training, Merian's ability to collect and preserve nature successfully was also dependent on a combination of firsthand observation and mastering a set of specialized skills. Merian's use of a darning needle in preparing specimens also suggests that working with fabric and thread and working with specimens were closely related practices for Merian and at times may have involved a shared set of instruments.

Another important lesson Merian would have learned from her visits to the collections, cabinets, and museums of Amsterdam was that collecting nature did not only involve preserving specimens but also the crafting of ingenious and sophisticated displays of these objects. Early modern collectors presented and stored their butterfly and moth specimens in boxes or flat panels, where they pinned the insects' wings flat in order to highlight their brilliant colors and intricate patterns. Some collectors created assemblages that were considered to rival the beauty and complexity of the insects themselves. Levinus Vincent and his wife Johanna Breda owned a well-known collection of rarities in Amsterdam, which was open to the public for an admission fee—an unusual practice for the time. The Vincents' collection is among those Merian mentions in the preface to *Metamorphosis,* and it was where she viewed examples of Surinamese insects before her trip to South America.[41] Aesthetic considerations played a central role in the Vincents' organization of their collection, and their techniques for displaying specimens were closely connected to the material

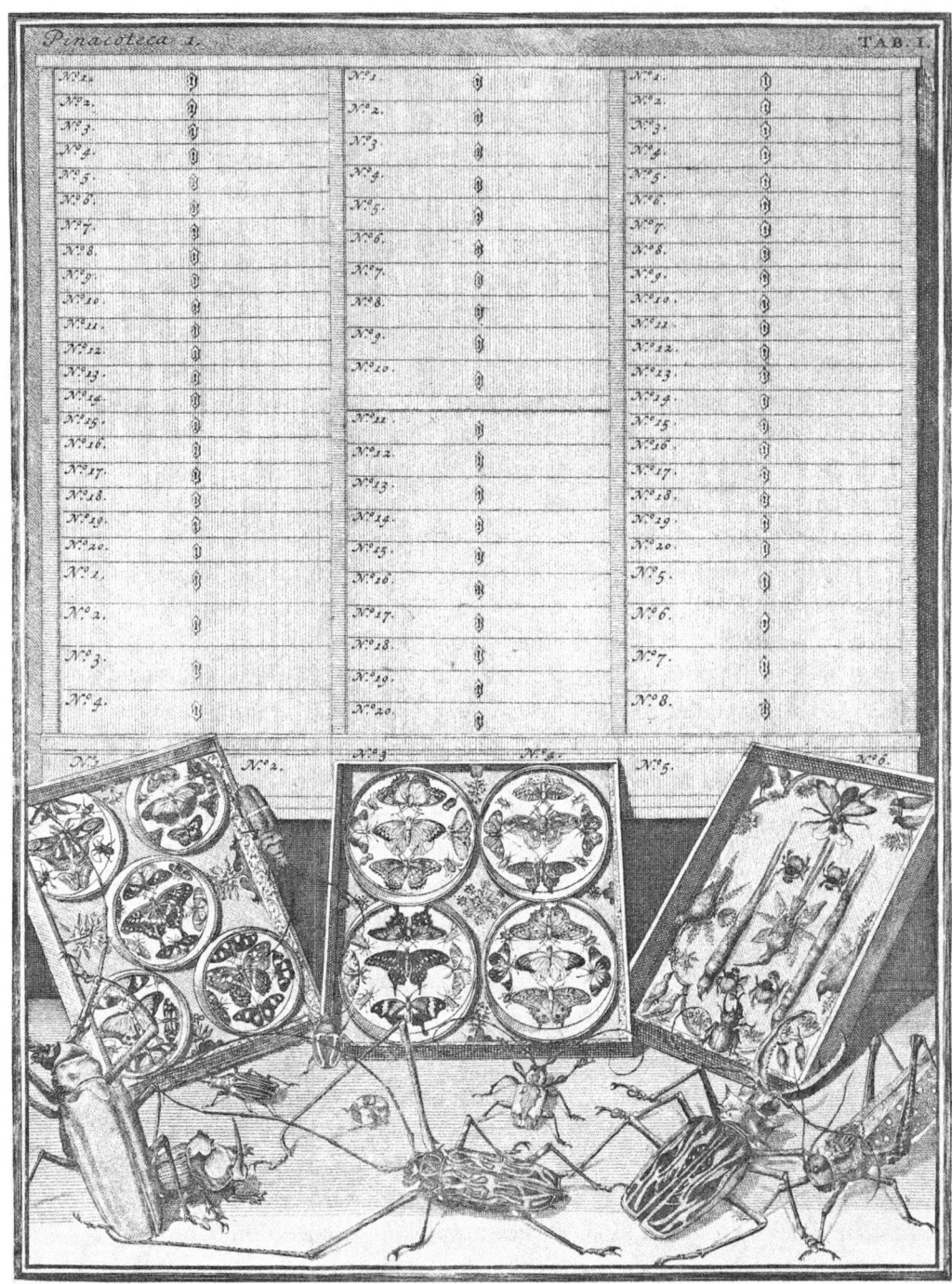

Figure 5.13. Levinus Vincent,
Wondertooneel der Nature . . ., volume 2
(Amsterdam, 1715), Table I. Research
Library, Getty Research Institute,
Los Angeles.

and decorative concerns of embroidery and needlework. Tomomi Kinukawa has shown that embroidery was one of the conceptual tools the Vincents used for ordering the natural world as well as a technique for creating dazzling displays of specimens. According to Kinukawa, the Vincents did not classify their specimens by name but instead "chose an alternative order: they organized them according to their sizes and colors into patterns of embroidery, formal gardens, and still lifes."[42]

The displays that attracted the most admiration from visitors to this collection were the arrangements of exotic butterflies and moths on embroidered silk panels. In 1706 Vincent began publishing a series of illustrated catalogs that provide valuable information about the content and appearance of the collection and include several illustrations of these insect displays.[43] In one of the illustrations three silk panels of insect specimens are arranged in front of a cabinet containing numbered drawers of insect specimens (Figure 5.13). The two panels on the left present specimens of butterflies and moths in circular frames surrounded by embroidered borders and floral motifs. The third panel contains specimens of small beetles and what appear to be two small birds, also surrounded by embroidered ornamentation. Specimens of larger beetles—whose bulky bodies were unsuited to display on flat panels—are scattered in the foreground like objects in a still life painting. One visitor to the collection, impressed by the scope and variety of its contents, wrote as follows: "What surprised me most was the never-ending display of all kinds of insects, perfectly preserved, whose variety of colors formed the most beautiful sight nature could offer."[44] Another visitor noted that from a distance the insect displays "resembled embroidery or tapestry from Brussels." It was said that Peter the Great, perhaps the most distinguished visitor to the Vincent's cabinet, was so overwhelmed that "he knelt in front of the insects to pay homage to their beauty and display."[45] To some, including Levinus Vincent himself, insects were themselves considered to possess the qualities of embroidery and were understood in terms of the colors and patterns of needlework. Vincent invited several poets to contribute works to his catalogue, one of whom described the insects as "a thousand little animals as embroidery, row by row, / so beautifully variegated like a painting."[46] According to Kinukawa, "Vincent himself emphasized that the insects were arranged in figures like embroidery with observation of light and shadow, magnification and reduction. It was impossible to describe them by pen or tongue, so that everyone stood in front of them in wonder as if stupefied."[47]

Embroidery and painting were two of the key practices used by the Vincents to organize and display their collection of exotic specimens and rarities. The engraved title page to their catalogue presents a view of the room that housed the collection framed by two human figures personifying the arts of painting and embroidery (Figure 5.14). The figure on the left, who represents embroidery, is shown stitching a design into fabric stretched over an embroiderer's

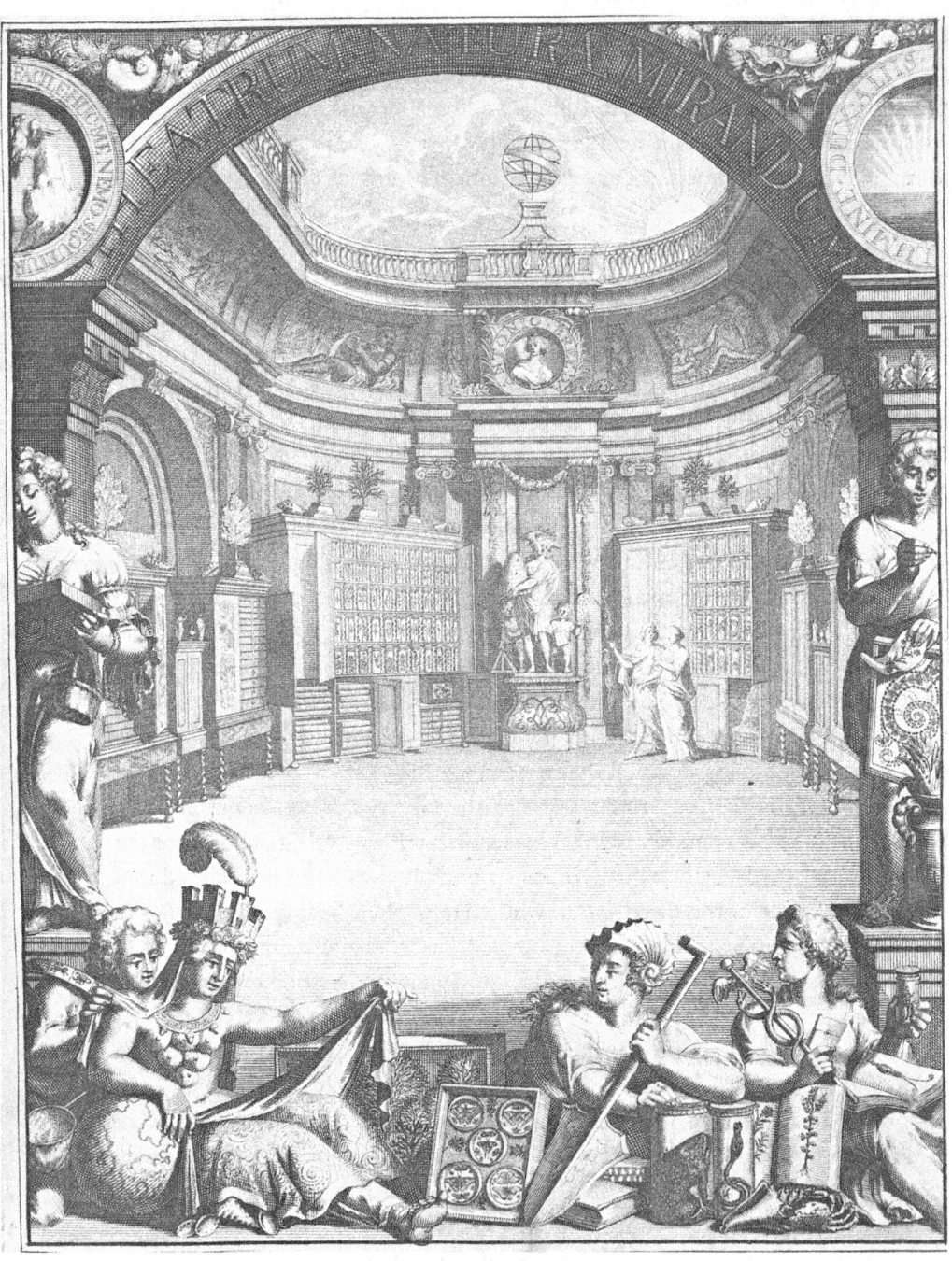

Figure 5.14. Romeyn de Hooghe,
frontispiece for Levinus Vincent,
Wondertooneel der Nature . . . (Amsterdam,
1706). Research Library, Getty Research
Institute, Los Angeles.

frame, while on the right side of the picture a figure representing the art of painting holds a paintbrush in one hand and in the other a drawing of a plant and a frame containing what appears to be an arrangement of shells. In her discussion of this image, Kinukawa has argued that Johanna Breda's role as the keeper of domestic order in the Vincent household was also one of the organizing principles of the museum, which was located inside the Vincents' home. Kinukawa writes that "the natural world cannot enter Vincent's neatly ordered room without being scrutinized and processed by these diligent guardians armed with needles and watercolors."[48] Johanna's work organizing and arranging the contents of the museum was also the subject of some of the poems included in Vincent's catalogue: "Poets assign to Johanna the essential role of standing at the threshold of the house and transforming the products of the dark womb of nature into the shining gems shown in their cabinets. It was an extension of the task conventionally assigned to the housemother, who oversees what her husband brings in, arranges it, and puts it in order . . . Johanna processes the natural wonder into gems / treasures of the house by choosing, embroidering, drawing, and neatly arranging them in the cabinets."[49]

Like Johanna Breda, Merian utilized the arts of embroidery and painting to process the natural world, not only to display it but also to sell it. In her visits to the collections of Witsen, Ruysch, and Vincent, Merian learned to see nature as composed of objects that could be collected and displayed in cabinets and panels. This understanding of the natural world was also shaped by the design principles of needlework, embroidery, and other artisanal and craft practices that sought to represent nature as a series of isolated pictorial elements. The process of preparing specimens for display and long-term preservation involved transforming nature into objects. Once their forms were fixed and stable, these objects could be bought and sold for profit. Before leaving for Surinam, Merian would have seen that collectors such as Ruysch, Vincent, and Witsen and others like them formed a lucrative market for exotic specimens from the East and West Indies. When she returned, she brought many specimens of insects, reptiles, and other animals that she offered for sale, and she continued to receive specimens from her contacts in Surinam in the years after her return to Amsterdam. Merian's daughter Johanna Helena returned to Surinam with her husband and served as Merian's supplier for some time. Merian also received specimens from other individuals in Surinam, and she seems to have had contact with potential suppliers in Barbados and other Caribbean islands from which she was pursuing plans to import specimens when hostilities between the Dutch and the English subsided.[50] Merian's correspondence shows that she sold exotic specimens to her friends Georg Volkammer and Clara Imhoff in Nuremberg. In one of her letters she asks Imhoff to tell other interested parties of the specimens she offered for sale, and that she would also be interested in receiving specimens from Germany as an exchange.[51]

Although Merian was sharply attuned to the commercial possibilities the natural world offered and was skilled in preparing attractive specimens, she was not always completely successful in negotiating the complex social interactions and cultural practices involved in the exchange of specimens in early modern Europe. In the Amsterdam cabinets and museums Merian frequented, nature provided an entertaining and elegant spectacle for owners and their visitors. However, even a collector who charged admission such as Levinus Vincent did not consider the museum to be a commercial establishment offering luxury goods for sale. Vincent and others like him collected and displayed nature as a leisure-time activity and not as a source of income. Wealthy collectors exchanged specimens with each other as gifts and of course purchased specimens from brokers, but they did not generally buy from each other or sell their specimens to museum visitors—although on occasion entire collections were sold at once, earning their owners large profits. Collectors did not part with their treasures piecemeal unless in great financial distress, and they routinely rejected offers from museum visitors to purchase individual items. As Kinukawa has pointed out, "The capital of the Republic [of Letters] was never money. Instead, service was returned by service, friendship by friendship. The exchanges were never supposed to be for profit."[52] Merian thus occupied an ambiguous and sometimes problematic position within these networks of exchange. As an entrepreneur who depended on the sale of her drawings, specimens, and books for her livelihood, Merian often violated the codes of the collecting community by seeking to sell rather than exchange items. In a letter to James Petiver, Vincent complained that Merian never kept any of the gifts of specimens that were sent to her, but "preferred instead to sell everything she obtained or produced . . . for the sole purpose of turning it into money."[53] Vincent, of course, speaks from the position of a retired merchant whose wealth allowed him the luxury of disdain for Merian's commercial activities. But as we have already seen in her rejection of Petiver's insect specimens, Merian's single-minded dedication to her research on insect life cycles at times interfered with her ability to engage in the genteel rituals of gift exchange. Merian earned the respect and admiration of this community of collectors and naturalists for her skills as a researcher and artist, but she was not always able to manage her multiple roles within that community smoothly. Unlike the insects she painted, preserved, circulated, and sold, Merian herself remained a somewhat unstable commodity within the cultural economy of natural history.

Processing Nature in the New World:
The *Metamorphosis insectorum Surinamensium*

When Merian and her daughter Johanna Helena set sail for Surinam in June 1699, Merian possessed a broad range of skills and interests that would shape her approach to studying, observing, and representing nature. In *Metamorphosis*, the publication that resulted from this trip, Merian would draw upon her experiences designing patterns for embroidery, her skills at observing and recording the life cycles of insects, and the techniques she developed for preparing specimens in order to process nature in the New World for European audiences. The vision of Surinam presented in *Metamorphosis* satisfied this audience's desire for exotic, elegant, and ordered spectacles of nature, and can thus be understood as part of the broader interest of Dutch authors and publishers in marketing the non-European world to European consumers. Benjamin Schmidt includes Merian's *Metamorphosis* among the enormous quantity of illustrated "geographical" materials published in Holland at the end of the seventeenth century that treated "exotic" subjects. Dutch "geography," as defined by Schmidt, comprised not only texts on geography and cosmology but also atlases, works of natural history, travel narratives and anthologies, and cartographic texts. Together these items "sold an *idea* of the world that appealed to readers, viewers, and consumers across Europe, and this idea marketed a world that was identifiably 'exotic.'"[54] Schmidt argues that the "resolutely disordered world" pictured in the frontispieces of these works erased distinctions and borders between non-European lands, thereby presenting European audiences with a world of curiosities, diversions, and delights in which commercial rivalries and colonial polemics had no presence: "Diffuse, digressive, often disorienting, sometimes recycled, purposely decontextualized: Dutch geography ended up being specific to none and thus palatable to all. The exotic world designed by the Dutch was a brand, ultimately, of very wide appeal."[55] Although Merian's *Metamorphosis* is resolutely specific to the local context of Surinam and does not engage in the type of "bric-a-brac" mixing of objects, curiosities, and people described by Schmidt, Merian's book did take part in constructing the non-European world as both exotic and accessible. In both their content and style, the images in *Metamorphosis* were designed to transform the natural world of Surinam into "purposely decontextualized" objects capable of entering into international exchange and circulation.

Merian's illustrations of insects and plants in *Metamorphosis* employed the same visual style and techniques she developed in the *Blumenbuch* and *Raupenbuch*, and she continued to rely on design principles of the decorative arts to structure her compositions. In plate 50 of *Metamorphosis*, for example, the leaves, stem, and blossom of the *Ipomoea alba* (white batatas) plant are arranged against the white ground of the page to form a pattern of positive and negative space, which in turn creates compartment-like areas that frame the insects (Figure 5.15). Merian frequently employed this framing technique in the illustrations for *Metamorphosis*, and it is conceptually linked to the "scrolling stem" embroidery patterns discussed earlier in this chapter, in which plant stems form compartments that frame and visually isolate pictorial elements. This technique resulted in orderly if somewhat improbable compositions. The beetle pupa shown at

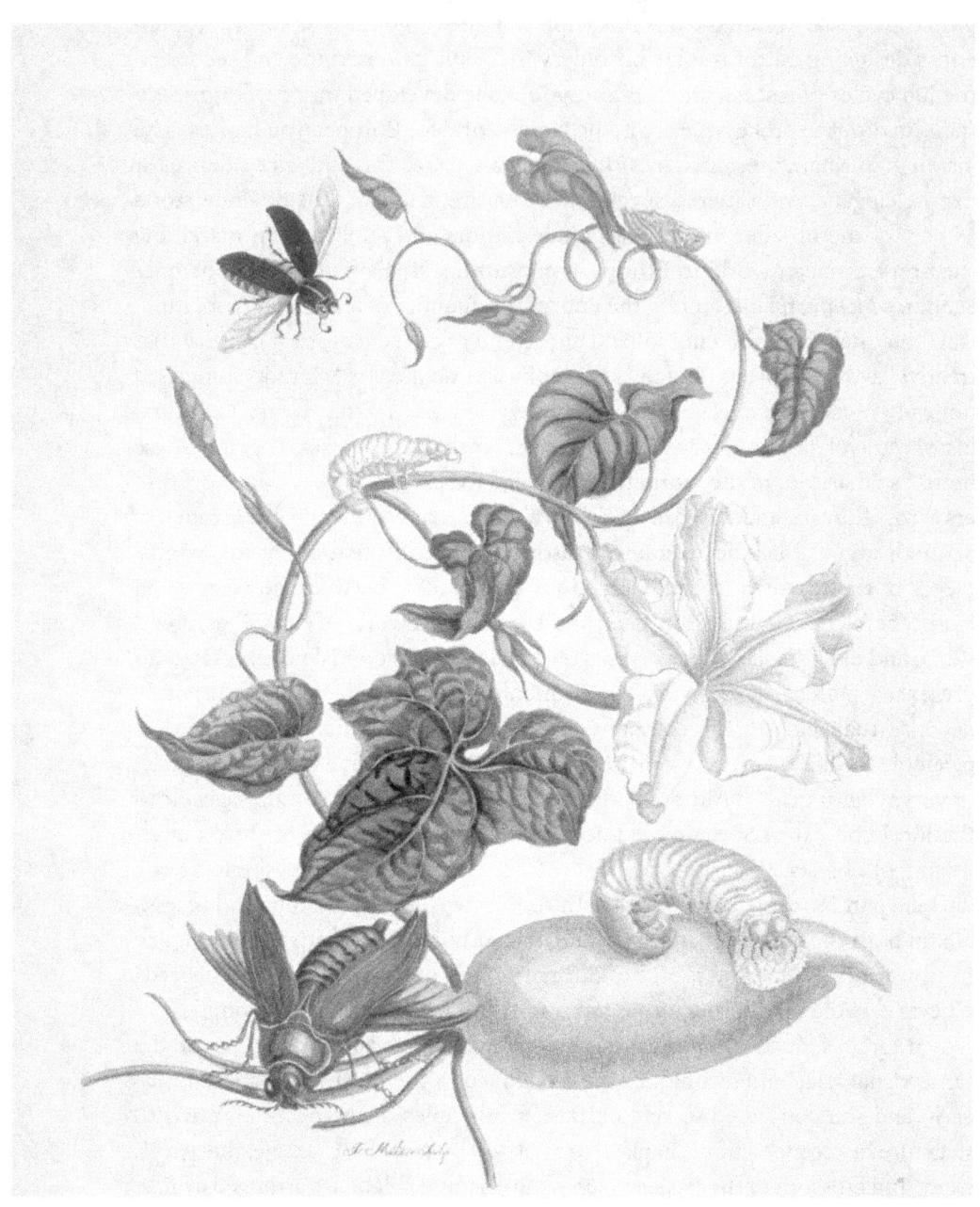

Figure 5.15. Maria Sibylla Merian,
Surinaamsche insecten (1726). First published
as *Metamorphosis insectorum Surinamensium*
(Amsterdam, 1705), plate 50. Photograph
copyright The Natural History Museum,
London.

the top of plate 50 rests upon a slender looping tendril that would not be able to support its weight. Below it, the larval instar clings to the stem of the plant, although Merian's text indicates that she actually found this specimen underground among the roots, where she also found another species of beetle larva that she has shown feeding on the root. The beetles and beetle larvae arranged on the leaves of the *Argemone mexicana* (Mexican pricklepoppy) in plate 24 of *Metamorphosis* are framed in a similar manner, and the composition is made up of pictorial elements that can be easily distinguished from one another, much like the elements of a pattern for embroidery or needlework (Figure 5.16). Although these configurations of insects and plants did not necessarily reflect the actual places where Merian found her specimens, they offered viewers a clear understanding of both the insects' appearance and their relationships to the plants they feed on.

Merian's written entries include information about the customs of the indigenous inhabitants of Surinam, which at times converge with her interest in the decorative and aesthetic aspects of the natural world. In one entry she describes how the red seeds of the annatto plant were used by indigenous peoples for making a paste "with which they paint various decorative patterns on their naked skin." In her observations of the musk flower, Merian notes that "the girls string them [seeds] on silk threads and wear them round their arms for decoration."[56] Merian considered information about the human uses of plants and insects for personal adornment to be as relevant in her observations of plants and insects as the information about life cycles, behavior, and habitat. Merian's observations of insects were also directed toward those aspects of their visual appearance that were most attractive and pleasing. She described the beetle shown in the lower-left corner of plate 50 (Figure 5.15) as a "beautiful golden beetle" having a "lovely gold-green color," and she has presented the beetle with its wings open in order to further highlight its colorful appearance.

In her illustrations for *Metamorphosis,* as with her prepared specimens, Merian preserved the appearance of insects by stabilizing their forms and fixing their positions. Once Merian established an insect's visual appearance in a drawing, she made very few modifications to its form. While in Surinam, Merian drew individual insects on small sheets of parchment and pasted these into her working journal, the *Studienbuch.* All but two of Merian's compositions for *Metamorphosis* were based on the studies gathered in the *Studienbuch,* which has been described as an "entomological archive."[57] Merian also executed several larger-format watercolor studies on parchment, some of which may have been made in Surinam. These watercolors represent an intermediate stage in the process of creating the final versions of the compositions that would appear in *Metamorphosis,* and they also incorporated motifs from the *Studienbuch.* In one of these studies several beetles are shown with one beetle larva (Figure 5.17). The insects are depicted in lively, dynamic positions but do not interact

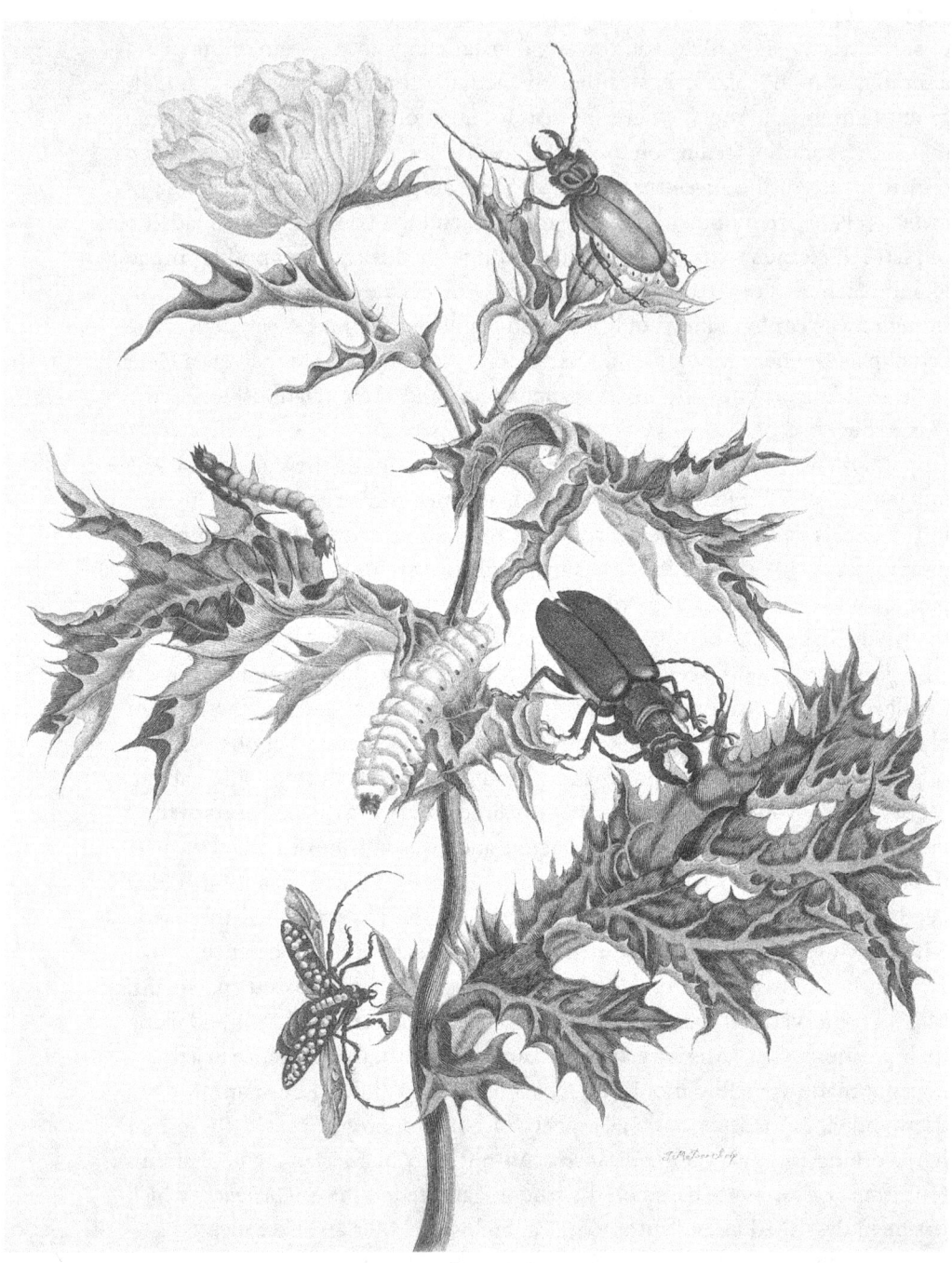

Figure 5.16. Maria Sibylla Merian,
Surinaamsche insecten (1726). First published
as *Metamorphosis insectorum Surinamensium*
(Amsterdam, 1705), plate 24. Photograph
copyright The Natural History Museum,
London.

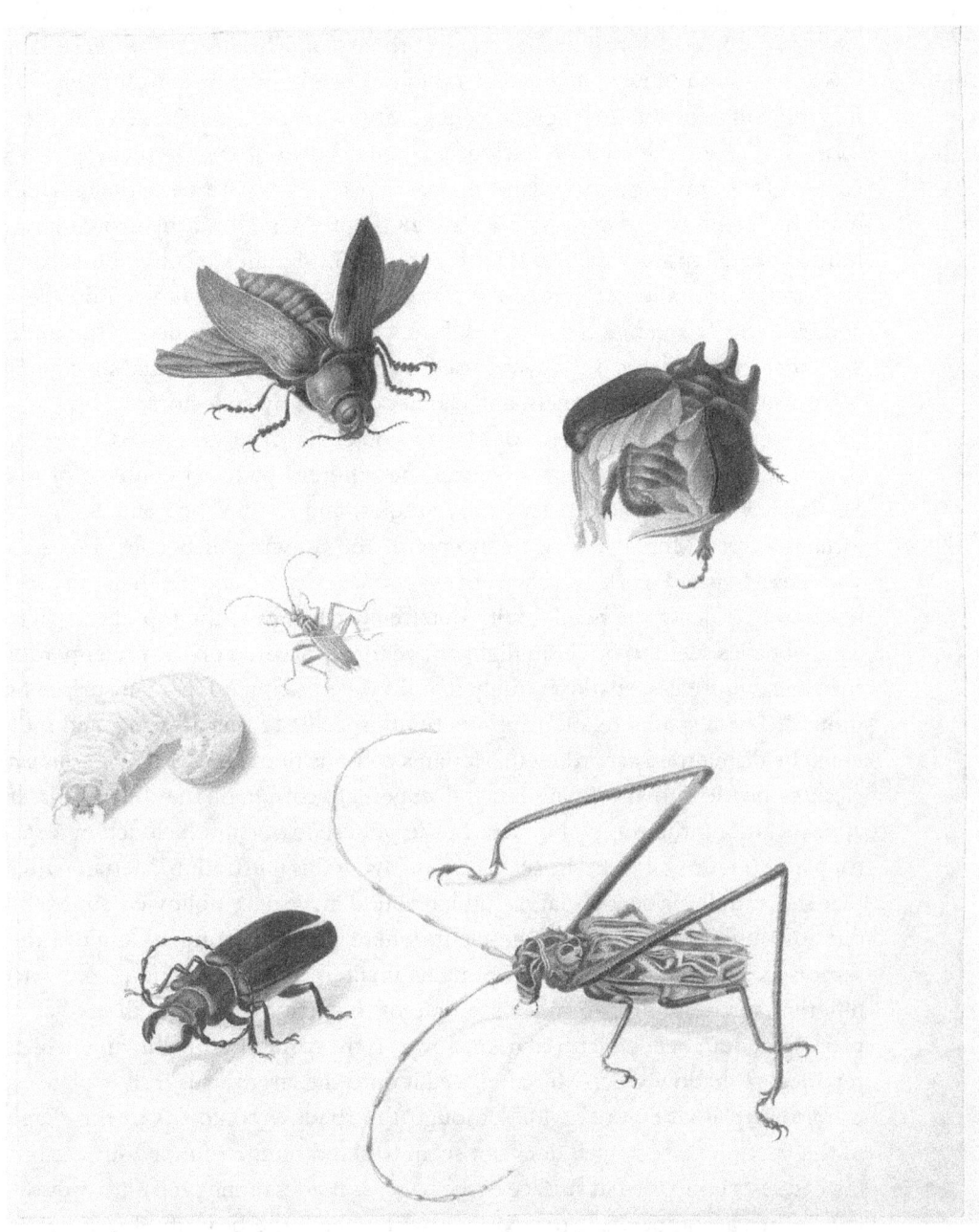

Figure 5.17. Maria Sibylla Merian,
watercolor on parchment, *Studienbuch*,
page IX. Copyright St. Petersburg Branch
of the Archive of the Russian Academy of
Sciences.

with one another and there is little to suggest any relationships between them. Within the empty space of the page, each insect exists as an independent entity, its crisply outlined forms and rich jewel tones contributing to the sense that it is a precious and rare object. Several of the beetles pictured in this preparatory drawing appear in different configurations in *Metamorphosis*, with their forms unaltered: the colorful *Euchroma gigantea* beetle shown in the upper-left corner of the drawing and the larva appear in plate 50 (Figure 5.15), the red-and-black harlequin beetle appears on plate 28 (Figure 5.18), and the brown long-horned beetle appears on plate 24 (Figure 5.16). Merian's technique of fixing an insect's form allowed her to circulate images of insects between different contexts and to arrange them into different configurations, much in the same way that embroidered slips were used by needleworkers or that specimens were arranged in boxes, panels, and cabinet drawers by collectors.

The stability of the forms of Merian's objectlike insects allowed her great flexibility in circulating them between the different pictorial contexts of the *Studienbuch*, intermediary preparatory studies, and the finished engravings of *Metamorphosis*. Merian did not conceive of the drawing of beetles discussed above as a finished work, but she conveys a sense of space and depth in the composition by placing the beetles with outstretched wings at the top of the page. These beetles seem to hover in flight above those pictured on the lower part of the page, whose cast shadows imply that they are resting upon a flat surface or ground. These spatial relationships were not specific to this drawing, and they could be dismantled according to Merian's compositional needs. The *Euchroma gigantea* beetle shown "flying" in the upper-left corner of the drawing later appears among the roots of a plant in the published version included in *Metamorphosis* (Figure 5.15). In the text accompanying this illustration, Merian writes that she found this beetle's larvae underground, in a space hollowed out of the dirt surrounding the roots of the plant, where she also found an adult of the same species. The beetle's form remains unaltered despite being placed into different surroundings. Another example of this fixity of form is the red-and-black harlequin beetle pictured in the lower-right corner of the drawing, which retained its shadow when Merian placed it onto the large citrus fruit in plate 28 of *Metamorphosis* (Figure 5.18). Although the shadow conforms to the shape of the fruit, the insect itself does not seem to stand on the rounded surface but instead rests upon the flat surface of the page, as it does in the preparatory drawing. In her accompanying text Merian writes that "the beautiful black beetle decorated with red and yellow flecks shown resting on the fruit was added on account of its rarity to complete and decorate the engraving although I do not know its origin."[58] This spectacularly colored beetle almost completely overshadows the presence of the moth pictured on one of the leaves, which is the insect Merian actually found on the plant and to which she devotes a lengthy account in the text. Merian's use of the harlequin beetle to "complete

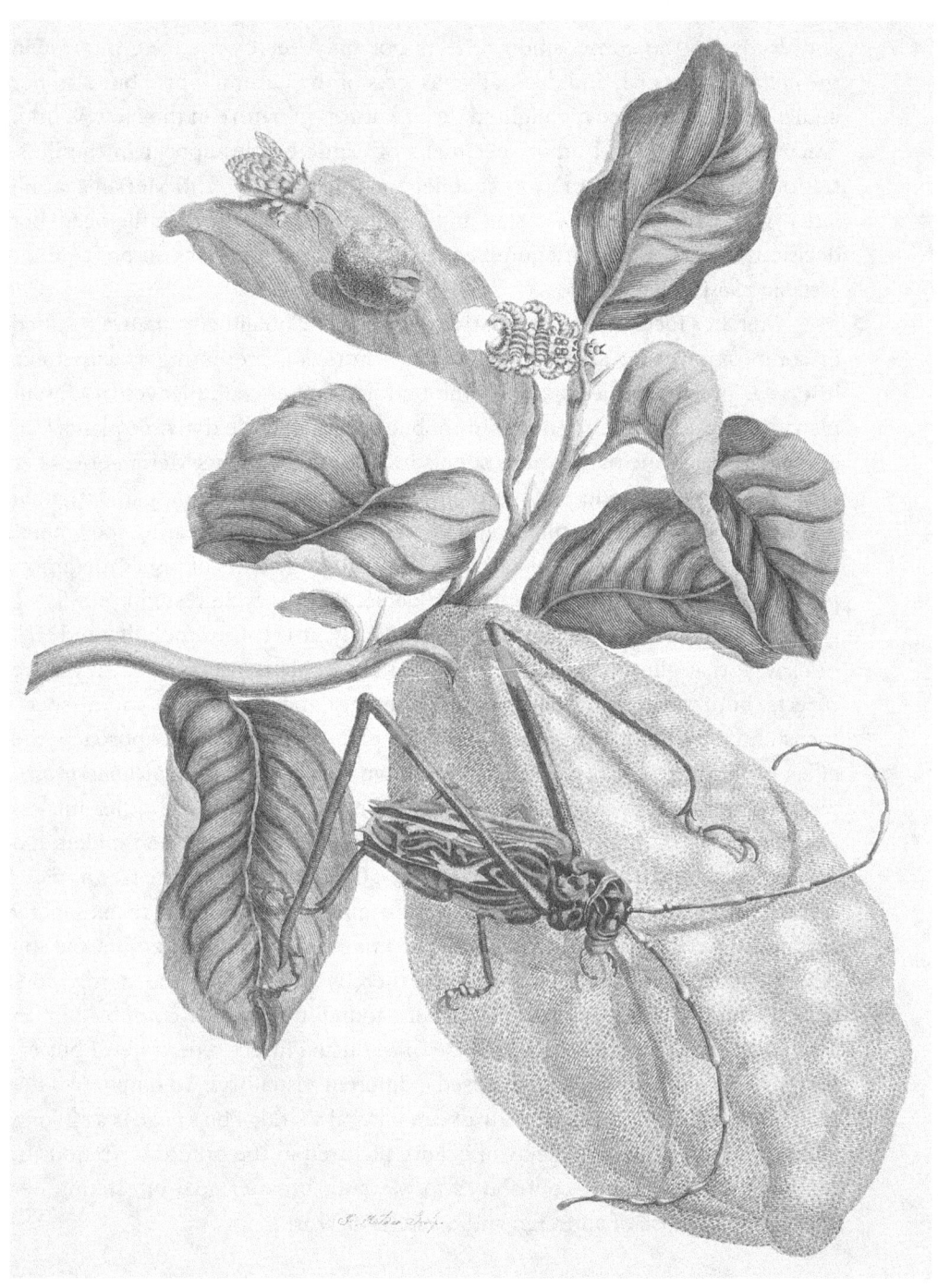

Figure 5.18. Maria Sibylla Merian,
Surinaamsche insecten (1726). First published
as *Metamorphosis insectorum Surinamensium*
(Amsterdam, 1705), plate 28. Photograph
copyright The Natural History Museum,
London.

and decorate" the composition reflects not only her own strong interest in the colorful, unusual, and decorative aspects of the natural world but also her audience's desire to see highlighted these features of nature in the New World. A harlequin beetle and other specimens of exotic beetles appear in the illustration of Levinus Vincent's insect collection (Figure 5.13), and Merian's familiarity with her audience's tastes and preferences may have influenced her decision to feature the harlequin beetle so prominently in her composition for *Metamorphosis*.

Merian's focus on the decorative and aesthetic qualities of nature resulted in compositions that are both orderly and intricate, presenting specimenlike insects as precious objects among the undulating stems and leaves of vibrant plants. The contrast between the immobile insects and the dynamic plants that entwine them is one of the many sophisticated visual delights Merian offers her viewers and is the product of her unique background and training in designing patterns for embroidery, still life and flower painting, and preparing specimens. Maria Sibylla Merian is associated with the changing forms of insect metamorphosis, but the visual style she used to depict the life cycle rested upon fixing insect forms in precise configurations. Ironically, it is this immobility and lack of change that allowed her great freedom to circulate insects as if they were objects, both within the spaces of her drawings and engravings and as specimens in the marketplace. When Merian departs from her usual approach, the effect is startling. One of Merian's best-known illustrations from *Metamorphosis*, plate 18, possesses a very different character and feel than the other images from the book. In this image, a violent struggle takes place between spiders and ants for control of a guava tree (Figure 5.19). On one branch of the tree a spider is shown in the process of devouring a hummingbird, and on the trunk several ants attack a cockroach.[59] Throughout the image, Merian presents ants and spiders in chaotic and deadly encounters, thereby abandoning the tightly controlled configurations of insects and plants found in her other compositions in *Metamorphosis*. This scene falls outside of her usual insect repertoire of butterflies, moths, and beetles, and she used a different visual style to emphasize the violence and drama of the non-European natural world. The image is a striking contrast to the peaceful objects of beauty pictured in the other illustrations in the book, but as such it contributes to Merian's broader goal of offering her audience a glimpse of a foreign and mysterious world.

Selecting, Framing, and Selling the Natural World of Surinam

Merian did not always succeed in her efforts to find and record the life cycles of unusual and beautiful species of Surinamese insects, but this did not hinder her ability to create fascinating visual images of these subjects. Merian sought out caterpillars that she believed would develop into large, brightly colored adults,

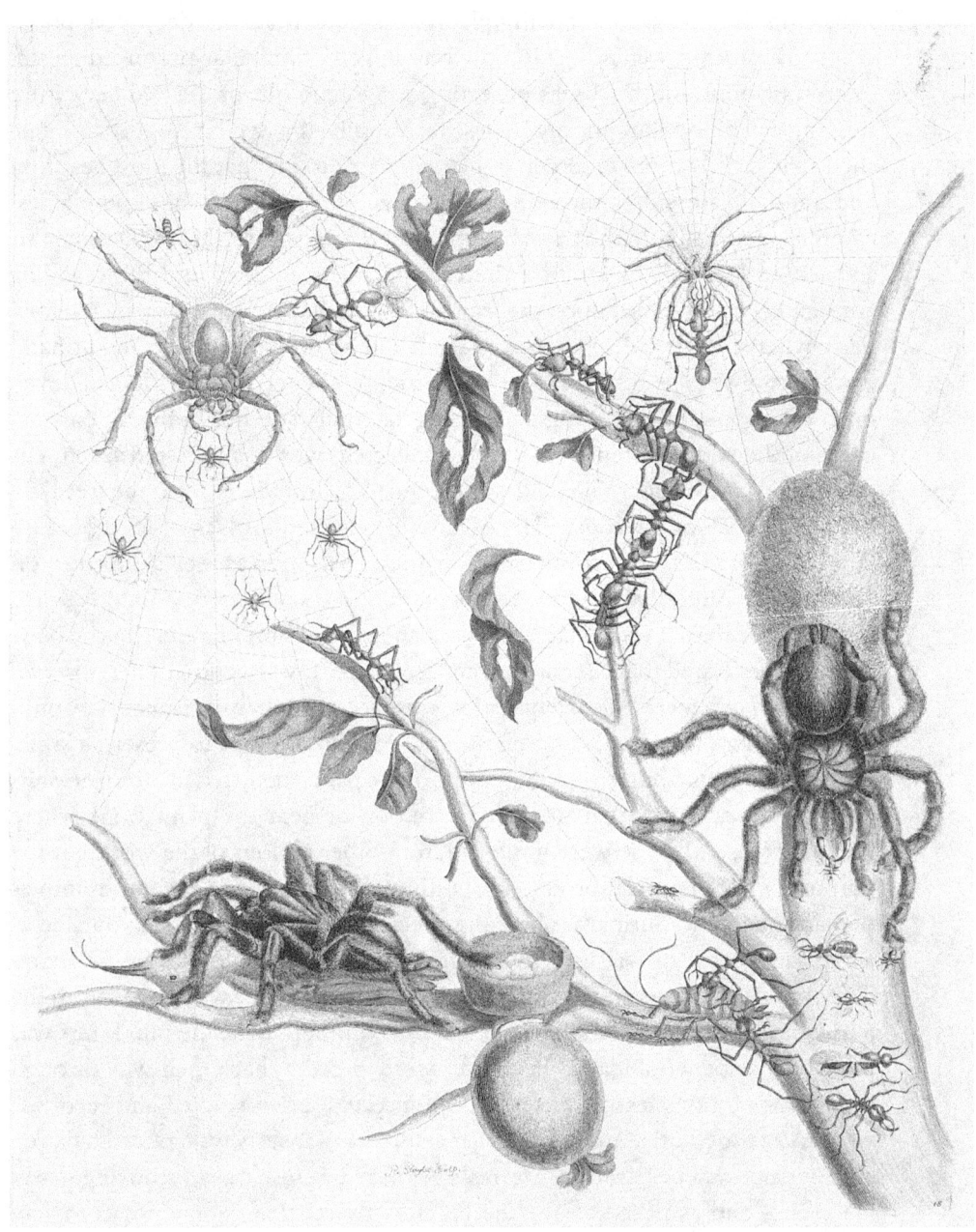

but she was sometimes disappointed in this search for continuity and beauty in nature. In the text accompanying plate 15 she describes her deep disappointment that an extravagantly colored "four-angled" caterpillar produced an uninteresting adult moth: "I was expecting something out of the ordinary from this unusual caterpillar, but my hopes were deceived when on August 10, 1700, such an unsightly moth emerged. It has happened to me that the most beautiful and unusual caterpillars turned into the plainest creatures, while the plainest caterpillars turned into the most beautiful moths and butterflies."[60] Despite this setback, in Merian's illustration the caterpillar, its cocoon, and the resulting "unsightly" moth are set upon the fruit and stems of a melon with bright yellow flesh, where together they form an elegant and harmonious image. In Surinam, it was not always possible for Merian to employ the techniques she had developed in Europe for observing and raising caterpillars in her home, in part because of the very different climate. She collected over one hundred specimens of a certain type of caterpillar and was careful to collect leaves from the tree they lived on, but all of the caterpillars except for one died for lack of food because "the leaves of this tree become hard and dried up as soon as they are broken off and then the caterpillars cannot eat them."[61]

Thus, offering European audiences an exotic vision of nature in the New World necessitated that Merian present only a narrow selection of her observations and experiences. *Metamorphosis* was not a comprehensive survey of South American flora and fauna, and neither did Merian attempt to present a study of every insect inhabiting Surinam. Most of the plants depicted in *Metamorphosis* are agricultural crops that were cultivated on or near the plantations where Merian lived, and some were imported from other regions of the world for the purposes of commercial production. Although she included several examples of plants native to Surinam, obtaining specimens of plants growing outside of agricultural contexts was extremely difficult. As Segal has noted, "her selection of motifs clearly suggests that Maria Sibylla Merian did not wander far from the Surinamese plantations. As she wrote in her commentaries, the rain forest was practically impenetrable, and her slaves were forced to hack their way through the thorns and thistles with axes."[62] The insects Merian studied and recorded were native to South America, but her selection of insects was guided by a relatively narrow set of criteria. Merian's primary interest was in studying insect life cycles, and as noted above she favored insects that experience complete metamorphosis, such as beetles, butterflies, and moths. Furthermore, Merian focused her attention on unusual and colorful insects that could serve as subjects of dazzling illustrations or sophisticated displays in collectors' cabinets. Although she included a great deal of information about the suffering caused by insects to the European and non-European inhabitants of Surinam, she did not choose to depict many of these insects in her images, with the exception of the cockroach. In one of her commentaries Merian describes cockroaches as "the most infamous of all insects in America on account of the great damage they cause to all the inhabitants by spoiling their wool, linen, food and drinks"; in the accompanying image a cockroach is shown perched upon a perfect, unspoiled pineapple rather than with any of its less visually appealing foods.[63]

Absent also from Merian's illustrations in *Metamorphosis* is evidence of the human labor involved in cultivating the plants she depicted. Merian herself relied on the labor of enslaved Africans to assist her in obtaining insect specimens from difficult locations, such as the tops of trees or the stalks of banana plants. They also were responsible for cultivating the garden she maintained, which was presumably a central source of food for her and her daughter. The indigenous inhabitants of Surinam were also important sources of specimens and information, and the European inhabitants also provided her with some information on the uses and growth cycles of plants. Merian was not completely unsympathetic to the plight of the enslaved Africans and indigenous inhabitants of Surinam whose labor and knowledge she depended upon. Her remarks in the commentary accompanying the illustration of the *Flos pavonis* plant have been extensively discussed by historians in relation to questions of European colonization and the history of female reproductive biology.[64] In one passage, Merian criticizes the Dutch plantation owners' harsh treatment of enslaved Africans and native people, but with an eye toward the economic benefits of more "benign" handling. "Indians, who are not well treated when in the service to the Dutch, use [the seeds of the plant] to abort their children so that their children should not become slaves as they are. The black slaves from Guinea and Angola must be treated benignly, otherwise they produce no children in this their state of slavery; nor do they have any; indeed they even kill themselves on account of the usual harsh treatment meted out to them."[65]

Intermixed with Merian's call for more humane treatment of slaves is a shrewd business sense that would have been appreciated by the Dutch plantation owners whose practices she criticized. Merian offers a number of other criticisms of the plantation owners in her commentaries—mostly directed toward the sugar monoculture—that reveal a similar concern with achieving economic self-sufficiency for the colony by reducing its reliance on expensive imported goods.[66] It was this sense of the commercial possibilities offered by the natural world of Surinam that made Merian adept at extracting the most spectacular and beautiful examples of insects and plants from their surroundings and including them in her illustrations. Merian's actual experiences of the natural and human world in Surinam were sometimes confusing, difficult, and dangerous, but the illustrations she produced convey only the exotic and beautiful aspects of that world, thereby giving her European audiences a safe and fully accessible vision of the New World on which to feast their eyes.[67] Collectors and other individuals who traded in exotic specimens utilized the infrastructure established by the much larger and economically significant movements of agricultural products, raw materials, and people conducted by European states and organizations such as the Dutch West India Company. With her images and specimens, Merian inserted the natural world into the overlapping networks of exchange that characterized both the practice of natural history and the commercial world of Europe at the end of the seventeenth century.

The 1719 Frontispiece to *Metamorphosis*

A second edition of Merian's *Metamorphosis* was published in 1719 by her daughters after the author's death. The 1719 edition differs from the first edition in that it contains an additional twelve illustrations as well as an illustrated frontispiece (Figure 5.20). The frontispiece tells us a great deal about Merian's legacy and how her descendents understood both the significance of the book and how they wanted readers to understand the contents and its author. In the foreground a finely dressed female figure—a personification of Merian—sits at a table preparing plant and insect specimens surrounded by putti assistants. She looks down to her left at a vignette of a book and a potted plant. The book shown is the *Metamorphosis* itself, and the potted plant is a living specimen of the pineapple plant pictured in the open book. The living specimen indicates to the reader that the contents of the book are based upon firsthand observations, and the open book functions as an advertisement of Merian's wares. The book is presented as a product made for the specific audience of wealthy collectors. The Merian figure resides in an elegant room decorated with classical motifs that make it clear that this is a book for an intellectual, educated audience, and that its author was also an educated person. This emphasis on Merian's learning was important, as is the emphasis on the putti as the workers who organize and assemble her specimens. By showing Merian as engaged with the materials of nature, but not expending too much physical effort on them, Merian's daughters sought to show her not as a businesswoman who collected specimens only for profit but as an upper-class person pursuing knowledge of nature as a serious but leisurely activity.

On the right side of the frontispiece one of Merian's putto assistants stands at the door of an elaborate specimen cabinet, and other putti are shown working on elegantly arranged trays of insect specimens. Although they do not feature embroidered decorations like those in Levinus Vincent and Johanna Breda's collection, these trays are meant to appeal to the same sensibilities and tastes for artfully presented natural objects. Following the gaze of the putto at the cabinet, the viewer's attention is drawn through the opening of the archway toward a view of Surinam. In the center of this scene is a female figure leaning close to the ground and holding a net, presumably Merian herself shown in the act of capturing one of the spectacular butterflies pictured in *Metamorphosis*. She is dressed in fine clothing, as are the two European men who observe her. Farther

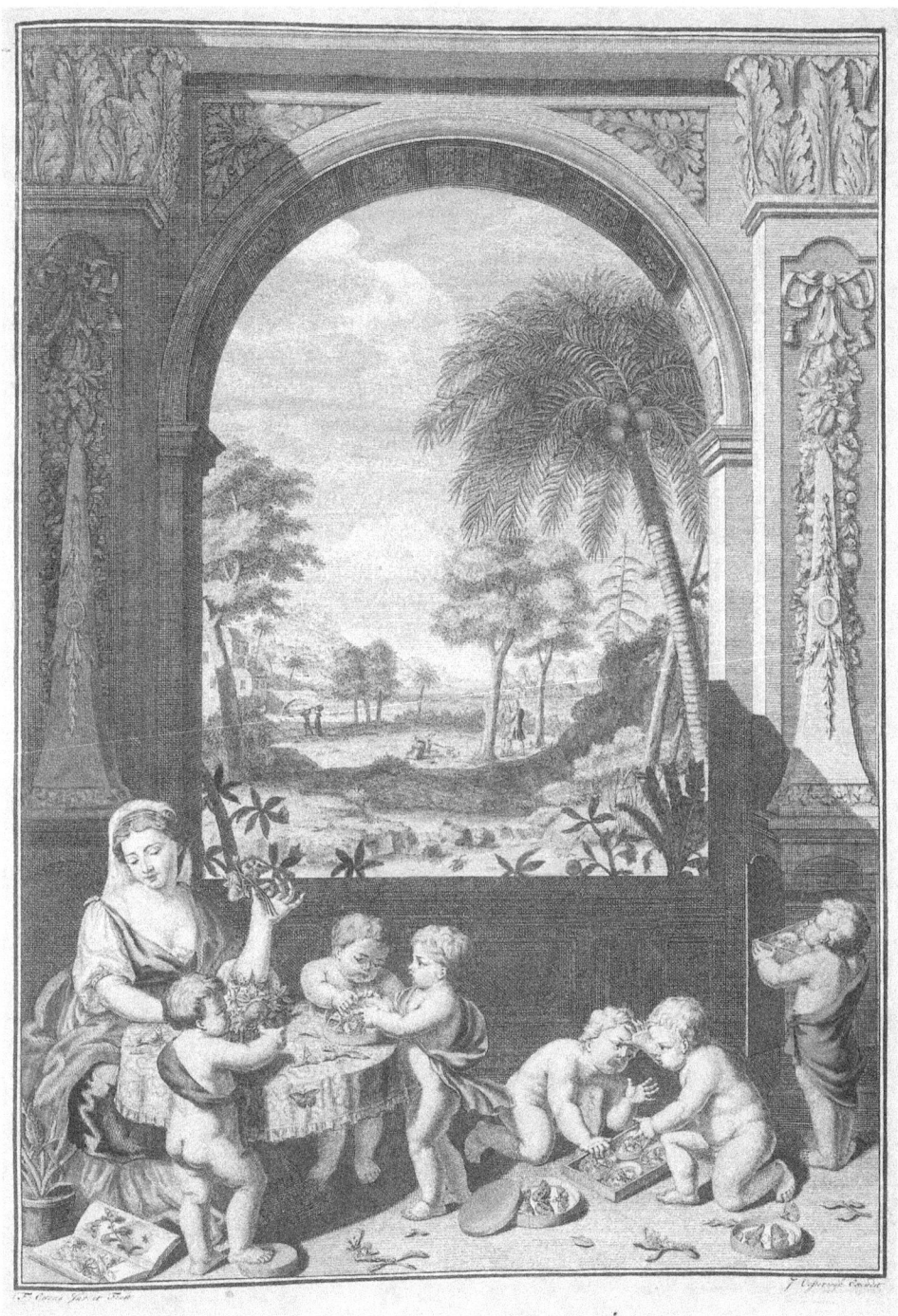

Figure 5.20. Maria Sibylla Merian,
Metamorphosis insectorum Surinamensium
(Amsterdam, 1719), frontispiece.
Photograph copyright The Natural History
Museum, London.

back on the left side are two additional figures, a female and a male, who are very likely the servants or slaves who accompanied Merian and her daughter on their collecting expeditions in Surinam. The cultivated grounds, the familiar-looking buildings, and the lack of dense vegetation present Surinam as a cultivated and controlled natural environment, exotic but unthreatening. The frontispiece also tells the reader that Merian's activities took place under the watchful eyes of European men who offer their protection and approval. Of course, the vision of Surinam presented in the frontispiece—a serene playground for European aristocratic living—is not maintained by either the illustrations or text of the *Metamorphosis*. Instead, the contents of Merian's book often hint at the instability and the difficulties posed by plants and insects to both European and non-European residents of Surinam despite Merian's efforts to present it as a world of immobile and timeless objects of beauty. However, in encapsulating many of the major themes of Merian's professional activities and publications, the frontispiece is an important example of the ways that visual images continued to shape both the European vision of nature and the construction of an artist's persona well into the eighteenth century and beyond.

DISCIPLINE AND SPECIMENIZE

*I*n this book I have traced the development of the insect as subject matter and the use of insects by artists and other practitioners to construct professional personae through making, circulating, and displaying insects as three-dimensional objects and as two-dimensional images. When the insect emerged as a subject of interest in the sixteenth century, Joris Hoefnagel utilized it as a vehicle for displaying his unrivalled talents as a miniaturist and to satisfy his courtly audiences' tastes for exotic and unusual entertainments. The naturalists Thomas Moffet and Ulisse Aldrovandi crafted identities around insects that showed them to be trustworthy organizers of the natural world, and early floral still life painters inserted insects into the rarefied spaces of cabinets and collections to generate interest in the new genre of still life painting. In all of these cases, the appeal of the insect was directly tied to its visual interest and its suitability to representation in a range of visual forms. By the later seventeenth century, when insects were well established as subject matter, Robert Hooke and Maria Sibylla Merian capitalized on the appeal of insects to feature them as spectacular illustrations in their publications—projects that were also intricately tied to the construction of their own complex professional identities.

Although this book demonstrates the increased visibility of insects over the course of the sixteenth and seventeenth centuries, the cases of Hooke and

Merian remind us that insects still retained their "marginal" status, and that this marginality was in some ways essential to maintaining interest in them as curiosities. Hooke and Merian each occupied ambiguous positions within their social, cultural, and professional worlds. Hooke's middling family background and work as an employee of Robert Boyle and later the Royal Society placed him on uncertain ground as an observer and witness when he commenced the research for *Micrographia*. Merian's situation as a divorced businesswoman who pursued her research on insects with a sometimes unsociably single-minded passion also placed her in an uncertain social and professional position in the natural history community. By associating themselves with the insect, usually overlooked or reviled but revealing itself as a quiet wonder to the patient and attentive observer, both Merian and Hooke took an opportunity to remake themselves as similarly hidden treasures.

Large illustrated books such as Hooke's and Merian's increased in popularity during the eighteenth century, both as a means of presenting new research and as a way of continuing to market the natural world to European consumers of natural history. The eighteenth and nineteenth centuries are often referred to as the "golden age" of natural history, and the illustrated works of natural history from this period reflect the parallel expansions of Europeans' knowledge of the natural world and the global reach of European political power along with the spread of international commercial and trade networks.[1] The desire to possess, control, and represent the natural world was but a smaller aspect of the efforts to establish and maintain far-flung political and commercial empires, and the study of insects in areas outside Europe occurred primarily in colonial holdings or areas of European influence. Works by Dru Drury, John Obadiah Westwood, and Edward Donovan followed in the tradition established by Maria Sibylla Merian of making foreign insects the subject of illustrated natural histories, and later works continued to fuel and satisfy the demand by Europeans for exotic and beautiful examples of insects from foreign lands.[2] At the same time that Europeans were studying and collecting insects from the rapidly expanding sphere of European global influence, other practitioners were producing studies of European insects that reflected a narrower scope of inquiry. Although major general studies of insects appeared during the eighteenth century, most notably John Ray's *Historia insectorum* and R. A. F. de Réaumur's *Mémoires pour servir de l'histoire des insectes,* a new development in the study of insects during this period was the rise of local studies of particular countries or geographical areas. No longer did those who studied insects purport to be able to encompass all insect life in a single volume. Instead, comprehensive local studies were undertaken that concentrated on identifying and describing local populations of insects.[3]

Visual images continued to play an important role in the study of insects, and artists of later periods continued to use the visual conventions and strategies for illustrating insects that had been developed in the sixteenth and seventeenth centuries. The specimen drawing gained importance as a mode of presenting insects, plants, and other productions of nature as discrete objects, but later practitioners sometimes combined the specimen drawing with other types of images to create multilayered descriptions of their experiences. Older pictorial techniques were reworked as naturalists used visual images to convey different types of information. An engraving by George Edwards (Figure C.1), for example, presents two rhinoceros beetles, insects whose visual pedigree had been established by Joris Hoefnagel in the sixteenth century. Unlike Hoefnagel's rhinoceros beetles, which always appear as specimens against a blank surface, Edwards's beetles are placed upon a map that outlines the artist's travels through northern Europe. The beetles are accompanied by a disembodied ibis beak and an inset of a hummingbird perched upon a tree stump. The image relies on the audience's familiarity with specimen logic—otherwise viewers might interpret the beetles as giant aquatic invaders launching simultaneous attacks on England and Scandinavia. By overlaying specimen drawings of exotic beetles with the record of his travels, Edwards associated himself with an appealing subject while also making it clear that his own collecting and picturing activities were founded on direct observations and knowledge of specific localities.[4]

Cutting and pasting images also continued to be an important method for organizing and conceptualizing knowledge about insects, and it was also used in new ways in the eighteenth century. The artist-naturalist Mark Catesby often created unexpected juxtapositions of plants and animals in his *Natural History of the Carolina, Florida, and the Bahama Islands* (1731–43), with compositions made through virtual cutting and pasting. For his illustrations of insects, Catesby also seems to have followed the visual practice established by Hooke and other microscopists in the seventeenth century of depicting magnified views of insects in monochrome. The illustrations of insects in Hooke's *Micrographia* were not meant to be colored, most likely because when insects were viewed through the microscope, line and light rather than color were the visual elements that combined to form an image. Microscopy became increasingly popular during the eighteenth century, and books of instruction were published for general audiences and microscopes and microscopists appeared in literature. Although the use of microscopes as scientific tools in the eighteenth century has been a question of debate among present-day researchers,[5] Hooke's images of insects were nevertheless widely copied in eighteenth-century publications, most notably in Diderot's *Encyclopédie*.

Naturalists who studied and collected insect specimens from far-flung regions of the earth continued to participate in international networks of correspondence and exchange. In the text accompanying Edwards's illustration of the rhinoceros beetles the author notes that the beetles had come from the East Indies via Dr. Matthew Lee, the hummingbird was drawn from a specimen

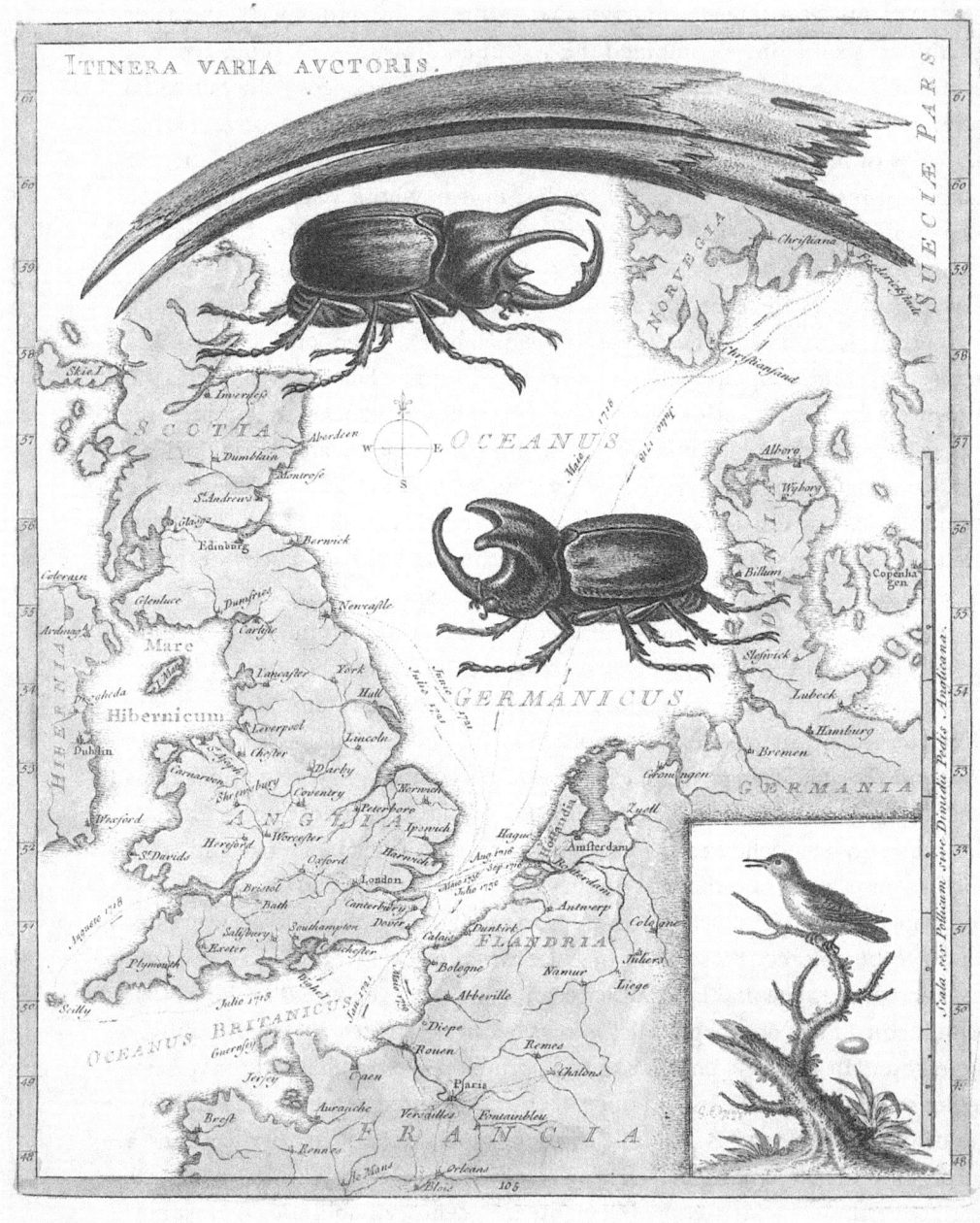

Figure C1. George Edwards, *A Natural History of Uncommon Birds* (London, 1747), volume 2, plate 105. Courtesy of University of Wisconsin Digital Collections.

brought from Jamaica, and the ibis beak came from an embalmed specimen from Egypt. Edwards's limited knowledge of the origins of his specimens is not betrayed by the image, which produces the appearance of knowledge of the natural world that encompasses Northern Europe, Africa, and the islands of the Pacific Ocean. Social networks remained an important aspect of the study of nature in the eighteenth century, but the new contours of knowledge resulting from the introduction of Linnaean classification brought about a shift in the social and cultural aspects of specimen and image exchange.

The work of the English entomologist Moses Harris is instructive for understanding the influence of Linnaeus's systems on studying and picturing insects. Harris published a number of illustrated books on insects during the eighteenth century, and the book for which he is best known is his work on English insects, *The Aurelian: A Natural History of English Moths and Butterflies* (London, 1766). In the frontispiece to *The Aurelian* a finely dressed gentleman rests in a forest after a collecting expedition, and he shows the reader his finds in neatly arranged oval specimen boxes (Figure C.2). In the plates that follow, butterflies and moths appear with the plants that they feed upon, in arrangements similar to those of Maria Sibylla Merian's *Metamorphosis* compositions. Butterflies and moths are elegantly arranged with plants and flowers, sometimes presented as specimens in small boxes (Figure C.3). Harris, however, is much more aggressive than Merian in his efforts to appeal to the tastes of wealthy collectors. Each plate is dedicated to a specific patron, among them some of the most prominent English supporters and practitioners of natural history of the day, including Dru Drury, Peter Collinson, John Fothergill, and several figures from the nobility. Harris no doubt hoped to secure subscriptions from the people he named in his plates, and he is careful to mention if specimens came from their collections, thereby paying compliments to his friends and colleagues who have shared information and specimens with him. The images were meant to appeal to the refined tastes of Harris's patrons and potential patrons, who were sophisticated consumers of natural history books and well familiar with lavishly illustrated works such as Merian's. Seventeen years later, Harris took a very different approach to illustrating insects in *An Exposition of English Insects* (London, 1783). The full title of this publication explains that the contents were "described, arranged, & named according to the Linnean system," a reflection of the growing popularity and influence of Linnaeus in England in the last half of the eighteenth century. *An Exposition of English Insects* has a different overall organization than *The Aurelian,* but the most striking change is in the way that Harris composed the illustrations for the later volume. Rather than showing a lively assortment of insects with plants, insects are instead organized into tight grid formations (Figure C.4). There are no dedications to friends, colleagues, or patrons, and there are no discussions of other collectors, their cabinets, or their contributions to Harris's research. The text mentions only one other person, Linnaeus.

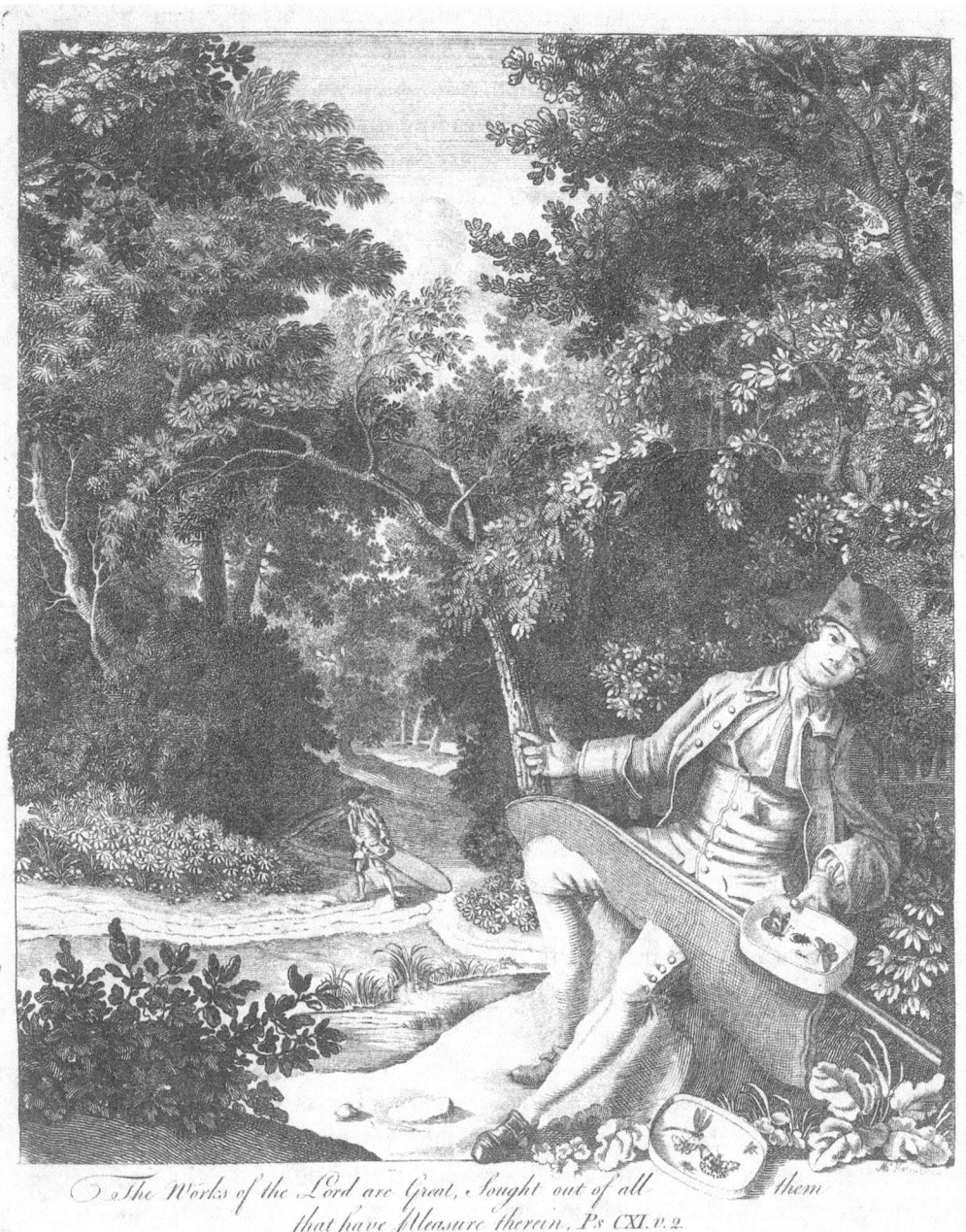

The Works of the Lord are Great, Sought out of all _____ them
that have pleasure therein. Ps CXI. v. 2.

Figure C.2. Moses Harris, *The Aurelian:
A Natural History of English Moths and
Butterflies* (London, 1766), frontispiece.
Reproduced by permission of The
Huntington Library, San Marino, California.

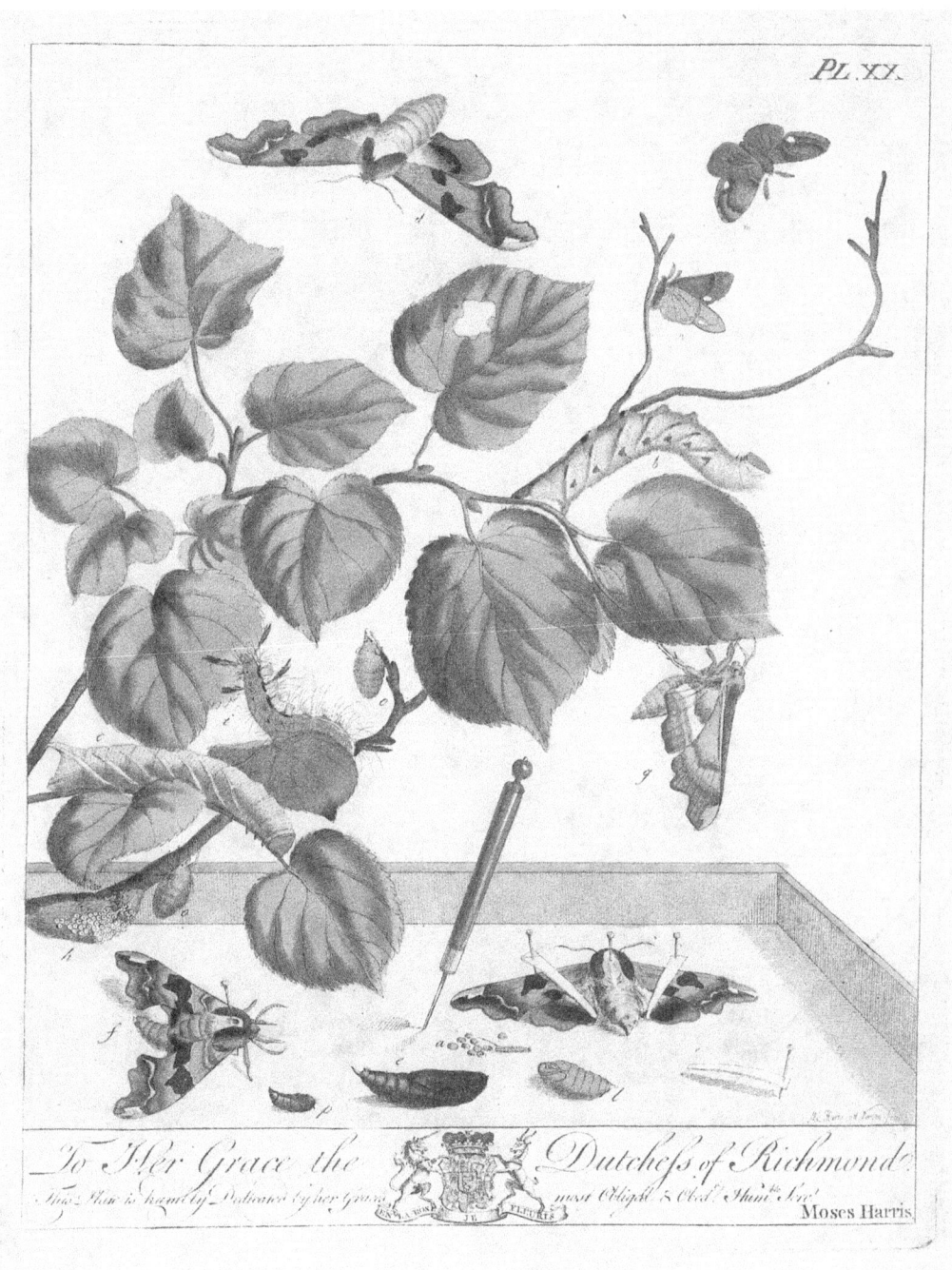

Figure C.3. Moses Harris, *The Aurelian: A Natural History of English Moths and Butterflies* (London, 1766), plate 20. Reproduced by permission of The Huntington Library, San Marino, California.

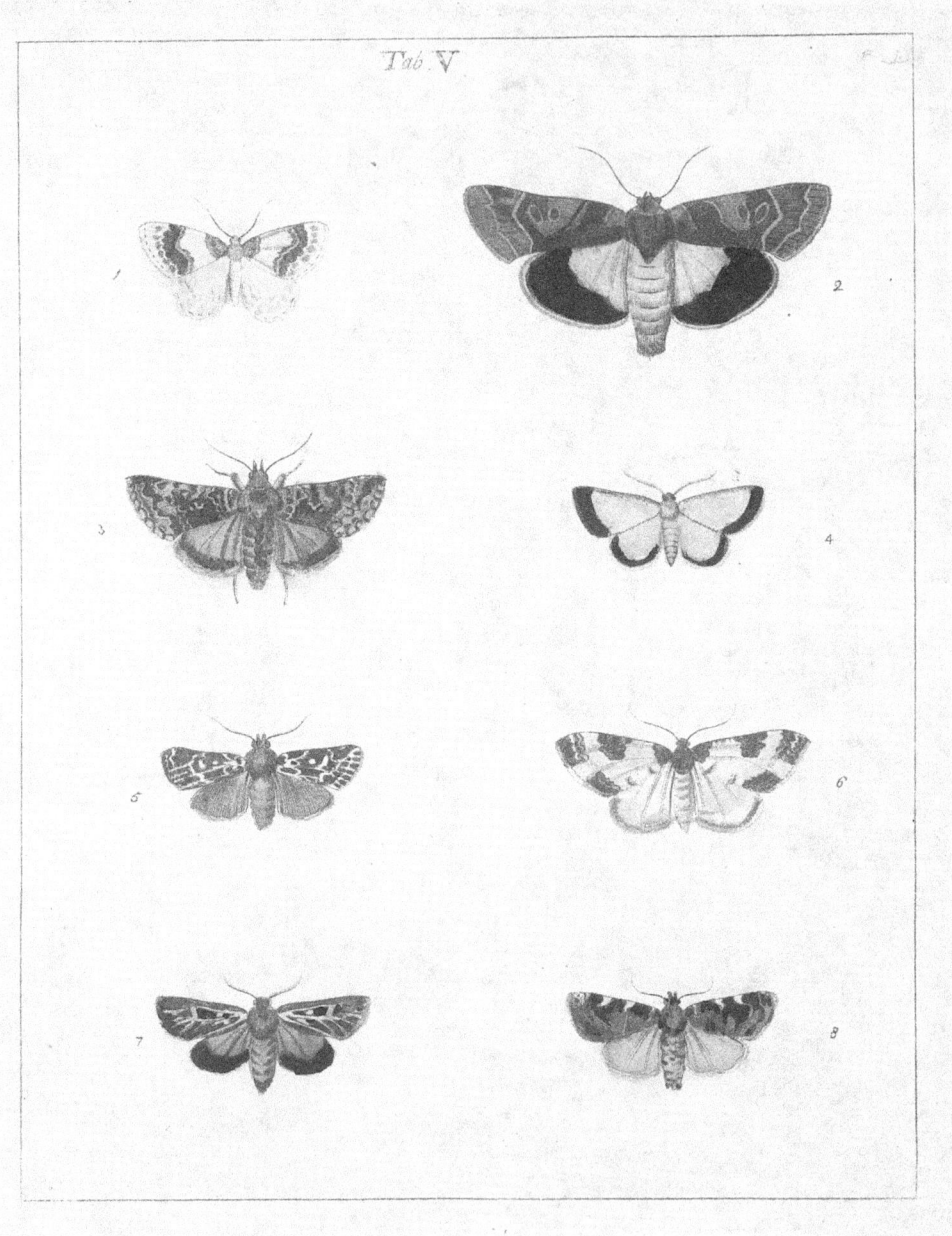

Figure C4. Moses Harris, *An Exposition of English Insects* (London, 1783), table V. Reproduced by permission of The Huntington Library, San Marino, California.

The system for naming and classifying species established by Linnaeus that would come to dominate the study of nature into the modern period was founded on narrowing the scope of inquiry to the anatomical structures involved in reproduction. The illustrations accompanying editions of Linnaeus's research on insects depicted only those parts of insects' bodies that were relevant to this system of classification, and thus left out other parts or rendered them only in general, schematic outline.[6] This narrowing process is what Michel Foucault refers to as the "screening" function of natural history in the eighteenth century. As I demonstrated in chapter 1, this selective approach to studying nature was already a part of sixteenth-century practices relating to picturing the natural world. It is in the eighteenth century that natural history texts would catch up with visual images, notably in their very narrow focus on a restricted number and type of structures.

Along with a narrowing of the range of inquiry, eighteenth-century naturalists also began to direct their efforts more and more toward the new, or the "nondescript"—that is, those species that had not yet appeared in print. Harris's contemporary Dru Drury is another example of this shift toward Linnaean classification in the eighteenth century, as well as the new emphasis on describing new species. The stated aim of Drury's *Illustrations of Natural History* was to present "figures of exotic insects, according to their genera; very few of which have hitherto been figured by any author."[7] In his preface Drury explains that at first he set out to include illustrations of any exotic insect, but later he changed his criteria because he "was soon sensible, that the figures, already known and published here could do no service to the study, or benefit of the reader . . . this would not be improving the reader's judgment, or increasing his knowledge . . . From that time I took care to delineate none that I was conscious had engaged the pencil of any preceding author."[8] The visual strategies that made nature visible according to the Linnaean method had already been in place for centuries, but they were now used in new ways. Specimen logic allowed eighteenth-century image makers to parse nature even further by using techniques that had been successfully deployed to transform nature into a cabinet of wonders to now focus attention on an even narrower view of nature's collection.

Eighteenth- and nineteenth-century practitioners who studied and illustrated insects still explored the visually appealing aspects of their subjects, as evidenced by titles that incorporate the words "exotic" and "cabinet." However, as the persona of the observer as a careful and faithful transcriber of nature rose in importance in the eighteenth century, older ways of studying insects that were embedded within multiple and overlapping cultural and social contexts began to wane. Or rather, the approach to nature as a collection of rare and precious objects moved out of the mainstream and into the living room. As the study of the natural world became increasingly professionalized and identified as "science," and the places where it was practiced restricted to museums, universities, and laboratories, the people associated with these new spaces were usually men. At the same time, interest in the aesthetic and decorative aspects of nature came to be associated with domestic spaces and practices, and the newly defined realm of the "amateur." The close connection between insects and the decorative arts evinced in the work of Maria Sibylla Merian could still be found in these later periods, but usually it would be found in domestic settings or in publications that lay outside of mainstream science. James Barbut's watercolor drawing for the title page to his edition of Linnaeus's *Genera insectorum,* which is called *A Collection of British Insects Drawn from Nature,* uses an embroidery pattern to frame the view of a landscape, and insects creep across the text of the title (Figure C.5).[9] However, Barbut's edition of Linnaeus was not widely known, and neither were his illustrations. Merian's own *Metamorphosis* illustrations were adapted for porcelain and faience designs, and thus were displayed alongside luxurious material objects that filled the homes of educated elites in Europe and North America in the eighteenth century. This blending of decorative and the aesthetic in the study of nature occurred in private domestic settings, mostly but not exclusively among women, and as leisure-time activities. Unlike Maria Sibylla Merian, whose professional career merged with domestic and decorative concerns, women who studied the natural world in the eighteenth and nineteenth centuries practiced mostly at home and were not welcomed into public scientific societies and institutions. The study of nature could also put women at risk, both socially and financially, as in the case of Lady Eleanor Glanvil, whose memory, according to Harris, "suffered for her curiosity. Some relations that was [sic] disappointed by her will, attempted to set it aside by acts of lunacy, for they suggested that none but those who were deprived of their senses, would go in pursuit of butterflies."[10] Women nevertheless played an important role in the study of nature in this later period, often as patrons but also as practitioners, and these contributions are receiving more attention from scholars as questions about disciplinarity and the categorization of knowledge move to the forefront of current research on the history of art and science.[11]

For early modern Europeans, insects were subjects around which they explored questions regarding relationships between the natural world, visual representation, and professional personae. The contexts and media through which these questions were examined were many and varied, including natural history but ranging beyond this field to include still life painting, manuscript illumination, and the cultures of collecting and display. It was through visual images and specimen logic that insects became objects of beauty and wonder. The consequences of this metamorphosis, of nature rendered into a collectable and a commodity, extend beyond the narrowly constructed borders of this book.

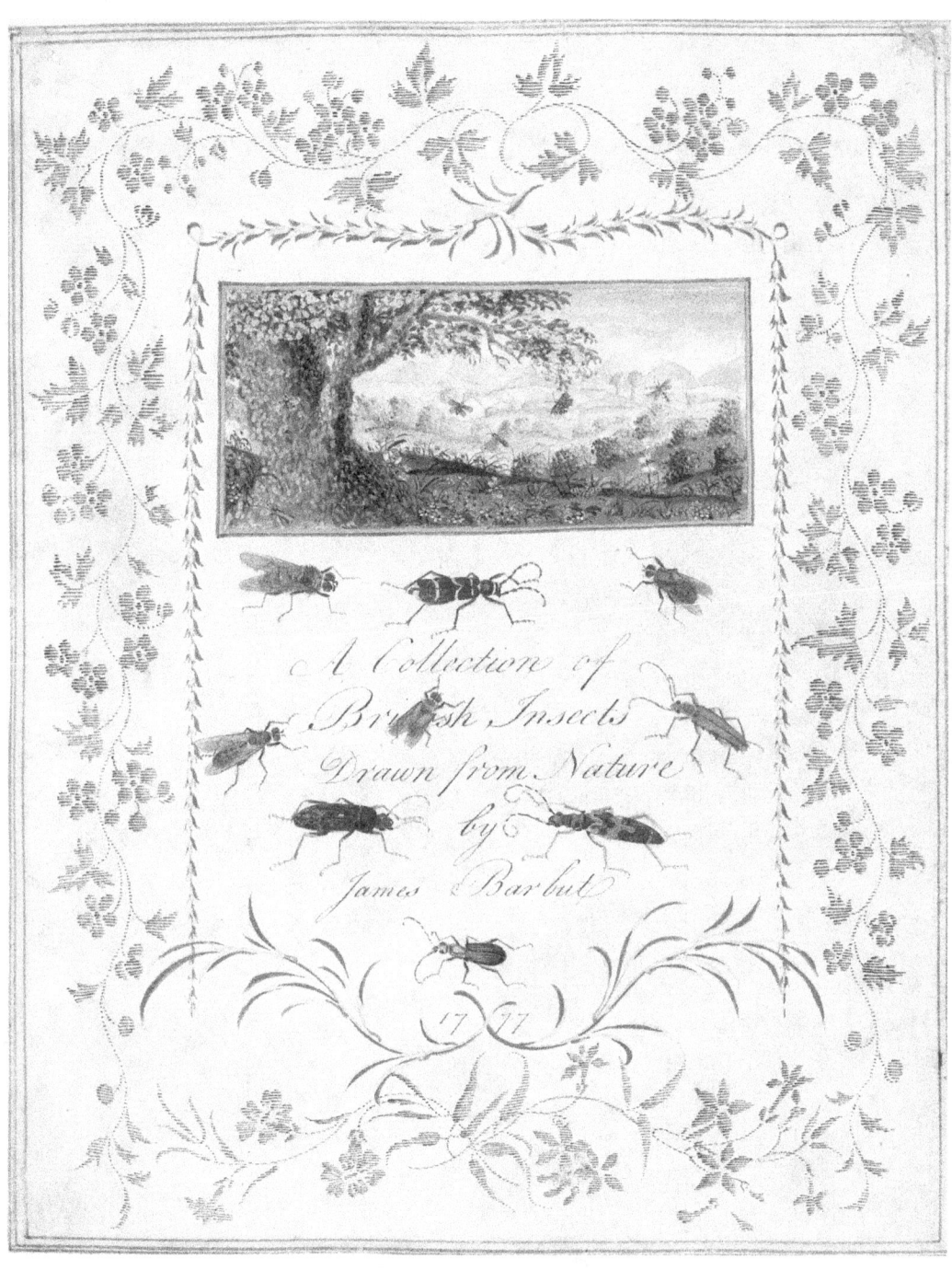

A Collection of
British Insects
Drawn from Nature
by
James Barbut

17 77

Figure C5. James Barbut, frontispiece from *A Collection of British Insects Drawn from Nature*. Photograph copyright The Natural History Museum, London.

ACKNOWLEDGMENTS

*M*y investigation of the networks of artists, naturalists, collectors, and other practitioners who made up the world of insect illustration in early modern Europe has been made possible by the encouragement, support, and assistance of an equally diverse and intellectually stimulating network of scholars and friends. I am grateful to George and Linda Bauer for their early encouragement for this project and their thoughtful reflection and clarity about how these materials could be understood in relation to the discipline of art history. Jane Newman suggested an early trip to Prague that resulted in my introduction to the extraordinary art of Joris Hoefnagel. Pamela Smith has been instrumental in shaping my interest in early modern Europe since my undergraduate days, and I am grateful for her sustained encouragement. Tara Nummedal's friendship has made research trips and conferences much more fun over the years, but I also appreciate her intellectual engagement with the central issues of this book and her expert readings and advice. Amy Meyers shared with me her extensive knowledge and her infectious enthusiasm for the visual culture of the natural sciences, and I continue to be inspired by her generosity, insights, and abilities to bring scholars together around this fascinating area of inquiry.

This book could not have been written without the generous support of the institutions and organizations that provided funding for travel, research, and writing. I received financial support from the School of Humanities at the University of California, Irvine; the Center for Austrian Studies at the University of Minnesota; the Samuel H. Kress Foundation; and the University of California Humanities Research Institute. Research fellowships at the Huntington Library and at the John Carter Brown Library gave me time to write and revise in congenial settings while having access to essential primary sources. Martin Schimpf of the College of Arts and Sciences and Mark Rudin of the Division of

Research at Boise State University provided crucial financial support for permissions and illustrations.

I wish to thank for their friendly assistance the staffs of the libraries, museums, and institutions where I conducted research for this project, in particular the personnel of the Department of Manuscripts at the British Library; Julie Harvey of the Natural History Museum, London; Henrietta McBurney of the Print Room at Windsor Castle; C. Danial Elliott of the Academy of Natural Sciences in Philadelphia; Dr. Helmut Trnek of Kunsthistorisches Museum in Vienna; and Laura Miani of Biblioteca Universitaria in Bologna. Chapter 4 benefited enormously from Monica Rumsey's close reading and suggestions. I very much appreciate the help I received from the staffs at the Huntington Library, Art Collections, and Botanical Gardens and at the John Carter Brown Library, where I was especially grateful to Edward Widmer for his excitement about this project. Beverly Howard, Carol Reagle, and Richard Young of the Art Department at Boise State University offered seemingly limitless help in administering grants for research and illustrations. Richard Morrison and the anonymous readers from the University of Minnesota Press gave important advice on shaping the argument of the book, and I am particularly indebted to "Reader number 2" for several close readings of the manuscript and astute insights about the intersections of artistic and scientific personae with new ways of studying and picturing insects. Jean Brady's expert copy editing greatly improved the manuscript.

I am grateful to Joel Floyd for sharing his expertise in entomology. I benefited from the diligent work of a number of research assistants over the years. Pomona College provided a research assistant, Lucy Meyer, and a faculty research grant from Boise State University funded another, Kimberly Cochrane, who did outstanding work on compiling bibliographic information. Jenaleigh Kiebert and Megan Pratt were tireless in the tasks of obtaining illustrations and permissions.

My thanks to Kathleen Keys and Jonathan Sadler for their friendly enthusiasm and support during the last stages of writing and revising. Joan and Richard Apel provided shelter, sustenance, and the wonderful chaos of their modern-day curiosity cabinet of a home during the lengthy writing of this book; I am also grateful for Richard's numerous readings of the manuscript and his assistance with Latin translations. My mother, Jane Neri, has always sustained me with her good humor and unconditional encouragement. Her unconventional outlook on life has in no small way inspired me to seek out the unusual and quirky aspects of my field.

There is no way that this book could have come into existence without the patience and sacrifice of Ted Apel. I dedicate this book to him and to our son, Abingdon, who napped sweetly in my lap while I revised, and revised, and revised again.

NOTES

Introduction

1. Early modern Europeans had a much more expansive understanding of what defined an insect than the criteria used by present-day entomologists; rather than a standard definition, the term functioned as a general category and included certain crustaceans as well as slugs and snails. I employ the term insect in the same broad manner, and for this reason I do not refer to this material as entomology or its practitioners as entomologists. Although insects have been studied since ancient times, entomology as a profession and an academic discipline dates only to the middle of the nineteenth century. For a discussion of these developments in relation to the study of ants, see Charlotte Sleigh, *Six Legs Better: A Cultural History of Myrmecology* (Baltimore: Johns Hopkins University Press, 2007), 3.

2. Yves Cambefort notes that the stag beetle appears in several drawings and paintings by Dürer. Yves Cambefort, "A Sacred Insect on the Margins: Emblematic Beetles in the Renaissance," in *Insect Poetics*, ed. Eric C. Brown (Minneapolis: University of Minnesota Press, 2006), 201. See also Fritz Koreny, *Albrecht Dürer and the Animal and Plant Studies of the Renaissance* (Boston: Little, Brown, 1988), 112–27.

3. On the Dürer Renaissance, see Koreny, *Albrecht Dürer and the Animal and Plant Studies of the Renaissance.*

4. These connections have been explored in a number of interdisciplinary collections of essays, including Londa Schiebinger and Claudia Swan, eds., *Colonial Botany: Science, Commerce, and Politics in the Early Modern World* (Philadelphia: University of Pennsylvania Press, 2005); Pamela H. Smith and Paula Findlen, eds., *Merchants and Marvels: Commerce, Science and Art in Early Modern Europe* (New York: Routledge, 2002); and Pamela H. Smith and Benjamin Schmidt, eds., *Making Knowledge in Early Modern Europe: Practices, Objects, and Texts, 1400–1800* (Chicago: University of Chicago Press, 2008).

5. Julie Hochstrasser, "The Conquest of Spice and the Dutch Colonial Imaginary: Seen and Unseen in the Visual Culture of Trade," in Schiebinger and Swan, eds., *Colonial Botany*. Similar points about still life and global trade are addressed in Elizabeth

Honig, "Making Sense of Things: On the Motives of Dutch Still Life," *Res* 34 (Fall 1998): 166–83. For a more extensive discussion of the material culture of Dutch still life, see Julie Hochstrasser, *Still Life and Trade in the Dutch Golden Age* (New Haven, Conn.: Yale University Press, 2007).

6. James Elkins has argued that art history should encompass such materials since "most images are not art." James Elkins, "Art History and Images That Are Not Art," *Art Bulletin* 77, no. 4 (1995): 533–71. For the importance of studying amateur art in art history, see Kim Sloan, *A Noble Art: Amateur Artists and Drawing Masters (c. 1600–1800)* (London: British Museum Press, 2000).

7. Cristopher Hollingsworth, *The Poetics of the Hive: The Insect Metaphor in Literature* (Iowa City: University of Iowa Press, 2001), ix.

8. Erika Olbricht, "'Made without Hands': The Representation of Labor in Early Modern Silkworm and Beekeeping Manuals," in Brown, ed., *Insect Poetics*.

9. Marc Olivier, "Through a Flea-Glass Darkly: Enlightened Entomologists and the Redemption of Aesthetics in Eighteenth-Century France," in Brown, ed., *Insect Poetics*.

10. Cambefort, "A Sacred Insect on the Margins." See also Eric C. Brown, "Insects, Colonies, and Idealization in the Early Americas," *Utopian Studies* 13, no. 2 (2002): 20–37.

11. Hollingsworth, *The Poetics of the Hive*; Juan Antonio Ramírez, *The Beehive Metaphor: From Gaudi to Le Corbusier* (London: Reaktion, 2000).

12. These visual processes of decontextualization parallel the construction of facts as argued in Lorraine J. Daston, "Baconian Facts, Academic Civility and the Prehistory of Objectivity," *Annals of Scholarship* 8 (1991): 337–63; and in Lorraine Daston and Katharine Park, eds., *Wonders and the Order of Nature, 1150–1750* (New York: Zone Books; distributed by MIT Press, 1998). On the construction of the fact in early modern Europe and its origins in the English legal system, see Barbara J. Shapiro, *A Culture of Fact: England, 1550–1720* (Ithaca: Cornell University Press, 2000).

13. Jan Swammerdam, for example, who made extensive use of the microscope to study insects, is not included here. The early microscopic images of bees made by the Accademia dei Lincei are also not discussed extensively. The political connotations and scientific content of the Lincei images have been astutely analyzed in David Freedberg, "The Microscope and the Vernacular," in *The Eye of the Lynx: Galileo, His Friends, and the Beginnings of Natural History* (Chicago: University of Chicago Press, 2002).

14. The essays in Hal Foster, ed., *Vision and Visuality* (San Francisco: Bay Press, 1988) were instrumental in establishing the idea of visuality. For a more recent discussion of the place of visuality within visual culture and its roots in imperial discourse, see Nicholas Mirzoeff, "On Visuality," *Journal of Visual Culture* 5, no. 1 (2006): 53–79. An extended analysis of the academic roots of visual culture and visual studies is available in Margaret Dikovitskaya, *Visual Culture: The Study of the Visual after the Cultural Turn* (Cambridge, Mass.: MIT Press, 2005).

15. Johnathan Crary, *Techniques of the Observer: On Vision and Modernity in the Nineteenth Century* (Cambridge, Mass.: MIT Press, 1990), 31.

16. The new (or renewed) attention to objects in the field of visual studies is discussed in Keith Moxey, "Visual Studies and the Iconic Turn," *Journal of Visual Culture* 7,

no. 2 (2008): 131–46. Examples of historical studies of objects influenced by this approach are found in Lorraine Daston, ed., *Things That Talk: Object Lessons from Art and Science* (New York: Zone Books; distributed by MIT Press, 2007).

17. Svetlana Alpers, *The Art of Describing: Dutch Art in the Seventeenth Century* (Chicago: University of Chicago Press, 1983). The historian of science David Topper has made similar arguments about extending the range of artifacts studied in order to better understand the processes of visualization that take place in scientific practice. David Topper, "Towards an Epistemology of Scientific Illustration," in *Picturing Knowledge: Historical and Philosophical Problems Concerning the Use of Art in Science*, ed. Brian S. Baigrie (Toronto: University of Toronto Press, 1996), 215–49.

18. See Freedberg, *Eye of the Lynx*; and Claudia Swan, *Art, Science, and Witchcraft in Early Modern Holland: Jacques de Gheyn II (1565–1629)* (Cambridge: Cambridge University Press, 2005). Freedberg showcases the project of the seventeenth-century Accademia dei Lincei and the many drawings they commissioned as part of their study of natural history, while Swan examines de Gheyn's intriguingly paradoxical images of nature and witches and argues that these images represent two different modes of thought, one based on science and the other on the imagination.

19. Pamela H. Smith, "Art, Science, and Visual Culture in Early Modern Europe," *Isis* 97 (2006): 84. Connections between knowledge, image making, and material practice have been examined in the context of images of the New World and from the point of view of art history by Michael Gaudio, *Engraving the Savage* (Minneapolis: University of Minnesota Press, 2008).

20. Foundational works that explore connections between art and science include Samuel Edgerton, *The Heritage of Giotto's Geometry: Art and Science on the Eve of the Scientific Revolution* (Ithaca: Cornell University Press, 1991); Brian Ford, *Images of Science: Art and Science on the Eve of the Scientific Revolution* (New York: Oxford University Press, 1993); Martin Kemp, *The Science of Art: Optical Themes in Western Art from Brunelleschi to Seurat* (New Haven, Conn.: Yale University Press, 1992); Erwin Panofsky, "Artist, Scientist, Genius: Notes on the 'Renaissance-Dämmerung,'" in *The Renaissance: Six Essays*, ed. Wallace Ferguson (New York: Harper and Row, 1953); Martin Rudwick, "The Emergence of a Visual Language for Geological Science, 1760–1830," *History of Science* 14 (1976): 149–95; Martin Rudwick, *Scenes from Deep Time: Early Pictorial Images of the Prehistoric World* (Chicago: University of Chicago Press, 1995); and John W. Shirley and David F. Hoeniger, eds., *Science and the Arts in the Renaissance* (Washington, D.C.: Folger Shakespeare Library, 1985).

21. Key early works on the history of botanical illustration are Agnes Arber, *Herbals: Their Origin and Evolution: A Chapter in the History of Botany* (1912; reprint Cambridge: Cambridge University Press, 1986); and Wilfrid Blunt and William Thomas Stearn, *The Art of Botanical Illustration* (London: Collins, 1950). For a discussion of issues of naturalism in botanical illustration, see Brian Ogilvie, "Illustrations in Renaissance Natural History," in *The Science of Describing: Natural History in Renaissance Europe* (Chicago: University of Chicago Press, 2006).

22. Clusius quoted in Ogilvie, *The Science of Describing*, 199–200.

23. Fuchs quoted in ibid., 195.

24. Robert Hooke, *Micrographia* (London, 1665), preface.

25. The contest for authority described in these accounts were part of the much broader construction of scientific personae and artistic identities in the early modern period. Some overlaps exist between these developments and the idea of self-fashioning described by Stephen Greenblatt, in *Renaissance Self-Fashioning: From More to Shakespeare* (Chicago: University of Chicago Press, 1983). On the construction of scientific personae, see Lorraine Daston and H. Otto Sibum, "Introduction: Scientific Personae and Their Histories," *Science in Context* 16 (2003): 1–8. For examples of the diverse venues in which contests over knowledge in the early modern period took place, see Antonio Barrera, "Local Herbs, Global Medicines: Commerce, Knowledge, and Commodities in Spanish America," in *Merchants and Marvels: Commerce, Science, and Art in Early Modern Europe,* ed. Pamela Smith and Paula Findlen (New York: Routledge, 2001); Lianne McTavish, *Childbirth and the Display of Authority in Early Modern France* (Burlington, Vt.: Ashgate, 2005); and Tara Nummedal, *Alchemy and Authority in the Holy Roman Empire* (Chicago: University of Chicago Press, 2007). The emergence of expertise and the trained expert in early modern England has been addressed by Eric H. Ash in *Power, Knowledge, and Expertise in Elizabethan England* (Baltimore: Johns Hopkins University Press, 2004); see also Eric H. Ash, "Queen v. Northumberland, and the Control of Technical Expertise," *History of Science* 39 (2001): 215–40. On the changing status of artists in the period, see Joanna Woods-Marsden, *Renaissance Self-Portraiture: The Visual Construction of Identity and the Social Status of the Artist* (New Haven, Conn.: Yale, 1998); and Thomas DaCosta Kaufmann, *Court, Cloister, and City: The Art and Culture of Central Europe, 1450–1800* (Chicago: University of Chicago Press, 1995).

26. This was a challenge for many early modern practitioners, no less for readers. See Ann Blair, "Reading Strategies for Coping with Information Overload, ca. 1550–1700," *Journal of the History of Ideas* 64, no. 1 (2003): 11–28.

27. Ogilvie, *The Science of Describing,* 194.

28. Numerous historians of science have addressed the problem of the "did they get it right?" approach by pointing out that this question does not take into account the motives, ideas, and circumstances of historical actors. A discussion of this problem in relation to historiographical issues can be found in Thomas Kuhn, "The History of Science," in *The Essential Tension: Selected Studies in Scientific Tradition and Change* (Chicago: University of Chicago Press, 1979). For a useful discussion of this question as it pertains to analyzing scientific illustrations, see Bert Hall, "The Didactic and the Elegant: Some Thoughts on Scientific and Technological Illustrations in the Middle Ages and Renaissance," in Baigrie, ed., *Picturing Knowledge.*

29. For further discussion of Foucault's epistemes as they relate to the study of nature, see Gary Gutting, *Michel Foucault's Archaeology of Scientific Reason* (Cambridge: Cambridge University Press, 1989). See also Ian MacLean, "Foucault's Renaissance Episteme Reassessed: An Aristotelian Counterblast," *Journal of the History of Ideas* 59, no. 1 (January 1998): 149–66.

30. Foucault does say more about the visual arts in other works. See Joseph A. Tanke, "Michel Foucault and Visual Culture: Toward a Genealogy of Modernity" (Ph.D. dissertation, Boston College, 2007).

31. Although some of Foucault's key examples of the Renaissance *episteme* do touch on material practices, as with the belief in the aconite plant as a treatment for ailments

of the eye due to the resemblance between the plant's seeds and the form of an eye. Gutting, *Michel Foucault's Archaeology of Scientific Reason*, 142.

32. Freedberg, *The Eye of the Lynx*, 166. For a fascinating discussion of these two broadsheets and the third, *Apes Dianiae*, see Freedberg, "The Chastity of Bees," in *The Eye of the Lynx*, 151–78.

33. Several collections of essays provide useful introductions to the diverse range of approaches to this topic, including Baigrie, ed., *Picturing Knowledge*; and Caroline A. Jones and Peter Galison, eds., *Picturing Science, Producing Art* (New York: Routledge, 1998).

1. Joris Hoefnagel's Imaginary Insects

1. Georg Bocskay and Joris Hoefnagel, *Mira calligraphiae monumenta*, The J. Paul Getty Museum, Los Angeles, Ms. 20. For an explanation of the circumstances of this manuscript's production, detailed analyses of its themes, and a biographical account of Joris Hoefnagel, see the essays by Lee Hendrix and Thea Vignau-Wilberg published in the facsimile edition *Mira calligraphiae monumenta: A Sixteenth-Century Calligraphic Manuscript Inscribed by Georg Bocskay and Illuminated by Joris Hoefnagel* (Los Angeles: J. Paul Getty Museum, 1992). Joris Hoefnagel, *Animalia Rationalia et Insecta (Ignis)*, National Gallery of Art, Washington, Gift of Mrs. Lessing J. Rosenwald, 1987.20.5.2. Single leaves from *Ignis* and other volumes of the *Four Elements* series are also found in collections in Berlin, Paris, Prague, and in private collections. For further details and a comprehensive analysis of the visual and textual sources and themes of the *Four Elements*, see Marjorie Lee Hendrix, "Joris Hoefnagel and the *Four Elements*: A Study in Sixteenth-Century Nature Painting" (Ph.D. dissertation, Princeton University, 1984). Jacob Hoefnagel and Joris Hoefnagel, *Archetypa studiaque patris Georgii Hoefnagelii . . .* (Frankfurt, 1592).

2. Paula Findlen, "Jokes of Nature and Jokes of Knowledge: The Playfulness of Scientific Discourse in Early Modern Europe," *Renaissance Quarterly* 43, no. 2 (1990): 292.

3. Michel Foucault, *The Order of Things: An Archaeology of the Human Sciences* (New York: Vintage Books, 1994), 132.

4. Ibid, 137.

5. Further explication of Foucault's ideas about natural history can be found in Gutting, *Michel Foucault's Archaeology of Scientific Reason*, 162–69. See also Dirk Stemerding, *Plants, Animals and Formulae: Natural History in the Light of Latour's Science in Action and Foucault's "The Order of Things"* (Enschede, Netherlands: University of Twente, 1991).

6. See the introduction for more discussion of Foucault's ideas from *The Order of Things* in relation to the broader themes and issues of this book.

7. The standard work on the Dürer Renaissance is Koreny, *Albrecht Dürer and the Animal and Plant Studies of the Renaissance.*

8. Museum of Fine Arts, Department of Prints and Drawings, Budapest, Inv. 184.

9. See, for example, the many stag beetles in still life paintings by Georg Flegel (1566–1638).

10. Hendrix, "Joris Hoefnagel and the *Four Elements*," chapter 1.

11. Lee Hendrix, "Of Hirsutes and Insects: Joris Hoefnagel and the Art of the Wondrous," *Word & Image* 11, no. 4 (1995): 385.

12. For a summary of the arguments for and against Dürer's authorship of the *Stag Beetle* drawing, see Koreny, *Albrecht Dürer and the Animal and Plant Studies of the Renaissance*, 120–21.

13. A useful discussion of this issue can be found in Hall, "The Didactic and the Elegant."

14. The open-wing stag beetle also appears in plate 6 of part 1 of the *Archetypa* series. Koreny discusses copies made by other artists, and he has determined that Hoefnagel's version of Dürer's *Stag Beetle* in *Ignis* was originally labeled with the Roman numeral "VI," and thus the two drawings originally appeared one after the other. Koreny also writes that Hoefnagel "has unmistakably expressed the Dürer connection he intended by placing his artistic flying beetle immediately after the Dürer crawling one." Although Koreny describes Hoefnagel's open-wing stag beetle as a flying beetle, the shadows under the insect's legs indicate that it rests upon a surface and is not depicted in flight. Koreny, *Albrecht Dürer and the Animal and Plant Studies of the Renaissance*, 126.

15. Thomas DaCosta Kaufmann and Virginia Roehrig Kaufmann have examined the relationship between Hoefnagel's illustrations and the tradition of naturalistic illumination in late fifteenth- and sixteenth-century Flemish book painting. They demonstrate the links between the development of trompe l'oeil illusionism in illuminated books of hours and the function of these books as religious objects in private devotional settings. In arguing that the objects represented in the borders of illuminated manuscripts should be understood in relation to actual objects collected by travelers to pilgrimage sites, they make crucial points about the relationship between image-making practices such as Hoefnagel's and the material culture of early modern Europe. See Thomas DaCosta Kaufmann and Virginia Roehrig Kaufmann, "The Sanctification of Nature: Observations on the Origins of Trompe l'oeil Netherlandish Book Painting of the Fifteenth and Sixteenth Centuries," in Thomas DaCosta Kaufmann, *The Mastery of Nature: Aspects of Art, Science, and Humanism in the Renaissance* (Princeton, N.J.: Princeton University Press, 1993).

16. These conventions for representing butterflies and moths were widely used until the late seventeenth and eighteenth centuries, when artists became interested in representing the positions of insects' bodies in flight.

17. Hendrix, "The Writing Model Book," in Hendrix and Vignau-Wilberg, *Mira calligraphiae monumenta*, 42. The illustrations in the Model Book and the *Archetypa* engravings, which predate the Model Book, are both believed to be based on a lost notebook of drawings kept by Hoefnagel that he used as models for compositions throughout his career. For a further discussion of the correlations between the *Archetypa* illustrations and those of the Model Book, see Thea Vignau-Wilberg, *Archetypa Studiaque Patris Georgii Hoefnagelii, 1592: Natur, Dichtung und Wissenschaft in der Kunst um 1600* (Munich: Staatliche Graphische Sammlung, 1994), 34.

18. This analysis is based on identifications provided in the facsimile edition of the manuscript (Hendrix and Vignau-Wilberg, *Mira calligraphiae monumenta*).

19. My understanding of these aspects of Hoefnagel's imagery owes much to Amy Meyers's discussion of the visual language of reflected form in relation to the eighteenth-century artist-naturalist Mark Catesby's images of New World flora and fauna. See Amy

R. W. Meyers, "Picturing a World in Flux: Mark Catesby's Response to Environmental Interchange and Colonial Expansion," in *Empire's Nature: Mark Catesby's New World Vision,* ed. Amy R. W. Meyers and Margaret Beck Pritchard (Chapel Hill: Published for the Omohundro Institute of Early American History and Culture by the University of North Carolina Press, 1998), esp. 230–38.

20. Again, this analysis is based on the identifications provided in the facsimile edition of the manuscript (Hendrix and Vignau-Wilberg, *Mira calligraphiae monumenta*).

21. Hendrix and Vignau-Wilberg, *Mira calligraphiae monumenta,* 82.

22. See also the damselflies pictured at the top of folio 99. Hendrix and Vignau-Wilberg, *Mira calligraphiae monumenta,* 250.

23. Thomas DaCosta Kaufmann, "Metamorphoses of Nature: Arcimboldo's Imperial Allegories," in Kaufmann, *The Mastery of Nature,* 100–28.

24. Hendrix, "*Mira calligraphiae monumenta*: An Overview," in Hendrix and Vignau-Wilberg, *Mira calligraphiae monumenta,* 4. For further details about the Constructed Alphabet and its contents, see Thea Vignau-Wilberg, "The Constructed Alphabet," in Hendrix and Vignau-Wilberg, *Mira calligraphiae monumenta,* 315–18.

25. Vignau-Wilberg, "The Constructed Alphabet," in Hendrix and Vignau-Wilberg, *Mira calligraphiae monumenta,* 318.

26. As, for example, in folio 53, where three plant stems recede back into a space that also serves as a flat surface on which a scorpion and a millipede confront one another, and on which rests a hazelnut plant.

27. Hendrix links this feature of Hoefnagel's imagery to the artist's favoring of vision over touch as a means of gaining knowledge of the natural world. "Notably, Hoefnagel's images do not evince an interest in texture, neither in those of diverse specimens nor in the workings of paint itself. Uniformly smooth and meticulously executed, inflated rather than sculpted in appearance, they suggest that vision, not touch, plays the critical role in gaining knowledge of nature. This process is not one of passively recording nature's surfaces, however; rather, vision acts aggressively on nature, prying into its recesses." Hendrix, "The Writing Model Book," in Hendrix and Vignau-Wilberg, *Mira calligraphiae monumenta,* 50.

28. White's one other insect drawing, that of a large swallowtail butterfly (also discussed in chapter 2), utilizes the flat view as well, thereby emphasizing the brilliant colors and bold patterns of the insect's wings. For details of White's life and the complex history of his drawings, see Paul Hulton, *America, 1585: The Complete Drawings of John White* (Chapel Hill: University of North Carolina Press; London: British Museum Publications, 1984); and Kim Sloan, *A New World: England's First View of America* (Chapel Hill: University of North Carolina Press, 2007). For an analysis of White's images in relation to the medium of engraving, see Gaudio, *Engraving the Savage.*

29. *Histoire naturelle des Indes,* The Pierpont Morgan Library, New York, folio 72.

30. For details about the manuscript, see the introduction by Verlyn Klinkenborg in the facsimile edition *Histoire Naturelle des Indes: The Drake Manuscript in the Pierpont Morgan Library* (New York: W. W. Norton, 1996).

31. The Drake Manuscript also contains an illustration of a firefly (folio 78), and like White's drawing the image contains no information about the insect's light-producing qualities and conveys little information about the insect's appearance. Indeed, the

creature portrayed by the artist bears some resemblance to a four-legged snail possessing wings in place of a shell.

32. Claudia Swan makes similar points with regard to the role of vision and visual images in the classification methods of the botanist Carolus Clusius. For Clusius, the characteristics most crucial to the classification of plants and animals were "those that can be observed in the immediate presence of the specimen, or . . . those that can be recorded pictorially." Clusius's analysis was dependent upon, and thus limited to, "visually apprehensible information." The examples by White and the artist of the Drake Manuscript discussed above show the difficulties that could arise when attempting to transfer a mode of representation developed to suit the needs and interests of a particular audience into a different context. However, Swan's discussion of Clusius also shows that such issues of translation were not always a problem for early modern European users and makers of naturalistic images. Although Clusius's use of botanical images was driven by taxonomic concerns, Swan points out that the images he relied on were not any different than those used by other botanists, such as Dodonaeus and Lobelius, whose concerns were directed toward more utilitarian, pharmaceutical purposes. See Claudia Swan, "From Blowfish to Flower Still Life Paintings: Classification and Its Images, Circa 1600," in *Merchants and Marvels: Commerce, Science, and Art in Early Modern Europe*, ed. Pamela H. Smith and Paula Findlen (New York: Routledge, 2002), 120–21.

2. Cutting and Pasting Nature into Print

1. Referred to hereafter as *Theatrum insectorum*.

2. Paula Findlen, *Possessing Nature: Museums, Collecting, and Scientific Culture in Early Modern Italy* (Berkeley: University of California Press, 1994), 245. The discussion above of Aldrovandi's teaching and medical background is based on Findlen's volume (pp. 243–57). General information on medical practice during this period can be found in Nancy G. Siraisi, *Medieval and Early Renaissance Medicine: An Introduction to Knowledge and Practice* (Chicago: University of Chicago Press, 1990). For a detailed discussion of Dioscorides and the *materia medica*, see John M. Riddle, *Dioscorides on Pharmacy and Medicine* (Austin: University of Texas Press, 1985).

3. Allen G. Debus, *The Chemical Philosophy: Paracelsian Science and Medicine in the Sixteenth and Seventeenth Centuries*, vol. 1 (New York: Science History Publications, 1977), 51.

4. Ibid., 55.

5. Allen G. Debus, *The English Paracelsians* (New York: F. Watts, 1966), 73.

6. Ibid., 80.

7. Georg Urdang, *Pharmacopoeia Londinensis of 1618* (Madison: State Historical Society of Wisconsin, 1944), 16–17. The initial project in the 1580s was not pursued, but the pharmacopoeia was completed in 1618. Theodor de Turquet de Mayerne was instrumental in its publication, as he was for Moffet's *Theatrum insectorum*. Mayerne is discussed in further detail below.

8. George Baker, *The newe jewell of health* (London: Henrie Denham, 1576), 174.

9. Quoted in Debus, *The English Paracelsians*, 72.

10. Moffet completed work on this book between 1595 and 1599 but, like the *Theatrum insectorum*, it was not published until well after his death. Raven cites several

sources supporting the idea that the chapters on birds were Moffet's work alone, but Hoeniger believes the bird material is drawn mostly from Gessner and William Turner. See David Hoeniger, *The Growth of Natural History in Stuart England: From Gerard to the Royal Society* (Charlottesville: Published for the Folger Shakespeare Library by the University Press of Virginia, 1969), 16; and Charles E. Raven, *English Naturalists from Neckam to Ray: A Study of the Making of the Modern World* (Cambridge: Cambridge University Press, 1947), 177.

11. The literature on Aldrovandi is extensive, but for treatments of his museum and collecting activities within a social and cultural context, see in particular Findlen, *Possessing Nature*; and Giuseppe Olmi and Paolo Prodi, "Art, Science, and Nature in Bologna Circa 1600," in *The Age of Correggio and the Carracci: Emilian Painting in the Sixteenth and Seventeenth Centuries* (Washington, D.C.: National Gallery of Art, 1986). Aldrovandi's illustration projects are discussed extensively in Giuseppe Olmi, *L'inventario del mondo: Catalogazione della natura e luoghi del sapere nella prima età moderna* (Bologna: Il Mulino, 1992); and Giuseppe Olmi, "Osservazioni della natura e raffigurazione in Ulisse Aldrovandi (1522–1605)," *Annali dell'Istituto Italo-Germanico in Trento* 3 (1977): 105–81.

12. Findlen, *Possessing Nature*, 3.

13. See ibid., 167–70; and Lucia Tongiorgi-Tomasi, "Gherardo Cibo: Visions of Landscape and the Botanical Sciences in a Sixteenth-Century Artist," *Journal of Garden History* 9 (1989): 199–216.

14. Olmi, *L'inventario del mondo*. Approximately three thousand drawings have survived and are now held at the Biblioteca Universitaria in Bologna. For short descriptions of the contents of the seven volumes on animals, the ten volumes of plant drawings, and the one volume of miscellaneous animal and plant drawings, see Lodovico Frati, *Catalogo dei manoscritti di Ulisse Aldrovandi* (Bologna: N. Zanichelli, 1907), 209–10. The remainder of Aldrovandi's natural history collections and the woodblock plates for the illustrations of his books are also housed at the University of Bologna.

15. The insect drawings are grouped together in the middle section of the volume (pages 70–103), with smaller sections of insect drawings dispersed throughout (pages 14–19, 105, 108, 110, 115–19, 122–23, 127).

16. Aldrovandi compiled a partial concordance between the insect drawings in Tomo VII and the woodcuts in *De animalibus insectis* in a document titled "Index insectorum in tabulis depictorum Ao. 1593," Biblioteca Universitaria, Bologna, Ms 136, XVIII c. 65–129. The contents of this document will be discussed in further detail later in this chapter.

17. In his notebooks, Aldrovandi records a catalogue of plants in the garden of Stephanus Springer of Heidelberg sent to him by Schwindt. Biblioteca Universitaria, Bologna, Ms. 136, XXXI, c. 73–77r, recorded in Frati, *Catalogo dei manoscritti di Ulisse Aldrovandi*, 169. For more on Schwindt (who is sometimes referred to as Cornelio) along with a transcription of Aldrovandi's records of payments to the artist, see Stefano de Rosa, "Cornelius Schwindt da Francoforte (1566–1632): 'Pictor et Designator' a Servizio di Ulisse Aldrovandi," *Mitteilungen des Kunsthistorischen Institutes in Florenz* 29 (1985): 401–9.

18. Aldrovandi writes, "Capta fuere hac inserta a fratre Gregorio Capuccino sub finem augusti ao. 1592." Biblioteca Universitaria, Bologna, Ms 136, XVIII c. 65.

19. See, for example, the hand-colored copy of *De animalibus insectis* (Biblioteca Universitaria, Bologna, A.IV.H.III.2) that Aldrovandi presented to the city of Bologna. Claudia Swan discusses the use of colored drawings and watercolors in the context of the study of botany in early modern Europe in her volume *The Clutius Botanical Watercolors: Plants and Flowers of the Renaissance* (New York: Harry N. Abrams, 1998).

20. There is very little information about the organization and layout of Aldrovandi's collections, though he did draw up a plan for a new museum in 1603. This plan called for the expansion of his museum from one room in his home to a gallery composed of four separate rooms in the Palazzo Publico in Bologna. Paula Findlen writes that Aldrovandi "placed the fossils, stones, gems, minerals, and seeds that filled the sixty-six boxes of his specimen cabinets, subdivided into 7000 compartments, in the last room of his imagined museum." For details and a diagram of Aldrovandi's proposed museum, see Findlen, *Possessing Nature*, 122–23.

21. The image can be viewed in the "Tavole acquarellate di Ulisse Aldrovandi" section of the "archivo on line," at www.filosofia.unibo.it/aldrovandi/.

22. Aldrovandi includes a table describing these divisions in the introductory pages of *De insectis*.

23. The silkworm images are discussed further below, in connection with Moffet's appropriation of them in his book.

24. Biblioteca Universitaria, Bologna, Ms 136, XVIII c. 151.

25. Aldrovandi noted these locations by assigning numbers to the insects. For example, the entry "Tomo Insect. f 76, no. 4 & 5. In Tabula no. 1 & 2" refers to the insects pictured on page 76 of the "Tomo insectorum" (the volume of insect drawings), in positions number 4 and 5. These same insects appear in the "Tabula" (the woodcut) in positions 1 and 2. The numbers were assigned from left to right, continuously across rows.

26. Paula Findlen, "The Museum: Its Classical Etymology and Renaissance Genealogy," *Journal of the History of Collecting* 1 (1989): 64.

27. The history of the *Theatrum insectorum* is recounted in a number of sources. The most comprehensive account, along with detailed biographies of Penny and Moffet, can be found in Raven, *English Naturalists from Neckam to Ray*. For details on Moffet's patrons and a reproduction of the 1590 title page, see A. Stuart Mason, "Little Miss Muffet's Father," *Journal of the Royal College of Physicians of London* 27, no. 3 (1993): 322–24. Concise accounts of the history of the *Theatrum insectorum* can also be found in Philip H. Swann, "Thomas Mouffet's *Theatrum Insectorum*, 1634," *Bulletin of the British Arachnological Society* 2, no. 8 (1973): 169–73. The history is also recounted in Willy Ley's introduction to the facsimile edition of the 1658 edition of the *Theatrum insectorum* included in Edward Topsell, *The History of Four-Footed Beasts and Serpents and Insects* (New York: Da Capo Press, 1967).

28. Raven speculates that Moffet's parents may have known Penny before he left England in 1565. Raven, *English Naturalists from Neckam to Ray*, 175.

29. Gessner's *Historia plantarum* did not appear in print until 1753, when it was published in Nuremberg by Christopher Trew and C. C. Schmiedel, who had obtained Gessner's collection of plant drawings. See Arber, *Herbals*, 111–13; and Raven, *English Naturalists from Neckam to Ray*, 155–57. Raven notes that additional botanical notes

by Penny appear in a later publication by Schmiedel from 1759 to 1760, which also contained selections from Gessner's *Historia plantarum*.

30. Raven, *English Naturalists from Neckam to Ray,* 156–57.

31. For details of Penny's botanical exchanges with these naturalists and others, see ibid., 155–64.

32. Ibid., 161.

33. Thomas Moffet, *Theater of Insects* (London: 1658; New York: Da Capo Press, 1967), 1009.

34. Ibid., 1008–9.

35. British Library, Sloane MS 4014.

36. Moffet, *Theater of Insects,* 1015. The "Quickelbergius" mentioned here by Moffet most likely refers to either Jacob or Pieter Quiccheberg, the sons of Samuel Quiccheberg (1529–1567), who was the author of the treatise on the organization of an ideal museum. Samuel Quiccheberg, *Inscriptiones Vel Tituli. Theatri Aplissimi, Complectentis Rerum Universitatis Singulas Materias Et Imagines Eximias . . .* (Munich, 1565).

37. Moffet, *Theater of Insects,* 959.

38. Ibid., 985.

39. Ibid., 1044.

40. For details on the Knyvets, and a description of their library, see Raven, *English Naturalists from Neckam to Ray,* 173–75.

41. Moffet, *Theater of Insects,* 948.

42. On the fly with silver wings, see ibid., 936, and on the horned beetle, see ibid., 1008.

43. On the report on scorpions, see ibid., 1052, and on the fly from Barber, see ibid., 934. The latter could be a reference to Thomas Barbar, a fellow of Trinity College, Cambridge, and a preacher active in London in the 1570s and 1580s.

44. On the earwigs, see Moffet, *Theater of Insects,* 1023. The water beetle drawing is found in Sloane MS 4014, p. 309v. Several other physicians are mentioned in the *Theatrum insectorum*. Dr. Roger Brown and Dr. John James are mentioned briefly in the preface; according to Raven the latter served as physician to Queen Elizabeth. Thomas Penny was a close friend of Peter Turner, the son of the eminent physician and natural historian William Turner. Raven, *English Naturalists from Neckam to Ray,* 173. Moffet makes several mentions of his former teachers at Basel, Felix Platter and Theodore Zwinger, although it does not seem that he maintained an active correspondence with them.

45. Moffet, *Theater of Insects,* 990.

46. Ibid., 1045.

47. On the cricket, see Moffet, *Theater of Insects,* 995, and on the insects that breed in stone, see ibid., 1081.

48. Ibid., 1060.

49. On the beetle, see ibid., 1014; on the caterpillars, see ibid., 1037; and on the first worm, see ibid., 1045, and the second 1046.

50. Ibid., 979.

51. Ibid., 955.

52. British Library, Sloane MS 4014, p. 103. Brewer is also recorded as having worked with Penny in Heidelberg, presumably sometime between 1565 and 1569 while

Penny was traveling in Europe. In Heidelberg, Brewer observed a goat beetle and conducted an experiment on a salamander with Penny, in which they tested the theory that salamanders live and breed in extremely hot places by placing one in a fire. The salamander perished, thus disproving the theory. Moffet, *Theater of Insects*, 1006 and 1020.

53. Moffet, *Theater of Insects*, preface.

54. Raven, *English Naturalists from Neckam to Ray*, 189. Philip Swann's slightly more recent assessment of Moffet is equally unforgiving; in referring to the passage in which Moffet alludes to the "tautologies and trivialities" he eliminated, Swann writes: "In so doing he obscured or destroyed a great deal of the work of Penny, the first great English entomologist." Swann, "Thomas Mouffet's Theatrum Insectorum, 1634."

55. Frances Dawbarn, "New Light on Dr Thomas Moffet: The Triple Roles of an Early Modern Physician, Client, and Patronage Broker," *Medical History* 47 (2003): 3–22.

56. The 1634 title page is based on a gouache drawing of a beehive and several images of individual insects pictured elsewhere in Moffet's manuscript; the 1658 title page differs slightly in that it contains an altered arrangement of insects, but for the most part it is based on the 1634 version. Mayerne makes no mention of the title page in his preface, but the reason for the change may have been that the plate for the original title page was lost by the time Mayerne acquired the manuscript.

57. The two vases come from the title pages of parts 3 and 4 of the *Archetypa*.

58. See the title pages of parts 1 and 2 (the grasshopper's form has been reversed).

59. W. J. Holland, "The First Picture of an American Butterfly," *Scientific Monthly* 29 (July 1929): 45. Holland proposes that the inscription on the drawing was written by Edward Wotton, the English naturalist whose book on insects was quoted extensively by Moffet, and whose portrait appears in the engraved title page for the *Theatrum insectorum*. However, Wotton could not have been the author of the inscription since he died in 1555. Charles Raven suggests that Penny, rather than Wotton, was the author of the inscription. The inscription was almost certainly written by Penny, who must have obtained the drawing shortly before his death. Holland seems to have confused the Edward Wotton whose portrait appears in the original title page of the *Theatrum insectorum* with a later Edward Wotton who died in 1626. The later Wotton does not seem to have had an interest in insects, and neither was he mentioned in the text by Moffet. See Charles E. Raven, "John White's Significance for Natural History," in *The American Drawings of John White*, ed. Paul Hulton and David Beers Quinn (Chapel Hill: University of North Carolina Press, 1964). White's drawings have a long and complicated history, the details of which are outlined in Paul Hulton, *America, 1585: The Complete Drawings of John White* (Chapel Hill: University of North Carolina Press; London: British Museum Publications, 1984). For a discussion of other drawings of insects by White, see chapter 1.

60. British Library, Sloane MS 4014, fol. 96.

61. Deborah E. Harkness, *The Jewel House: Elizabethan London and the Scientific Revolution* (New Haven, Conn.: Yale University Press, 2007), 27.

62. Harkness discusses this issue in detail in "Living on Lime Street," in *The Jewel House*, 15–56.

63. Dawbarn, "New Light on Dr Thomas Moffet," 9.

64. Extensive treatment of Mayerne is found in H. R Trevor-Roper, *Europe's Physician: The Various Life of Sir Theodore De Mayerne* (New Haven, Conn.: Yale University Press, 2006).

65. Hoeniger, *The Growth of Natural History in Stuart England*, 17.

66. Théodore Turquet de Mayerne et al., *Lost Secrets of Flemish Painting: Including the First Complete English Translation of the De Mayerne Manuscript, B.M. Sloane 2052* (Hillsville, Va.: Alchemist, 2001).

67. Ibid., 210–11.

68. Moffet, *Theater of Insects,* preface.

69. He is clear about this in Thomas Moffet, *The Silkewormes, and Their Flies* (London: V.S. for Nicholas Ling, 1599).

70. British Library, shelfmark 444f.3.

71. Dubois's natural history notebook is reproduced in facsimile. Charles DuBois, "Notes on Insects, 1692 & 1695," *Bulletin of the British Museum (Natural History),* Historical Series 17, no. 1 (1989).

72. Thomas Moffet, *Theater of Insects* (London, 1658). British Library, shelfmark 444f.3, p. 97.

73. These occur on pages 94 and 103.

74. The Moffet manuscript came to the British Library from Hans Sloane, who acquired Courteen's collection after his death.

75. Moffet, *Theater of Insects,* British Library, shelfmark 444f.3, p. 99.

76. Marshal's other known botanical paintings, thirty-three sheets of floral arrangements, are in the collection of the British Museum, Department of Prints and Drawings. For reproductions of the flower album, biographical details on Marshal, and a listing of all of the artist's known works with their locations, see Prudence Leith-Ross and Henrietta McBurney, *The Florilegium of Alexander Marshal in the Collection of Her Majesty the Queen at Windsor Castle* (London: Royal Collection, 2000). Selected reproductions from Marshal's flower album can also be found in John Fisher, *Mr. Marshal's Flower Album from the Royal Library at Windsor Castle* (London: V. Gollancz, 1985).

77. Leith-Ross and McBurney, *The Florilegium of Alexander Marshal,* 19.

78. Alexander Marshal, Marshal Album, Philadelphia, Academy of Natural Sciences, Coll. 941.

79. The catalogue was written by Dr. William Freind (1715–1766), who inherited many of Marshal's drawings and written materials. Freind's catalogue is transcribed in Leith-Ross and McBurney, *The Florilegium of Alexander Marshal,* 364–68.

80. Freind catalogue in ibid., 367.

81. Marshal's network of contacts are further discussed in James A. Mears, "An Analysis of Information Preserved in a Recently Identified Collection of Insect Drawings by Alexander Marshall [sic] (1639?–1682)," in *History in the Service of Systematics: Papers from the Conference to Celebrate the Centenary of the British Museum (Natural History) 3–16 April 1981,* ed. Alwyne C. Wheeler and J. H. Price (London: Society for the Bibliography of Natural History, 1981).

82. Freind catalogue, quoted in Leith-Ross and McBurney, *The Florilegium of Alexander Marshal,* 367.

83. "This locust was given me by one Captaine Stains whoe brought it from hispagnola, on that expedition that was sent by Cromwell." Alexander Marshal, Marshal Album, Philadelphia, Academy of Natural Sciences, Coll. 941, p. 25. It is possible that the note refers to another locust, as it is not written on the same paper as the drawing.

84. Moffet, *Theater of Insects*, 983.

85. Freind catalogue, in Leith-Ross and McBurney, *The Florilegium of Alexander Marshal*, 367.

86. Ulisse Aldrovandi, *De animalibus insectis libri septem* (Bologna, 1602), preface. Translation my own, with assistance from Richard Apel.

87. Harkness, *The Jewel House*, 44.

3. Suitable for Framing

1. For a related discussion of early still life painting within the context of sixteenth-century collecting practices, see Thomas DaCosta Kaufmann and Virginia Roehrig Kaufmann's argument in "The Sanctification of Nature" that the origins of still life lie in the collecting practices of religious pilgrims and their use of books of hours and other illuminated manuscripts as repositories for coins, medals, and plants.

2. Most of the literature on still life painting in early modern Europe focuses on seventeenth-century Dutch still life. For treatments of the early formative period of the late sixteenth and early seventeenth centuries, and the influence of Flemish painters on the development of the genre of European still life, see Laurens J. Bol, *The Bosschaert Dynasty: Painters of Flowers and Fruit* (Leigh-on-Sea, U.K.: F. Lewis, 1960); Alan Chong and Wouter Kloek, eds., *Still-Life Paintings from the Netherlands, 1550–1720* (Zwolle, Netherlands: Waanders Uitgevers, 1999); Wilfried Seipel, *Das flämische Stillleben, 1550–1680: Eine Ausstellung des Kunsthistorischen Museums Wien und der Kulturstiftung Ruhr Essen* (Lingen, Germany: Luca, 2002); Ger Luijten, *Dawn of the Golden Age: Northern Netherlandish Art, 1580–1620* (Amsterdam: Rijksmuseum Waanders; New Haven, Conn.: Yale University Press, 1993); and Peter Mitchell, *European Flower Painters* (London: A. and C. Black, 1973). For an overview of historiographical issues in Netherlandish art that includes useful discussions of still life, see also Mariet Westermann, "After Iconography and Iconoclasm: Current Research in Netherlandish Art, 1566–1700," *Art Bulletin* 84, no. 2 (2002): 351–72.

3. The moth is possibly a great tiger moth or garden tiger moth *(Arctia caja)*, both of which are common species in Belgium.

4. The stag beetle images by Dürer and other artists have also been interpreted as a symbol of Christ: see Koreny, *Albrecht Dürer and the Animal and Plant Studies of the Renaissance*, 112; and Hope B. Werness, *The Continuum Encyclopedia of Animal Symbolism in Art* (New York: Continuum, 2004), 392. An example of an interpretation of floral symbolism as connected to Christ is Robert Koch, "Flower Symbolism in the Portinari Altar," *Art Bulletin* 46, no. 1 (1964): 70–77.

5. F. Lugt Collection, Paris. Bosschaert may have modeled his dead frog on a similar image of a dead frog published in Hoefnagel's *Archetypa* (part 2, plate 5).

6. Joris Hoefnagel may have maintained a personal collection of insects, from which he could have selected the dragonfly wings affixed to plate 54 of the *Ignis* volume

from his *Four Elements* series of illuminated manuscripts. For further discussion of Hoefnagel, see chapter 1, and for insects in natural history collections, see chapter 2.

7. S. Peter Dance, *A History of Shell Collecting* (Leiden, Netherlands: Brill, 1986), 33.

8. J.-A. Goris and G. Marlier, eds., *Albrecht Dürer: Diary of His Journey to the Netherlands, 1520–1521* (London: Lund Humphries; Greenwich, Conn.: New York Graphic Society, 1971), 79.

9. Ibid., 58, 83, 84.

10. Pierre Belon's *L'histoire naturelle des estrange poissons . . .* (Lyons, 1551) was the earliest printed book devoted to fish, although it concentrates mostly on unusual or rare nonfish species such as dolphins, hippos, and nautilus shells. Belon's second book on fish, *De aquatilibus* (Lyons, 1553), expanded on his earlier work by including a wider range of species, and it appeared in ten editions before 1620. Around the same time that Belon was publishing his books, another French naturalist, Guillaume Rondelet, published a book on fish, *Libri de piscibus marinis* (Lyons, 1554), and Ulisse Aldrovandi was at work preparing *De piscibus*. Aldrovandi treated shells in a separate volume, *De reliquis animalibus*, which would be published posthumously in 1606. Further discussion of these books is found in E. W. Gudger, "The Five Great Naturalists of the Sixteenth Century: Belon, Rondelet, Salviani, Gessner and Aldrovandi. A Chapter in the History of Ichthyology," *Isis* 22 (1934): 21–40.

11. The undated painting by Van der Ast is illustrated in Bol, *The Bosschaert Dynasty*, plate 36.

12. Guillaume Rondelet, *L'histoire entiere des poissons* (Lyons, 1558), 16, 34. This was the later French edition of Rondelet's 1554 book.

13. Bocskay and Hoefnagel, *Mira calligraphiae monumenta*, fol. 151v. Mussel shells were used by painters as far back as the medieval period for mixing and storing pigments as well as for mixing and storing gold paint, which for this reason is sometimes referred to as "shell gold."

14. Shells appear throughout the pages of the *Mira calligraphiae monumenta*, and Hoefnagel depicts another mussel shell on folio 37. See chapter 1 for an extended discussion of the insects in this manuscript.

15. "It is in light of the natural world as a collection of valuable documents that conceal and reveal the handwriting of the invisible Creator that collecting shells becomes a legitimate source of action for the attainment of wisdom. In fact, gathering shells proved to be a major argument in the defence of humanist piety as a cultural practice and a way of life." Leopoldine van Hogendorp Prosperetti, "'Conchas Legere': Shells as Trophies of Repose in Northern European Humanism," *Art History* 29, no. 3 (2006): 395.

16. Shell identified by Bol, *The Bosschaert Dynasty*, 71. Bol notes that the same shell appears in another painting by Van der Ast in the Boymans Museum.

17. My thanks to Joel Floyd for identifying this insect.

18. Aldrovandi, *Miscellanea di animali e piante depinte, Tavole di animali*, Tomo VII. This volume also includes drawings of frogs, fish, and other subjects. See chapter 2 for an extended discussion of the insect drawings in Tomo VII.

19. De Gheyn the Elder, *Vanitas Still Life* (1603), New York, Metropolitan Museum of Art. In De Gheyn's painting numerous gold and silver coins lie upon a stone ledge in the foreground, along with two medals in the lower corners of the painting depicting

Charles V and his mother Joanna of Aragon and Castile. (Identification of medals from Heilbrunn Timeline of Art History, The Metropolitan Museum of Art, New York, http://www.metmuseum.org/toah/hd/nstl/hod_1974.1.htm [accessed October 2006]).

20. The painting described by Brueghel in this letter is believed to be the painting of a large bouquet of flowers in the collection of the Pinacoteca Ambrosiana (Milan) that features a small collection of shells in the lower-right corner along with a piece of jewelry and coins.

21. Prosperetti, "'Conchas Legere,'" 397. See also Pamela M. Jones, *Federico Borromeo and the Ambrosiana: Art Patronage and Reform in Seventeenth-Century Milan* (Cambridge: Cambridge University Press, 1992).

22. Martha McCrory, "Coins at the Courts of Innsbruck and Florence: The Numismatic Cabinets of Archduke Ferdinand II of Tyrol and Grand Duke Francesco I De'Medici," *Journal of the History of Collections* 6, no. 2 (1994): 153–72. McCrory provides much useful information about the contents and methods of displaying these coin collections.

23. John Cunnally, "Ancient Coins as Gifts and Tokens of Friendship during the Renaissance," *Journal of the History of Collections* 6, no. 2 (1994): 129.

24. John Cunnally, *Images of the Illustrious: The Numismatic Presence in the Renaissance* (Princeton, N.J.: Princeton University Press, 1999), 3.

25. John Cunnally, "Ancient Coins as Gifts and Tokens of Friendship during the Renaissance," 129.

26. Obadiah Walker, *The Greek and Roman History Illustrated by Coins and Medals* (London, 1692).

27. John Cunnally, drawing on ideas from Francis Haskell and Susan Sontag, argues that practices associated with collecting coins encouraged an approach to history as composed of a series of pictures or events. "The original models for this kind of classification and storage of visual data were the coin cabinets of the Renaissance and their literary counterpart, the illustrated numismatic book." Cunnally, *Images of the Illustrious,* 14.

28. McCrory, "Coins at the Courts of Innsbruck and Florence," 162.

29. Glyn Davies and Julian Hodge Bank, *A History of Money: From Ancient Times to the Present Day,* 3rd ed. (Cardiff: University of Wales Press, 2002), 249–51.

30. Thomas J. Sargent and François R. Velde, *The Big Problem of Small Change* (Princeton, N.J.: Princeton University Press, 2002), 221.

31. Insects and coins also appear together in a still life painting of circa 1620 by Clara Peeters in the Ashmolean Museum.

32. Thomas DaCosta Kaufmann, "From Mastery of the World to Mastery of Nature: The *Kunstkammer,* Politics and Science," in *The Mastery of Nature,* 176.

33. Lorraine Daston and Katharine Park, "Wonders of Art, Wonders of Nature," in *Wonders and the Order of Nature, 1150–1750.*

34. On the shifting attitudes toward curiosity in the early modern period, see Peter Harrison, "Curiosity, Forbidden Knowledge, and the Reformation of Natural Philosophy in Early Modern England," *Isis* 92, no. 2 (2001): 265–90; and on its relationship to early modern science, see Lorraine Daston, "Curiosity in Early Modern Science," *Word and Image* 11 (1995): 391–404.

35. Kaufmann, "From Mastery of the World to Mastery of Nature," 181.

36. Daston, "Curiosity in Early Modern Science," 397.

37. Krzysztof Pomian, *Collectors and Curiosities: Paris and Venice, 1500–1800* (Cambridge, U.K.: Polity Press; Cambridge, Mass.: Basil Blackwell, 1990), 81.

38. Paul Taylor, *Dutch Flower Painting, 1600–1720* (New Haven, Conn.: Yale University Press, 1995), 122. The motto has also been translated as "There are powers in herbs, words, and stones," Piero Camporesi, *Bread of Dreams: Food and Fantasy in Early Modern Europe* (Chicago: University of Chicago Press, 1989), 144. The motto appears in a sixteenth-century text as part of a magical charm to cure a toothache; see Keith Thomas, *Religion and the Decline of Magic* (London: Weidenfeld and Nicholson, 1971), 180. For further discussion of this motto and the relationship between words and healing, see Claire Fanger "Things Done Wisely by a Wise Enchanter: Negotiating the Power of Words in the Thirteenth Century," *Esoterica* 1 (1999): 97–132.

39. Taylor, *Dutch Flower Painting*, 122.

40. Alabaster was a popular material for sculpture during this time, but alabaster containers were also associated with Mary Magdalen as the vessel that contained the ointment with which she anointed Christ. Katherine Ludwig Jansen, *The Making of the Magdalen: Preaching and Popular Devotion in the Later Middle Ages* (Princeton, N.J.: Princeton University Press, 2000), 109. European collectors of this period could also have associated alabaster with ancient Egypt, where it was used to make jars and other objects.

41. One of the most spectacular examples is the piece known as the Augsburg Art Cabinet in the collection of the Museum Gustavianum, Uppsala University.

42. Amin Jaffer, "Furniture Made in Asia for Export West," in *Encounters: The Meeting of Asia and Europe, 1500–1800*, ed. Anna Jackson and Amin Jaffer (London: Victoria and Albert Museum; distributed in North America by Harry N. Abrams, 2004), 256.

43. Amin Jaffer, *Luxury Goods from India: The Art of the Indian Cabinet-Maker* (London: Victoria and Albert Museum; distributed in North America by Harry N. Abrams, 2004), 10.

44. Pyrard de Laval (1619) quoted in Jaffer, "Furniture Made in Asia for Export West," 252.

45. Mark Meadow, "Samuel Quiccheberg, the Wunderkammer and the Copious Object," in *The Lure of the Object*, ed. Stephen W. Melville (Williamstown, Mass.: Sterling and Francine Clark Art Institute; distributed by Yale University Press, 2005), 41.

46. Fred G. Meijer, *The Collection of Dutch and Flemish Still-Life Paintings Bequeathed by Daisy Linda Ward: Catalogue of the Collection of Paintings* (Zwolle, Netherlands: Waanders, 2003), 230.

4. Between Observation and Image

1. *Micrographia; or, Some Physiological Descriptions of Minute Bodies Made by Magnifying Glasses, with Observations and Inquiries Thereupon* (London, 1665). On Robert Hooke's life and work, see J. A. Bennett et al., eds., *London's Leonardo: The Life and Work of Robert Hooke* (Oxford: Oxford University Press, 2003); Stephen Inwood, *The Man Who Knew Too Much: The Strange and Inventive Life of Robert Hooke, 1635–1703* (London: Pan Books,

2002); and Lisa Jardine, *The Curious Life of Robert Hooke: The Man Who Measured London* (New York: HarperCollins, 2004).

2. According to Inwood, *The Man Who Knew Too Much*, 8, Hooke's inheritance was between fifty and one hundred pounds. These details of Hooke's biography are taken from Michael Cooper, "Hooke's Career," in Bennett et al., eds., *London's Leonardo*, 1–8; Inwood, *The Man Who Knew Too Much*, 6–17; and Jardine, *The Curious Life of Robert Hooke*, 21–56.

3. Balthasar de Monconys (1611–1665), a French diplomat and art connoisseur, related that on his visit to the king's cabinet he saw these drawings, along with Wren's lunar globe (see note 5): "Je vis la Lune, que M. Rene à faite de relief en carton, suivant le dessein d'Hevelius, et des desseins à la plume d'un Poû, d'une Pûce de la teste, et d'une Aille d'une Mouche, fait par le Microscope" (I saw the Moon that Monsieur Wren had made in cardboard relief, following the design of [the astronomer Johannes] Hevelius, and some pen-and-ink drawings of a louse, a flea, and the wing of a fly, made by the use of a microscope). Balthasar de Monconys, *Journal des voyages de Monsieur de Monconys . . . publié par le sieur de Liergues son fils*, 3 vols. (Lyon, 1665–66), 2:82. Christiaan Huygens (1629–1695), a Dutch mathematician and physicist, also reported seeing Wren's insect drawings in the king's cabinet; see J. A. Bennett, *The Mathematical Science of Christopher Wren* (Cambridge: Cambridge University Press, 1982), 73. The location of the insect drawings Wren presented to Charles II is not presently known. Harwood reports searching for them without success in the collections of the Royal Library, Windsor; the Natural History Museum, London; and the British Library. See John T. Harwood, "Rhetoric and Graphics in *Micrographia*," in *Robert Hooke: New Studies*, ed. Michael Hunter and Simon Schaffer (Woodbridge, U.K.: Boydell Press, 1989), 122 n. 15.

4. The letter by Henry Powle (1630–1692) to Christopher Wren is published in Wren, *Parentalia; or, Memoirs of the Family of the Wrens; Viz., of Mathew, Bishop of Ely, Christopher, Dean of Windsor, Etc. But Chiefly of Sir Christopher Wren . . . in Which Is Contained, Besides His Works, a Great Number of Original Papers and Records . . . Compiled by His Son Christopher* (Farnborough, U.K.: Gregg Press, 1965 [1750]), 210.

5. Wren, *Parentalia*, 210. Along with this request for drawings, the king also asked that Wren construct a lunar globe for him. Wren's activities involving the lunar globe and the problems it raised for the Royal Society are discussed in Adrian Tinniswood, *His Invention So Fertile: A Life of Christopher Wren* (Oxford: Oxford University Press, 2001), 77–79; and in Bennett, *The Mathematical Science of Christopher Wren*, 40–41. Wren would eventually complete the lunar globe, again presenting the gift directly to the king, in August 1661.

6. Thomas Birch, *The History of the Royal Society of London for Improving of Natural Knowledge from Its First Rise*, 4 vols. (New York: Johnson Reprint Corp., 1968 [1756–57]), 1:33.

7. Wren, *Parentalia*, 211.

8. In July 1663 the Royal Society was preparing for a visit from Charles II, and Hooke was directed to prepare a "handsome book" of his microscopic drawings for the king's entertainment. Hooke was in the midst of his work on *Micrographia* at this time, and the drawings he was asked to present were probably those he had been showing at

meetings since March of that year. There is no record, however, that the visit from Charles II actually occurred. See Birch, *The History of the Royal Society of London*, 1:272.

9. Hooke did not receive a salary from the Royal Society until 1664, when he was appointed professor of "the history of the trades" at Gresham College. Birch, *The History of the Royal Society of London*, 1:453.

10. Ibid., 1:21. The publication referred to at this meeting is Robert Hooke, *An Attempt for the Explication of the Phaenomena, Observable in an Experiment Published by the Honourable Robert Boyle* (London, 1661).

11. Shapin and Schaffer point out that Hooke was appointed curator in November 1662 and "from this moment, significantly, the pump in London worked better. Within six months it had been completely rebuilt along the lines of Boyle's machine in Oxford, a machine whose existence was partly due to Hooke himself. Thus it was not until Hooke was given responsibility for this pump . . . that the London pump, now reconstructed, could work reliably." Steven Shapin and Simon Schaffer, *Leviathan and the Air-Pump: Hobbes, Boyle, and the Experimental Life* (Princeton, N.J.: Princeton University Press, 1985), 235. On Hooke's work with Robert Boyle, see esp. 36–38, 231, and 250.

12. For further discussion of Hooke's status within the Royal Society, see J. A. Bennett, "Robert Hooke as Mechanic and Natural Philosopher," *Notes and Records of the Royal Society of London* 35 (1980): 33–48, esp. 34; and Steven Shapin, "Who Was Robert Hooke?" 253–85, esp. 253–54, 256, 263, and 283–85.

13. Steven Shapin, "The House of Experiment in Seventeenth-Century England," *Isis* 79 (1988): 396.

14. Birch, *The History of the Royal Society of London*, 1:213.

15. The image of moss would later be published as Schema 13 in *Micrographia*. In Harwood's chronology of Hooke's observations, the moss is listed as Hooke's fourth observation. However, the first three observations listed in Harwood's chronology did not involve presentations of microscopic drawings. Hooke later made drawings of the first three items listed by Harwood and published illustrations of them in *Micrographia*, but the initial presentations of these items involved verbal rather than visual reports. See Harwood, "Rhetoric and Graphics in *Micrographia*," 124–25.

16. Harwood's chronology lists one further presentation, of petrified wood, on August 24, 1664. This presentation did not involve a drawing; see also note 15 above.

17. Birch, *The History of the Royal Society of London*, 1:231. Harwood's chronology states that the six-eyed spider was illustrated in Schema 31 of *Micrographia*, but the spider pictured in that plate has only two eyes.

18. Birch, *The History of the Royal Society of London*, 1:250, 255, 262, 270.

19. On Wren's early work with the microscope, see Bennett, *The Mathematical Science of Christopher Wren*, 73. The historian A. Rupert Hall dates Hooke's early work with the microscope to this period as well and speculates that Hooke and Wren worked together on microscopy in Oxford during the 1650s. Hall, *Hooke's "Micrographia," 1665–1965* (London: Athlone Press, 1966), 6–7.

20. On later collaborations between Wren and Hooke, see Margaret 'Espinasse, "Hooke as Surveyor and Architect," in Hunter and Schaffer, eds., *Robert Hooke*, 83–105; Bennett, *The Mathematical Science of Christopher Wren*, esp. 19, 41–43, 50–56, 61–70, 73–74, 87–88; Tinniswood, *His Invention So Fertile*, 98–99, 150–51; and Lisa Jardine,

"Monuments and Microscopes: Scientific Thinking on a Grand Scale in the Early Royal Society," *Notes and Records of the Royal Society of London* 55 (2001): 289–308.

21. Recorded in Hartlib's *Ephemerides* for 1655, as quoted in H. W. Turnbull, "Samuel Hartlib's Influence on the Early History of the Royal Society," *Notes and Records of the Royal Society of London* 10 (1952): 114. Bennett mistakenly gives the year for this conversation as 1665, the year in which *Micrographia* was published (and three years after Hartlib's death). Bennett, *The Mathematical Science of Christopher Wren*, 73.

22. Henry Power, *Experimental Philosophy* (London, 1664). See also Harwood, "Rhetoric and Graphics in *Micrographia*," 122–29.

23. Birch, *The History of the Royal Society of London*, 1:266. John Wilkins (1614–1672) was one of the founders of the Royal Society and warden of Wadham College, Oxford. While Hooke was studying at Oxford he attended scientific meetings organized by Wilkins and later praised him in the preface to *Micrographia*.

24. Maxwell E. Power, "Sir Christopher Wren and the *Micrographia*," *Transactions of the Connecticut Academy of Arts and Sciences* 36 (1945): 37–44.

25. Birch, *The History of the Royal Society of London*, 1:442.

26. Hooke, *Micrographia*, preface, n.p.

27. Ibid.

28. Hooke's apprenticeship to Lely is reported by Richard Waller in *The Posthumous Works of Robert Hooke* (New York: Johnson Reprint Corp., 1969 [1705]), iii. Hooke's diary entries in later years record purchases of paintings, prints, and drawing materials, as well as his acquaintanceship with a number of artists, including Lely, his former painting master. See Leona Rostenberg, *The Library of Robert Hooke: The Scientific Book Trade of Restoration England* (Santa Monica, Calif.: Modoc Press, 1989), 8–9.

29. Hooke, *Micrographia*, 153.

30. Svetlana Alpers also notes the relationship between Hooke's illustration of thyme seeds and Dutch still life paintings of lemons and other objects. Alpers argues that both types of images exhibit a "fragmenting approach," one that aims to draw the viewer's attention to the individual features of an object. Although the isolation of individual features is characteristic of Hooke's images, his overall pictorial aim was to build unified compositions and self-contained worlds from these fragments, rather than to create a series of fragmented, isolated views. See Alpers, *The Art of Describing*, 85.

31. Hooke, *Micrographia*, Observation 31, 156.

32. Hooke, *Micrographia*, 152.

33. Francesco Redi published his illustrations of insects based on microscopic investigations shortly after Hooke's *Micrographia* was published. See Francesco Redi, *Esperienze intorno alla generazione degli insetti* (Florence, 1668). The original drawings for Redi's publication have been identified in the collections of the Biblioteca Nazionale Centrale in Florence. See Lucia Tongiorgi Tomasi and Paolo Tongiorgi, "Il naturalista e il cappellano: Osservazione della natura e immagini 'dal naturale,' in Francesco Redi," in *Natura e immagine: Il manoscritto di Franceso Redi sugli insetti delle galle*, ed. Walter Bernardi et al. (Pisa, Italy: Edizioni ETS, 1997); see also Lucia Tongiorgi Tomasi, "L'infinitamente piccolo: Immagini al microscopio di Redi e al tempo di Redi," in *Franceso Redi: Un protagonista della scienza moderna*, ed. Walter Bernardi and Lucia Guerrini

(Florence: Leo S. Olschki, 1999). For further discussion of insects as subject matter in the sixteenth century, see chapters 1–3.

34. Harwood, "Rhetoric and Graphics in *Micrographia*," 128.

35. Hooke, *Micrographia*, 203.

36. Ibid., 203.

37. Ibid., 204.

38. Willem Piso (1611–1678) et al., *Historia naturalis Brasiliae* . . . (Leiden and Amsterdam, 1648); Richard Ligon (1585?–1662), *A True and Exact History of the Island of Barbados* (London, 1657). Harwood has shown that Hooke copied the illustration of an aquatic beehive that appears as figure 3 of Schema 27 in *Micrographia* from another work by Piso, *Indiae utriusque re naturali et medici* (Amsterdam, 1658), 113. Harwood, "Rhetoric and Graphics in *Micrographia*," 142, n. 52.

39. Hooke, *Micrographia*, 204.

40. Moffet's and Aldrovandi's books and illustrations are the subject of chapter 2.

41. Hooke, *Micrographia*, 184.

42. Harwood, "Rhetoric and Graphics in *Micrographia*," 121.

43. Ibid., 131.

44. Ibid., 142.

45. Michael Aaron Dennis, "Graphic Understanding: Instruments and Interpretation in Robert Hooke's 'Micrographia,'" *Science in Context* 3 (1989): 312.

46. Ibid., 313.

47. Ibid., 324.

48. The sale catalogue listing is quoted from Geoffrey Keynes, *A Bibliography of Dr. Robert Hooke* (Oxford: Clarendon Press, 1960), 23. The sale catalogue is reprinted in Rostenberg, *The Library of Robert Hooke*, 143–221.

49. Royal Society Classified Papers, vol. XX, fol. 6. The drawing is reproduced in Harwood, "Rhetoric and Graphics in *Micrographia*," 126, plate 9a. Hooke discusses figure 1 on page 88 of *Micrographia* and discusses figure 6 on page 93.

50. John Covel, Natural History Notebook and Commonplace Book: 1660–1713, British Library, Add. 57495. Hereafter referred to as BL Add. 57495.

51. On Covel, see *Extracts from the Diaries of Dr. John Covel, 1670–1679*, vol. 2 of *Early Voyages and Travels in the Levant*, ed. J. Theodore Bent (New York: B. Franklin, 1964 [1893]); and Jean-Pierre Grélois, ed., *Dr. John Covel: Voyages en Turquie, 1675–1677* (Paris: Lethielleux, 1998).

52. On folio 114 of the notebook, Petiver refers to his own publication, *The Monthly Miscellany; or, Memoirs for the Curious* (London, 1707–9), and on folio 2 he refers to having seen a specimen, while in Holland, of an insect drawn by Covel. Petiver's only known visit to Holland occurred in 1711. For more on Petiver's life and collecting activities, see Raymond Phineas Stearns, "James Petiver: Promoter of Natural Science, c. 1663–1718," *Proceedings of the American Antiquarian Society* 62 (1952): 243–365.

53. BL Add. 57495, fol. 58v.

54. Ibid., fols. 32v and 2.

55. Ibid., fol. 96.

56. Identifications of the insects and other animals pictured in this folio, along with full transcriptions of the annotations, can be found in Janice L. Neri, "Some Early Drawings by Robert Hooke," *Archives of Natural History* 32 (2005): 41–47.

57. Hooke's lowercase *e* is particularly distinctive.

58. A lengthy written description of this drawing on a separate page of the notebook describes the steps taken to immobilize, dissect, and observe the insect. The treatment of the legs is described as follows: "I layd his leggs in the posture you see there drawn, but befour long they were crumpled as you may see prickt." BL Add. 57495, fol. 118v.

59. For Hooke's written description of this illustration, see Hooke, *Micrographia*, 187.

60. Here, Hooke seems to have used the word "sad" to mean "dark" or "deep," referring to the color green.

61. Hooke, *Micrographia*, 208.

62. Piso, *Historia naturalis Brasiliae . . .* , 185.

63. Shapin, "Who Was Robert Hooke?" 259.

64. On the work of Greatorex and Hooke on the air pump, see Shapin and Schaffer, *Leviathan and the Air-Pump*, 379. On Greatorex, see A. V. Simcock, "An Equinoctal Ring Dial by Ralph Greatorex," in *Making Instruments Count: Essays on Historical and Scientific Instruments Presented to Gerard L'Estrange Turner*, ed. R. G. W. Anderson, J. A. Bennett, and W. F. Ryan (Aldershot, U.K.: Variorum, 1993), 201–15.

65. Michael Hunter, *The Royal Society and Its Fellows, 1660–1700: The Morphology of an Early Scientific Institution* (Chalfont St. Giles, U.K.: British Society for the History of Science, 1994), 11.

66. Steven Shapin, "Pump and Circumstance: Robert Boyle's Literary Technology," *Social Studies of Science* 14 (1984): 492.

67. In Hooke's case, experiments took place in his rooms. For a discussion of the domestic setting of laboratories and of the physical layout of the Royal Society's meeting rooms, see Shapin, "The House of Experiment in Seventeenth-Century England," 391.

68. Steven Shapin, *A Social History of Truth: Civility and Science in Seventeenth-Century England, Science and Its Conceptual Foundations* (Chicago: University of Chicago Press, 1994), 381.

69. Hooke, *Micrographia*, preface, n.p.

70. Ibid., 175.

71. Hooke described the experiment on a live dog, and in a letter to Robert Boyle dated November 10, 1664, he confessed his unwillingness to perform the experiment again. See "Letter from Hooke to Boyle, 10 Nov. 1664," in Gunther, *Early Science in Oxford*, 6:216–18, 217, as cited in Andreas-Holger Maehle and Ulrich Tröhler, "Animal Experimentation from Antiquity to the End of the Eighteenth Century: Attitudes and Arguments," in *Vivisection in Historical Perspective*, ed. Nicholas Rupke (London: Routledge, 1987), 43 n. 76. Nevertheless, Hooke did perform the experiment again for the Royal Society, with Richard Lower, on October 10, 1667. For a discussion of the 1664 experiment and Hooke's reaction to it, see Maehle and Tröhler, "Animal Experimentation from Antiquity to the End of the Eighteenth Century," 23. For a brief discussion of the experiment done with Lower, see Andreas-Holger Maehle, "Literary Responses to Animal Experimentation in Seventeenth- and Eighteenth-Century Britain," *Medical History* 34 (1990): 27–51, 31.

72. Hooke, *Micrographia*, 186.

5. Stitches, Specimens, and Pictures

1. Elisabeth Rücker, ed., *Maria Sibylla Merian, 1647–1717* (Nuremberg, Germany: Germanisches Nationalmuseum, 1967); as quoted in Kurt Wettengl, "Maria Sibylla Merian: Artist and Naturalist between Frankfurt and Surinam," in *Maria Sibylla Merian, 1647–1717: Artist and Naturalist*, ed. Kurt Wettengl (Ostfildern, Germany: G. Hatje, 1998), 18.

2. This account of Merian's life is based on the biography of Merian found in Natalie Zemon Davis, *Women on the Margins: Three Seventeenth-Century Lives* (Cambridge, Mass.: Harvard University Press, 1997). For a shorter biography and an introduction to the literature on Merian, see Wettengl, "Maria Sibylla Merian," which also includes important essays by Merian scholars and provides excellent reproductions of many unpublished drawings. Both of these publications contain exhaustive bibliographical information on Merian. Additional biographical information and useful analyses of Merian's illustrations can be found in Ella Reitsma, *Merian and Daughters: Women of Art and Science* (Los Angeles: J. Paul Getty Museum, 2008), and Kim Todd, *Chrysalis: Maria Sibylla Merian and the Secrets of Metamorphosis* (Orlando, Fla.: Harcourt, 2007). Invaluable facsimiles of Merian's major works have been published in recent years; for a useful explanation and summary of these, see Sharon Valiant, "Maria Sibylla Merian: Recovering an Eighteenth-Century Legend," *Eighteenth-Century Studies* 26, no. 3 (1993): 467–79. David Freedberg's essay on the relationship between Merian's *Metamorphosis* and seventeenth-century Dutch art, commerce, natural history, and the New World provides an important basis for understanding the historical and historiographical issues explored in this chapter. Freedberg, "Science, Commerce, and Art: Neglected Topics at the Junction of History and Art History," in *Art in History, History in Art: Studies in Seventeenth-Century Dutch Culture, Issues and Debates* (Santa Monica, Calif.: Getty Center for the History of Art and the Humanities; distributed by University of Chicago Press, 1991), 377–428.

3. Davis, *Women on the Margins*, 151. Davis's discussion of the topos of the remarkable woman appears on 154–56.

4. The problem of spontaneous generation was discussed by many natural philosophers. The two major published works on the topic from this period are Redi, *Esperienze intorno alla generazione degl'insetti*; and Jan Swammerdam, *Historia insectorum generalis* (Amsterdam, 1669). The topic is discussed in relation to the development of the microscope in Edward G. Ruestow, *The Microscope in the Dutch Republic: The Shaping of Discovery* (Cambridge: Cambridge University Press, 1996); and in Catherine Wilson, *The Invisible World: Early Modern Philosophy and the Invention of the Microscope* (Princeton, N.J.: Princeton University Press, 1995). On Swammerdam's research and experiments on the question of spontaneous generation, see Matthew Cobb, "Reading and Writing *The Book of Nature*: Jan Swammerdam (1637–1680)," *Endeavour* 24, no. 3 (2000): 122–28. On Antony van Leeuwenhoek's work in this area, see Edward Ruestow, "Images and Ideas: Leeuwenhoek's Perception of the Spermatozoa," *Journal of the History of Biology* 16 (1983): 185–224. For an extensive bibliography on the topic, see also Tomomi Kinukawa, "Art Competes with Nature: Maria Sibylla Merian (1647–1717) and the Culture of Natural History" (Ph.D. dissertation, University of Wisconsin, 2001), 14 n. 32.

5. Wettengl, "Maria Sibylla Merian," 33.

6. First editions of the *Blumenbuch* series are now exceedingly rare. For publication details and information on these early editions, see Thomas Bürger, epilogue to Maria Sibylla Merian, *Neues Blumenbuch* (Munich: 1999 [1680]), 81–95.

7. As quoted in Wettengl, ed., *Maria Sibylla Merian*, 98.

8. General information on early modern pattern books can be found in Janet S. Byrne, *Renaissance Ornament Prints and Drawings* (New York: Metropolitan Museum of Art, 1981); and Edward F. Strange, "Early Pattern-Books of Lace, Embroidery, and Needlework," *Transactions of the Bibliographical Society* [London] 7 (October 1902–March 1904): 209–46. For the history of European embroidery, see Mary Eirwen Jones, *A History of Western Embroidery* (London: Studio Vista, 1969); Marie Schuette and Sigrid Müller-Christensen, *A Pictorial History of Embroidery* (New York: Praeger, 1964); and Pamela Warner, *Embroidery: A History* (London: B. T. Batsford, 1991). The relationship between embroidery and gardening has been extensively treated in Thomasina Beck, *Gardening with Silk and Gold: A History of Gardens in Embroidery* (London: David and Charles, 1997). Analyses of pattern books and embroidery in the construction of gender roles can be found in Ruth Geuter, "'The Silver Hand': Needlework in Early Modern Wales," in *Women and Gender in Early Modern Wales*, ed. Simone Clarke and Michael Roberts (Cardiff: University of Wales Press, 2000); Rozsika Parker, *The Subversive Stitch: Embroidery and the Making of the Feminine* (London: Women's Press, 1984); and Stacey Shimizu, "The Pattern of Perfect Womanhood: Feminine Virtue, Pattern Books and the Fiction of the Clothworking Woman," in *Women's Education in Early Modern Europe: A History, 1500–1800*, ed. Barbara J. Whitehead (New York: Garland, 1999). These works concentrate primarily on the rhetoric of the written portions of pattern books rather than on the visual qualities of the images. The work of the Italian printmaker Isabella Parasole forms an interesting comparison to Merian's work in this area. Parasole published pattern books for lace design and was also involved in creating botanical illustrations. See Evelyn Lincoln, "Models for Science and Craft: Isabella Parasole's Botanical and Lace Illustrations," *Visual Resources* 17 (2001): 1–35.

9. Sam Segal, "Maria Sibylla Merian as a Flower Painter," in Wettengl, ed., *Maria Sibylla Merian*, 74.

10. Other sources include Merian's maternal grandfather Jan Theodor de Bry's *Florilegium novum*; Jan Jonston's *Historia naturalis de insectis*; and Joris and Jacob Hoefnagel's *Archetypa* series. See Segal, "Maria Sibylla Merian as a Flower Painter," 69–87.

11. Merian's copies of Robert's iris engraving have also been discussed by Segal and Bürger. See Bürger's epilogue in Merian, *Neues Blumenbuch*, 90; and Segal, "Maria Sibylla Merian as a Flower Painter," 70.

12. Another copy of Robert's composition, a drawing attributed to Merian, is in the collection of the Senckenbergische Bibliothek, Frankfurt. It presents a reversed version of the Robert composition illustrated here. An illustration of the drawing is available in Wettengl, ed., *Maria Sibylla Merian*, 70.

13. Merian, *Neues Blumenbuch*, 90; Segal, "Maria Sibylla Merian as a Flower Painter," 70.

14. Beck, *Gardening with Silk and Gold*, 36. For a discussion of the technique of slipwork, see also Warner, *Embroidery*, 73–77.

15. Warner lists Gessner's *Catalogue plantarum* (1546) and Gerard's *Herbal* (1597) among the widely used sources for embroidery design in the sixteenth and seventeenth centuries, and Thomas Moffet's *Theatrum insectorum* (1634) and Edward Topsell's *Historie of Four-Footed Beasts* (1607) as sources for the seventeenth century. Warner, *Embroidery*, 67, 94.

16. Warner, *Embroidery*, 95.

17. Jacob and Joris Hoefnagel, *Archetypa studiaque patris Georgii Hoefnagelii* . . . . Merian based the damselfly in the pansy illustration and the swallowtail butterfly in the iris illustration on insects appearing in plate 7 of part 1 and plate 12 of part 3, respectively.

18. Segal has shown that several of these garlands were adapted from Robert's designs, and he has noted that Merian's garlands were also influenced by the work of her stepfather Jacob Marrel. See Segal, "Maria Sibylla Merian as a Flower Painter," 72–73.

19. Dirck van Rijswijck was a Dutch goldsmith active in Amsterdam during the third quarter of the seventeenth century. For further information on Van Rijswijck and a complete catalogue of his inlay work, see Danielle Kisluk-Grosheide, "Dirck van Rijswijck (1596–1679): A Master of Mother-of-Pearl," *Oud Holland* 111, no. 2 (1997): 77–152. Information on Merian's friendship with Koerten, along with further references to Koerten's life and work, can be found in Kinukawa, "Art Competes with Nature," 157, n. 69.

20. Wettengl, ed., *Maria Sibylla Merian*, 98.

21. As quoted and translated in ibid.

22. Translated in ibid., 103.

23. The first *Raupenbuch* was published in 1679 while Merian was living in Nuremberg, the second volume was published in 1683, and the third was published posthumously in 1717. For further publication details about the *Raupenbuch* series, see Heidrun Ludwig, "The Raupenbuch, a Popular Natural History," in Wettengl, ed., *Maria Sibylla Merian*, 52–67, esp. 53–54.

24. Segal, "Maria Sibylla Merian as a Flower Painter," 70.

25. Ludwig, "The Raupenbuch, a Popular Natural History," in Wettengl, ed., *Maria Sibylla Merian*, 59.

26. Johannes Goedaert, *Metamorphosis et historia naturalis insectorum*, 3 vols. (Middleburg, 1662–69).

27. Merian's preparatory watercolors utilizing this format are found in the artist's *Studienbuch*, as noted by Wettengl, "Maria Sibylla Merian," 24–25. On Goedaert's influence on Merian, see also Davis, *Women on the Margins*, 152–53.

28. George McGavin, *Insects, Spiders, and other Terrestrial Arthropods* (New York: Dorling Kindersley, 2000), 158.

29. Wettengl, "Maria Sibylla Merian," 21.

30. The *Studienbuch* contains entries from Merian's childhood years, but these were added later by Merian when she began assembling the journal in the 1680s. A facsimile edition of the *Studienbuch* was published as Maria Sibylla Merian, *Schmetterlinge, Käfer und andere Insekten: Leningrader Studienbuch*, ed. Wolf-Dietrich Beer, 2 vols. (Leipzig: Edition Leipzig, 1976). For details on the discovery of the *Studienbuch*, see also Valiant, "Maria Sibylla Merian," 469.

31. As quoted and translated in Wettengl, "Maria Sibylla Merian," 25. Merian made other analogies between insects and fabric; in one passage in the *Raupenbuch* (1:32) she compares the color in moths' wings to dyed wool. See Kinukawa, "Art Competes with Nature," 134.

32. Wettengl, "Maria Sibylla Merian," 21.

33. British Library, Sloane MS 4064, fol. 70. Translated by Elizabeth Rücker and published in Wettengl, ed., *Maria Sibylla Merian*, 268.

34. For information on Cornelis van Aerssen van Sommelsdijk's activities in Surinam and a detailed account of the Labadist presence there, see Trevor J. Saxby, "Disaster in the Jungle: Labadist Colonial Enterprise in Surinam, 1683–1719," in *The Quest for the New Jerusalem: Jean de Labadie and the Labadists, 1610–1744* (Dordrecht, Netherlands: M. Nijhoff Publishers, 1987).

35. Saxby believes that this collection of butterflies awakened Merian's interest in the insects of Surinam, but he provides no further discussion of this point. The collection of Surinamese butterflies at Waltha Castle has also been noted by Elisabeth Rücker, who states that the butterflies were brought back to Waltha Castle by Labadists who had traveled to Surinam. Although it is not possible to determine for certain the origin or fate of this collection of butterflies, it is true that they must have played a role in Merian's later decision to travel to South America. See Saxby, *The Quest for the New Jerusalem*, 277, 384; and Elisabeth Rücker, "Life and Personality of Maria Sibylla Merian," in *Maria Sibylla Merian in Surinam*, 10–11.

36. The English were the first European presence in Surinam, dating from Francis Willoughby's claim of the land for England in 1650. Control over the area was contested for a number of years until 1667, when the Dutch reached a settlement with the English. On the early history of the European presence in Surinam, see Rudolf Asveer Jacob van Lier, *Frontier Society: A Social Analysis of the History of Surinam* (The Hague: Martinus Nijhoof, 1971), 1–37.

37. Maria Sibylla Merian, *Metamorphosis insectorum Surinamensium*, ed. Elisabeth Rücker and William T. Stearn, 2 vols. (London: Pion, 1980), 1:85. Nicolaas Witsen and his nephew Jonas Witsen had close contacts with Surinam; Nicolaas served as a director of the Dutch East India Company and Jonas's wife was the daughter of plantation owners in Surinam who maintained a collection of rarities at their estate there. This collection was inherited by Jonas and was incorporated into Nicolaas's collection in Amsterdam, where Merian is believed to have encountered more specimens of Surinamese insects during the 1690s. See Roelof van Gelder, "Art, Commerce, Passion and Science," in Wettengl, ed., *Maria Sibylla Merian*, 143. Van Gelder's essay provides a useful introduction to the community of collectors in Amsterdam in the late seventeenth century.

38. Elisabeth Rücker, "Maria Sibylla Merian: Businesswoman and Publisher," in Wettengl, ed. *Maria Sibylla Merian*, 259.

39. Stadtbibliothek Nürnberg, Manuscript no. 167. Translated by Elisabeth Rücker and published in Wettengl, ed., *Maria Sibylla Merian*, 264.

40. Merian's technique of coating insect specimens with turpentine oil is described by her in a letter accompanying a shipment of specimens from Surinam to Georg Volkammer, another friend from Nuremberg. See Universitätsbibliothek Erlangen, Trew-Bibliothek, Brief-Sammlung Ms. 1834, Merian No. 2, translated by Elisabeth

Rücker and published in Wettengl, ed., *Maria Sibylla Merian*, 265. Turpentine oil, also known as terebinth, was a key ingredient in the preparation of specimens in early modern Europe. For a discussion of the substance in this context, see also Harold J. Cook, "Time's Bodies: Crafting the Preparation and Preservation of Naturalia," in Smith and Findlen, eds., *Merchants and Marvels*, 238–40.

41. Merian and Levinus Vincent were personally acquainted, and in later years they shared some of the same contacts and correspondents. Vincent, a Mennonite silk merchant specializing in brocade fabric, came from a family of merchants specializing in overseas trade. Vincent's business earned him much wealth, which allowed him to spend most of his time on leisurely pursuits. Vincent and his wife Johanna Breda worked together on developing, organizing, and displaying their collection of rarities, the nucleus of which was formed by the collection of Johanna's brother Anthony Breda. The material discussed here on their collection is based on Kinukawa's invaluable study of Levinus Vincent and the culture of collecting in Amsterdam. See Kinukawa, "Art Competes with Nature," 167–216.

42. Ibid., 183.

43. Romyn de Hooghe completed the engravings for the catalogues; see Levinus Vincent, *Wondertooneel der Nature . . .* (Amsterdam, 1706). For a complete bibliography of Vincent's publications, see Kinukawa, "Art Competes with Nature," 177 n. 133.

44. Edmé-François Gersaint, *Catalogue raisonée de coquilles et autres curiosités naturelles* (Paris: Flahault et Prault, 1736). As quoted and translated in Barbara Maria Stafford and Frances Terpak, *Devices of Wonder: From the World in a Box to Images on a Screen* (Los Angeles: Getty Research Institute, 2001), 148.

45. Stafford and Terpak, *Devices of Wonder*, 150.

46. As quoted in Kinukawa, "Art Competes with Nature," 200.

47. Kinukawa, "Art Competes with Nature," 200.

48. Ibid., 185.

49. Ibid., 195. Kinukawa provides evidence that other women also participated in ordering and arranging cabinets in this manner in the Netherlands during this time. A collection in Harlem belonging to a man named Dorville contained shells arranged in displays that were embroidered with silk thread by his wife. See Kinukawa, "Art Competes with Nature," 199.

50. Letter to Volkammer, Universitätsbibliothek Erlangen, Trew-Bibliothek, Brief-Sammlung Ms. 1834, Merian No. 1, as quoted and translated by Rücker in Wettengl, ed., *Maria Sibylla Merian*, 264–65.

51. Stadtbibliothek Nürnberg, Manuscript no. 167. As quoted and translated by Rücker in ibid., 264.

52. Kinukawa offers a detailed discussion of the culture of specimen exchange in Amsterdam during Merian's time. See Kinukawa, "Art Competes with Nature," 217–46.

53. British Library, Sloane MS 4064, fol. 3, quoted and translated in Gelder, "Art, Commerce, Passion and Science," 148.

54. Benjamin Schmidt, "The Project of Dutch Geography and the Marketing of the World, circa 1700," in Findlen and Smith, eds., *Merchants and Marvels*, 349, 354.

55. Ibid., 361–62.

56. Merian, *Metamorphosis insectorum Surinamensium*, 1:124.

57. For further details on the contents of the Studienbuch, see Wettengl, ed., *Maria Sibylla Merian*, 134, 224.

58. Merian, *Metamorphosis insectorum Surinamensium*, 1:111.

59. This illustration was cause for controversy during the eighteenth century, since many people did not accept Merian's account as truthful due to its unusual subject matter. See Valiant, "Maria Sibylla Merian," 474; and Davis, *Women on the Margins*, 198.

60. Merian, *Metamorphosis insectorum Surinamensium*, 98.

61. Plate 51. Ibid., 1:130.

62. Segal, "Maria Sibylla Merian as a Flower Painter," 78. For further discussion of the plants in Merian's book, see William T. Stearn, "The Plants, the Insects and Other Animals of Merian's *Metamorphosis insectorum Surinamensium*," in Rücker and Stearn, eds. *Metamorphosis insectorum Surinamensium*.

63. The cockroach and pineapple appear on plate 1 of *Metamorphosis*.

64. The two major discussions of Merian's remarks on the *Flos pavonis* are found in Davis, *Women on the Margins*, and in Londa Schiebinger, "Lost Knowledge, Bodies of Ignorance, and the Poverty of Taxonomy as Illustrated by the Curious Fate of *Flos Pavonis*, an Abortifacient," in *Picturing Science, Producing Art*, ed. Caroline A. Jones and Peter Louis Galison (New York: Routledge, 1998). For an analysis and critique of Davis's and Schiebinger's arguments, see Viktoria Schmidt-Linsenhoff, "Metamorphosis of Perspective: 'Merian' as a Subject of Feminist Discourse," in Wettengl, ed., *Maria Sibylla Merian*.

65. Merian, *Metamorphosis insectorum Surinamensium*, 125–26.

66. For plantation owners and their financial backers in Amsterdam, however, the enormous profits generated by the sugar trade far outweighed the costs of importing food, livestock, manufactured goods, and laborers.

67. When Merian returned to Amsterdam from Surinam, she brought with her the indigenous woman who had worked either as her servant or her slave. This person, with whom Merian may have had a more complex relationship than she had with any of the other people she met in Surinam, received the least amount of attention in Merian's accounts, and her presence is known only from the manifest of the ship on which they traveled home. Natalie Zemon Davis discusses the relationship between Merian and this woman in her biographical account of Merian in *Women on the Margins*, 194. Elizabeth Honig's discussion in "Making Sense of Things" of the question of human labor, and its absence, in the context of seventeenth-century Dutch still life painting and collecting practices provides a valuable model for understanding these aspects of Merian's approach. For a discussion of the ambiguous relationships between identity and property in representations of enslaved Africans on Dutch plantations in Surinam, see the discussion of the work of the Dutch painter Dirk Valkenburg in Charles Ford, "People as Property," *Oxford Art Journal* 25, no. 1 (2002): 1–16.

Conclusion

1. The connection between European colonialism and the study of the natural world has been the focus of a number of studies that have concentrated on the history

of botany. Examples include Daniela Bleichmar, Paula De Vos, Kristin Huffine, and Kevin Sheehan, eds., *Science in the Spanish and Portuguese Empires, 1500–1800* (Stanford: Stanford University Press, 2009); David Philip Miller and Peter Hanns Reill, eds., *Visions of Empire: Voyages, Botany, and Representations of Nature* (Cambridge: Cambridge University Press, 1996); Margaret Beck Pritchard and Amy R. W. Meyers, eds., *Empire's Nature: Mark Catesby's New World Vision* (Chapel Hill: Published for the Omohundro Institute of Early American History and Culture by the University of North Carolina Press, 1998); Bernard Smith, *European Vision and the South Pacific*, 2nd ed. (New Haven, Conn.: Yale University Press, 1985); and Londa L. Schiebinger and Claudia Swan, *Colonial Botany: Science, Commerce, and Politics in the Early Modern World* (Philadelphia: University of Pennsylvania Press, 2005). For discussions of these issues in the context of gender, see also Carolyn Merchant, *Ecological Revolutions: Nature, Gender, and Science in New England* (Chapel Hill: University of North Carolina Press, 1989); and Londa L. Schiebinger, *Nature's Body: Gender in the Making of Modern Science* (Boston: Beacon Press, 1993).

2. Dru Drury, *Illustrations of Natural History: Wherein Are Exhibited Upwards of Two Hundred and Forty Figures of Exotic Insects, According to Their Different Genera . . .* , 3 vols. (London, 1770); John Obadiah Westwood, *The Cabinet of Oriental Entomology: Being a Selection of Some of the Rarer and More Beautiful Species of Insects, Natives of India and the Adjacent Islands* (London, 1848); Edward Donovan, *An Epitome of the Natural History of the Insects of China* (London, 1789); *An Epitome of the Natural History of the Insects of India* (London, 1800); and *An Epitome of the Natural History of the Insects of New Holland, New Zealand, New Guinea, Otaheite, and Other Islands in the Indian, Southern, and Pacific Oceans* (London, 1805).

3. Some examples of local studies of insects are Eleazar Albin, *A Natural History of English Insects* (London, 1720); Moses Harris, *The Aurelian: A Natural History of English Moths and Butterflies* (London, 1766); and Jacob Christian Schäffer, *Icones insectorum circa Ratisbonam indigenorum coloribus naturam referentibus expressae / Natürlich ausgemahlte Abbildungen Regensburgischer Insecten* (Regensburg, Germany, 1766).

4. George Edwards, *A Natural History of Uncommon Birds* (London, 1743).

5. Tita Chico, "Minute Particulars: Microscopy and Eighteenth-Century Narrative," *Mosaic: A Journal for the Interdisciplinary Study of Literature* 39, no. 2 (2006): 143–61; and Mary B. Campbell, *Wonder and Science: Imagining Worlds in Early Modern Europe* (Ithaca, N.Y.: Cornell University Press, 1999).

6. See, for example, the plates accompanying William Curtis's translation of Linnaeus's work on insects. Carl von Linné, *Fundamenta entomologiæ; or, An Introduction to the Knowledge of Insects, by William Curtis* (London, 1772).

7. Drury, *Illustrations of Natural History*, title page.

8. Ibid., preface, xvii.

9. The drawing appears in an album of insect drawings by Barbut in the Entomology Library of the Natural History Museum, London.

10. Harris, *The Aurelian*, 34.

11. A recent example is the exhibition Mrs. Delany and Her Circle, which featured the eighteenth-century Englishwoman Mary Delany and her extraordinary paper mosaics and botanical embroideries. On the broader context of her activities see the exhibition catalogue and essays in Mark Laird and Alicia Weisberg-Roberts, eds., *Mrs. Delany and*

Her Circle (New Haven, Conn.: Yale University Press, 2009), including my essay addressing her work in natural history and zoology, "'A Beautiful Mixture of Pretty Objects': Mrs. Delany's Natural History and Zoological Activities." For further information on Delany, see Ruth Hayden, *Mrs. Delany: Her Life and Her Flowers* (London: British Museum Press, 2000). Delany's close friend, Margaret Cavendish Bentinck, the second Duchess of Portland, was a major patron and collector of natural history in England and has also been the subject of renewed scholarly interest.

INDEX

Academy of Linceans, xxii, 115–16, 196n13, 197n18

accuracy, xv, xvii, 9, 105

Africa, 52, 65

Africans, 177, 222n67

alabaster, 92, 211n40

alchemy, 61–62

Aldrovandi, Ulisse: on birds, 33; collections of, 32, 44, 84–85, 203n14, 204n20; education of, 29; museum of, 32, 203n11, 204n20; on shells, 81, 84, 209n10. *See also De animalibus insectis; Tavole di animali*

Alpers, Svetlana, xvi, 214n30

amateurs, xiv, 63, 143, 190. *See also* anonymous artists

Amsterdam, 160–66

anonymous artists, 24, 84, 126, 131. *See also* amateurs

ants, 117, 174–75, 195n1

Apiarum (Cesi, Stelluti), xxiii, xxv

apothecaries, 94

Archetypa studiaque patriis Georgii Hoefnagelii . . . (Hoefnagel), 3, 11, 47, 99, 208n5; illustrations, 13, 49; as source for Merian, 150, 219n17; as source for Moffet, 47, 55; sources for, 200n17; stag beetle image, 9, 200n14

Arcimboldo, Giuseppe, 18

art and science: debate, xix, xviii, xvi; in Merian's work, 141, 155

Augsburg, 95, 211n41

Aurelian: A Natural History of English Moths and Butterflies, The (Harris), 185–87

authority, 30, 198n25

Barbut, James, 190–91

bees, xiv, 115–16, 196n13. *See also Apiarum* (Cesi, Stelluti)

beetles, 170, 171, 99–100, 169; cockchafer, 89; and embroidery, 191; *Euchroma gigantea*, 172; harlequin, 172–74; life cycles of, 157, 167–69, 172; long-horned, 172; rhinoceros, 183–85; specimens, 163; worn as jewelry, 124. *See also* stag beetles

Belon, Pierre, 81, 209n10

Bennett, J. A., 108

birds, 33, 51, 95, 202n10; culinary uses, 31; specimens, 163. *See also* hummingbirds

Blumenbuch series (Merian), 142–54; as pattern book, 139; and *Raupenbuch* (Merian), 158; second edition, 155

Bocskay, Georg, 13, 15, 16, 19, 199n1; calligraphy, 14–15, 23. *See also Mira*

insect world, 5, 28; service to Rudolf II, 5; stag beetle images by, 5–10, 18, 23, 68, 200n14; and vision, 201n27. *See also Archetypa studiaque patriis Georgii Hoefnagelii . . .* (Hoefnagel); *Mira calligraphiae monumenta* (Bocskay, Hoefnagel)

Hoffmann, Hans, 6

Honig, Elizabeth, 222n67

Hooghe, Romeyn de, 164, 221n43

Hooke, Robert: assignment to draw insects, 107, 212n8; biographies of, 211n1; collaborations with unnamed colleagues, 131–35; collaborations with Wren, 110–12, 213n19, 213n20; early life, 106, 212n2; as employee of Royal Society, 108–10, 213n9; experiments with vivisection, 138, 216n71; persona as observer, xviii, xix, 105–6, 112, 131–33, 134–35; persona as scholar, 130–31; presentations to Royal Society, 109, 111, 134, 213n15, 216n67; sketches by, 124–31, 215n56; social status, 182; status in Royal Society, 112, 133, 134–35, 213n12; work on Robert Boyle's air pump, 213n11. *See also Micrographia*

Hours of Catherine of Cleves, 11, 12, 20, 118, 119

hummingbirds, 174, 183. *See also* birds

hybrid insects, 44. *See also* imaginary insects

illuminated manuscripts, xi, 6, 11, 200n15, 208n1; and Hooke, 113; and still life, 76–77. *See also* Hoefnagel, Joris: *Ignis;* Hours of Catherine of Cleves; *Mira calligraphiae monumenta* (Bocskay, Hoefnagel)

Illustrations of Natural History (Drury), 189

imaginary insects, 17–22

Imhoff, Clara, 165

India: cabinets from, 95, 98; shells from, 86; source for insect specimens, 53, 65, 68

insect: as collectable object, xiii, xv, 18, 28, 77; definition, xi, 195n1; symbolic meanings, xiv, 78, 208n4

Insectorum sive minimorum animalium theatrum (Moffet). *See Theatrum insectorum* (Moffet)

iris, 143–45, 150–51. *See also* flowers

Jamnitzer, Wenzel, 95–98, 99

Kaufmann, Thomas DaCosta, 89, 91, 200n15, 208n1

Kessel, Jan van, the Elder, 99–101

Kinukawa, Tomomi, 141, 163–65, 166

Knyvet, Edmund and Thomas, 51

Koerten, Johanna, 152–53

Koreny, Fritz, 199n7, 200n14

kunstkammer, 18, 76, 91, 92

Labadists, 140, 160, 220n34, 220n35

lantern fly, 68

Lely, Peter, 113, 214n28

lice. *See* louse

life casting, 95–98, 99

lifelike appearances: in Hoefnagel's imagery, 7, 11; in Merian's *Metamorphosis,* 169–72; skill in depicting, 89; in still life, 78–80, 82, 100–101

Ligon, Richard, 118

Linnaean classification, 185–89. *See also* Linnaeus

Linnaeus, 185, 190, 223n6. *See also Genera insectorum*

locusts, xi, 44, 51, 65–66, 208n83

louse: drawing by Wren, 107, 212n3; in Hooke's *Micrographia,* 111, 128, 129, 135; in Moffet's *Theatrum insectorum,* 136

Ludwig, Heidrun, 156

lusus naturae, 4. *See also* visual jokes

Marseus van Schrieck, Otto, 158

Marshal, Alexander, 64–68, 207n76

Marshal Album, 64–68, 69

nature painting. *See* Dürer Renaissance

nautilus shells, 113, 209n10. *See also* shells

needlework. *See* embroidery

networks, xiv, 55, 178, 182–85, 189

Neues Blumenbuch (Merian). *See* *Blumenbuch* series (Merian)

Newe jewell of health, The (Gessner), 31, 61

New World: Drake manuscript's drawings of, 25–26; and Hooke, 118; White's drawings of, 23–24

Nine Panels of Insects and Flowers (Kessel, attributed), 100

Ogilvie, Brian, xx

Olmi, Giuseppe, 32

Order of Things, The (Foucault), xxi–xxii. *See also* epistemes; Foucault, Michel

Paracelsus, 29–31, 61. *See also* chemical medicine

Park, Katharine, 91

patrons: courtly, 6, 26, 28, 61; Moffet's, 45, 60, 204n27; of natural history, 185; of still life, 78, 86, 91, 98; women, 190, 224n11

pattern books. *See* embroidery: patterns for

Penny, Thomas, 45–49, 205n29, 206n52

Peter the Great, 163

Petiver, James, 124, 128–29, 159–60, 166, 215n52

pigments, 61, 81–82

Piso, Willem, 118, 131, 132

plants, 156, 158, 168, 169. *See also* botanical illustration; botanists; botany; flowers

Pomian, Krzysztof, 92

Power, Henry, 111, 135

Powle, Henry, 107

Raupenbuch (Merian), 140, 154–58, 159, 220n31; editions of, 219n23

Raven, Charles, 46, 47, 54

Ray, John, 64, 182

recipes, 31, 61. *See also* medicinal preparations

Redi, Francesco, 214n33, 217n4

rhinoceros beetle, 14, 16, 47–49, 183–85. *See also* beetles

Rijswijck, Dirck van, 152–54, 219n19

Ring, Tom, 93

Robert, Nicolas, 143–45, 152, 155

Rogers, William, 55

Rondelet, Guillaume, 81–82, 209n10

Royal College of Physicians. *See* College of Physicians

Royal Society: and Charles II, 107; experiments with vivisection, 138; and instrument makers, 133; interest in insects, 116; public image of, 122; social and political concerns of, 108, 121–23

Rücker, Elizabeth, 160

Rudolf II, 5, 13, 18, 20, 23

Ruysch, Frederick, 160, 165

salamanders, 206n52

Savery, Roelant, 78, 79, 80, 95

Saxby, Trevor, 160

scale, 118, 131; in embroidery, 146; juxtapositions, 22, 146, 155; and microscope, 127–28

Schaffer, Simon, 108, 213n11

Schmidt, Benjamin, 167

Schweyfbuch (Ebelmann and Guckeisen), 95–97

Schwindt, Cornelius, 33–34, 36, 37, 44, 203n17

Scientific Revolution, xvii

screening, 4, 26, 189; for courtly audiences, xxiii; and Dürer's *Stag beetle* (drawing), 10; Hoefnagel's use of, 11, 13, 15, 22. *See also* epistemes; Foucault, Michel

Segal, Sam, 144–45, 155, 176

Shapin, Steven, 108–9, 133–34, 213n11

shells, 33, 81, 118; artists' use of, 81, 82, 209n13; collecting and displaying, 221n49; and humanist learning, 82,

209n15; nautilus, 113, 209n10; and still life, 77, 80, 81–82, 210n20

silkworms, 71, 72; Aldrovandi's research on, 36, 70; economic benefits, xi; and labor relations, xiv; Merian's research on, 139; Moffet's illustration of, 70

slips. *See* embroidery

Small Flower Bouquet in Earthenware Vase (Brueghel), 87

Smith, Pamela, xvii

Sommelsdijk, Cornelis van Aerssen van, 160, 220n34

space, 22, 100, 155, 172

specimen drawings, 82–85, 100, 182. *See also* botanical illustration; natural history illustration; specimen logic; specimens

specimen logic: in eighteenth-century images, 183, 189; and failure to convey information, 23–24; in Hoefnagel's work, 4, 7; in *Micrographia,* 114; and still life, 85

specimens: displaying, 161–66, 178; exchanging, 46, 221n52; Hoefnagel's, 208n6; images as substitutes for, 33–34, 36, 47, 77, 88; Merian's, 158–66, 169, 220n40; as models for illustrations, 9, 11, 65; preparing, 22, 34, 116–17, 120, 221n40; selling, 165–66, 141; storing, 185, 204n20. *See also* specimen drawings; specimen logic

spiders, 14, 57, 109–10, 174–75. *See also* tarantula

spontaneous generation, 141, 217n4

Stag Beetle (Dürer), xi–xii, 6–10, 18, 118, 200n12, 208n4. *See also* Dürer Renaissance; Hoefnagel, Joris; stag beetles

stag beetles: and Dürer Renaissance, 5–6; Hoefnagel's depictions of, 5–10, 23, 68, 200n14; Hoffmann's depiction of, 6; specimens, 7; as symbol of Christ, 208n4. *See also* Dürer Renaissance; Hoefnagel, Joris; *Stag Beetle* (Dürer)

Stelluti, Francesco, 116. *See also Apiarum* (Cesi, Stelluti)

Still Life with Flowers (Flegel), 90

Studienbuch (Merian), 159, 169, 171, 172, 219n30

Surinaamsche insecten (Merian). See *Metamorphosis insectorum Surinamensium* (Merian)

Surinam, 160, 167, 179, 180; agriculture, 176–77; colonization of, 220n36; indigenous inhabitants of, 169, 177, 222n67; as source for specimens, 160–61

swallowtail butterfly: depictions by Merian, 143, 146, 150–52, 219n17; White's drawing of, 57, 64, 201n28, 206n59. *See also* butterflies

Swammerdam, Jan, 196n13, 217n4

Swan, Claudia, 197n18, 202n32

tarantula, 52

Tavole di animali (Aldrovandi), 32–37, 39, 40, 85

Theater of Insects (Moffet), 48, 65, 66, 67, 68. *See also Theatrum insectorum* (Moffet)

Theatrum insectorum (Moffet): and Hooke's *Micrographia,* 135; illustrations, 58, 59, 71, 136; manuscript, 45–46, 50; as medical treatise, 61; title page, 45, 55–57; uses in later seventeenth century, 62–68. *See also Theater of Insects* (Moffet)

tiger moth, 78, 208n3

title pages, 45, 55–57, 163, 190, 204n27, 206n56. *See also* frontispieces

Tom Ring, Ludger, the Younger, 92, 93

Topsell, Edward, 46, 219n15

trompe l'oeil, 14, 200n15

True and Exact History of the Island of Barbados, A (Ligon), 118

value: artistic and scientific, 141; of coins, 89; of images, 101; of objects, 77, 98, 101; in still life, 89

Van der Ast, Bartolomeus, 81, 82, 84

Variae ac multiformes florum species (Robert), 145

Vase with Irises (Tom Ring), 93

Vase with Lilies and Irises (Tom Ring), 93

Vignau-Wilberg, Thea, 20

Vincent, Levinus: collections of, 160, 161–66, 221n41; insect collection of, 162, 174. *See also* Amsterdam; Breda, Johanna

virtual collections, 34, 60, 77, 88, 98. *See also* virtual specimens

virtual specimens, 28, 32–45, 68; and formation of community, 55; images as, 28, 36, 44. *See also* specimens; virtual collections

virtual witnessing, 133–34

visual jokes, 4, 18

vivisection, 138, 216n71

Wettengl, Kurt, 141, 159

White, John: drawing of fireflies, 23, 24; drawing of swallowtail butterfly, 57, 63–64, 201n28, 206n59; history of drawings by, 206n59

Witsen, Nicolaas, 160, 165, 220n37

women: education of, 143; and eighteenth-century natural history, 190, 223n11; and embroidery, 143, 155, 218n8; role in organizing collections, 221n49. *See also* embroidery; patrons

Wondertooneel der Nature (Vincent), 162, 164

Wotton, Edward, 55, 206n59

Wren, Christopher: collaborations with Hooke, 110–12, 213n19, 213n20; drawings of insects, 106–8, 112, 212n3; plans for publishing microscopical observations, 111–12; and Royal Society, 107, 212n5. *See also* Charles II; Royal Society

wunderkammer, 97–98

Janice Neri is associate professor of art history and visual culture at Boise State University.

www.ingramcontent.com/pod-product-compliance
Lightning Source LLC
Chambersburg PA
CBHW080955170526
45158CB00010B/2813